EARTH

IN **100** GROUNDBREAKING DISCOVERIES

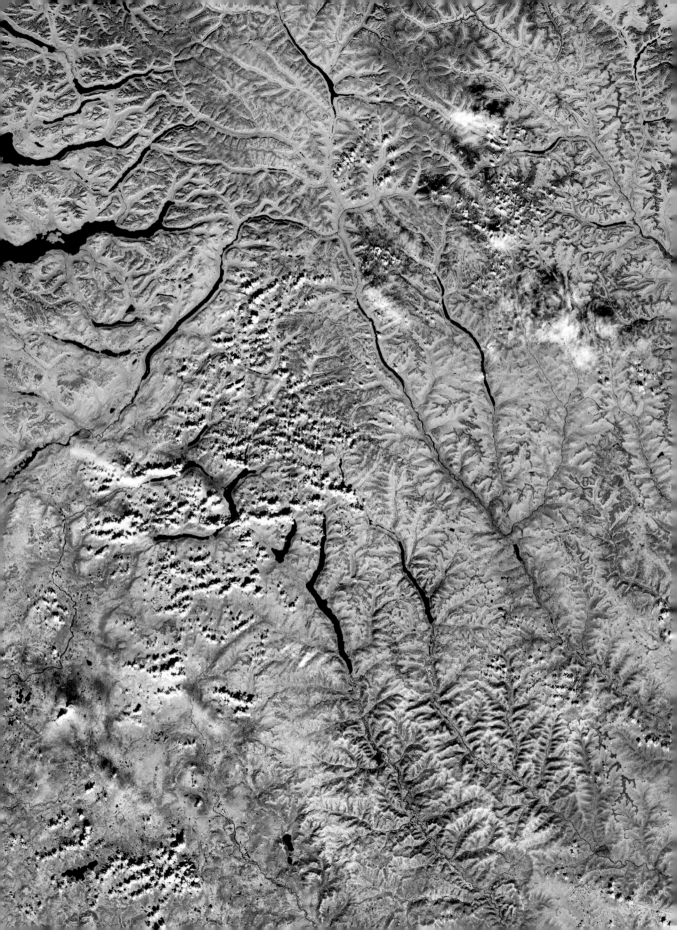

EARTH

IN **100** GROUNDBREAKING DISCOVERIES

DOUGLAS PALMER

FIREFLY BOOKS

Contents

Natural disasters are always news. In the course of writing this book, earthquakes, landslides, storms, floods, tsunamis and volcanic eruptions have all made their presence felt around the world, frequently making the headlines due to their catastrophic effects on both human life and the wider environment.

But as the saying goes, 'You ain't seen nothing yet.' The geological record of our planet is replete with evidence of similar catastrophic events scattered through billions of years of the deep past, often on a much greater scale than anything experienced within the relatively brief span of historical time. Ultimately, however uncomfortable and unmanageable these events and processes might be, they are also manifestations of Earth's internal dynamic. Without these problematic surface expressions of the energy within Earth, our planet would be as lifeless as the Moon. There would be no atmosphere, water or habitable environments, and no life.

A changing planet

Over the last 200 years and more, studies of Earth's history, the development of its environments and the evolution of its life have gradually revealed a complex and dynamic system. Our planet's 4.5-billion-year story has involved continued interaction between the organic world of living organisms and the inorganic world of materials that range from rocks and minerals, through the waters of the oceans and rivers, to the gases of the atmosphere. For example, one of the earliest and most important of these 'co-evolutionary' events was the increase in oxygenation of the atmosphere and oceans some 2.4 billion years ago, known as the Great Oxidation Event. Triggered by a rise in the abundance and efficiency of photosynthetic micro-organisms, this boom in oxygen made the evolution of aerobic organisms and ultimately humans possible.

Much more recently, within the relatively brief lifespan of our own *Homo sapiens* species, we have seen similar co-evolutions of Earth and life. A very different kind of perturbation of the atmosphere caused by the Quaternary Ice Age, while subsequent human influence on the climate has had a range of complex impacts on life and the environment.

Another great lesson from the geological record of Earth and its history is that, despite the extraordinary range of often-cataclysmic events that have had an impact on life, the flexibility offered by the evolutionary process has

ensured that it has survived and thrived. Despite mass extinctions that, at their most extreme, have cut down the majority of living organisms, both in the sea and on land, enough have always survived to restock and replenish Earth's environments.

A dynamic science

The aim of this book is to illustrate some of the most interesting current scientific studies across the entire range of Earth Sciences. These range from the investigations of the inner workings of Earth's metallic core, which ultimately drives many of the dynamic processes that impact upon the thin surface layer we inhabit, to discussions of our planet's future prospects in both the short and long term. Almost without exception, the areas covered involve rapidly changing fields of research, and every effort has been made to include the most recent research. But, this is not a textbook, and despite its scope it is simply not possible to be inclusive across the ever-widening range of Earth Science topics.

It is ironic that humans are the first species to be able to gain such an insight into the complexity of Earth's entire 'geosystem', but are also the first to have radically altered the global environment and affected the fate of its other inhabitants. Despite our awareness of our actions and our relationship with the surrounding environment on which we depend, we seem incapable of comprehending the results, both real and potential, of our actions. As this book and recent experience shows, while there may be many natural processes that we can do little about, there are others that we can hope to affect, to the benefit of future generations. This has been cruelly reinforced by the March 2011 earthquake and tsunami in Japan. We will never be able to prevent such natural events, but the crippling blow to nuclear reactors and the cost in human lives, both present and future, can and should be prevented. We may need nuclear reactors but as has been said before, they should not be built near major fault zones or within reach of tsunami waves.

Finally, I hope that this book shows just how much there is still to be discovered, learned and understood about the Earth. The study and pursuit of the Earth Sciences is an ever-expanding and increasingly complex field that today extends its reach far beyond the traditional limitations of physical geology. Hopefully this book will encourage enquiring minds to take up the fascinating quest for a better understanding of our planet.

DOUGLAS PALMER

Birth of a planet

1

DEFINITION THE EARTH FORMED FROM DEBRIS LEFT ORBITING THE NEWBORN SUN, THROUGH A PROCESS CALLED ACCRETION

DISCOVERY THE SOLAR NEBULAR DISC MODEL WAS DEVELOPED IN THE 1970S BY US ASTRONOMER GEORGE WETHERILL

KEY BREAKTHROUGH ROCK SAMPLES RETURNED FROM THE MOON BY THE APOLLO LUNAR MISSIONS ALLOWED PRECISE DATING OF KEY EVENTS IN EARTH'S EARLY HISTORY

IMPORTANCE MODELLING THE WAY IN WHICH EARTH FORMED IS VITAL TO UNDERSTANDING ITS SUBSEQUENT DEVELOPMENT

Earth was formed, along with the other planets of the solar system, in orbit around the newborn Sun more than 4.5 billion years ago. But how much can we really know about this remote and turbulent period of our planet's history?

The most widely accepted explanation for the origin of Earth and its solar system, known as the nebular hypothesis, was originally developed by Swedish scientist Emanuel Swedenborg (1688–1722) in 1734. In almost three centuries since, it has remained essentially intact despite enormous scientific advances. Its current manifestation takes into account our understanding of the true age of the Earth (derived from radiometric dating – see page 19), and modern ideas about the raw materials from which the Sun and its planets formed. This 'solar nebular disc' model was developed in the 1970s by US scientist George Wetherill (1925–2006) from the ideas of Russian astronomer Viktor Safronov (1917–99). It describes events leading to Earth's formation through a process of gradual build-up or 'accretion'.

The solar system was born around 4.56 billion years ago from a dense solar nebula considerably larger than the present Solar System. This nebula was dominated by molecular hydrogen gas, the most abundant and lightest chemical element in the Universe, and dust grains containing heavier elements. As the giant molecular cloud coalesced, it became gravitationally unstable and collapsed to form a slowly spinning 'protoplanetary disc' surrounding a young star. Condensing and contracting through a process known as cold accretion, the disc began to rotate faster, feeding material to the increasingly hot and luminous protosun at its centre. As the disc spun ever faster, the icy gas and dust clouds in the outer parts of the disc collided

OPPOSITE In the beginning: our solar system, including planet Earth, began life as a dense solar nebula, dominated by molecular hydrogen and dust. Over millions of years, this nebula coalesced into a slowly spinning disc illuminated by the youthful but increasingly hot protosun at its centre.

and clumped together. Their increasing gravity attracted ever more material into what are known as planetesimals, which were around 1km (0.6 miles) in diameter.

As each planetesimal grew larger and its gravity increased, there was a process of runaway accretion with each planetesimal becoming denser and drawing in more rock material from its immediate surroundings. They then merged to form the primordial Earth and three other rocky protoplanets (Mercury, Venus and Mars) closest to the increasingly hot Sun. During Earth's early accretion the first solid rocks were created by the condensation of gaseous silica (silicon dioxide), leaving light elements such as hydrogen and helium to form a proto-atmosphere, but this was driven away by the solar wind, Earth's internal heating and the impact that formed the Moon.

Origins of the Moon

Earth is unusual in having a large natural satellite – none of the other rocky planets has anything to compare with our Moon, and its origins have long been a subject for debate. Traditionally, there were three rival hypotheses, in which the Moon either formed independently in orbit around Earth, was flung out from a considerably larger and rapidly spinning 'proto-Earth', or was captured from elsewhere in the solar system. However, rocks collected by the Apollo astronauts and dated to around 4.5 billion years ago, show an overall composition similar to the rocks of Earth's crust and mantle, while measurements of the Moon's overall density shows that it is similar to that of Earth's upper mantle and therefore cannot have a large metallic core. This puzzling information has led to a new theory of the Moon's formation – the impact hypothesis, also known as the Big Splash. Around 4.5 billion years ago, it seems, Earth was struck by a Mars-sized planet in an impact so immense that it tore away and fragmented a sizeable chunk of the primordial crust and mantle. This debris formed a dense ring in orbit around Earth, which rapidly accreted to become the Moon. The energy from the collision also raised Earth's surface temperatures to around 6,000°C (11,000°F).

'High-energy impacts melted Earth's surface, and continued until most of the planetesimals in Earth-crossing orbits had been "swept up". The Moon's cratered surface still preserves evidence of this bombardment in the form of lunar "seas". '

Differentiating Earth

At first, the primordial Earth was a more or less uniform mass of rock, but within 10 million years or so, the release of heat from radioactive materials raised temperatures so much that the interior became mobile, triggering the formation of Earth's layered structure – a process called differentiation. Iron and other heavy metals such as nickel separated out and sank towards the centre of the planet, where they formed the intensely hot metallic core whose internal motion created Earth's magnetic field. The remaining, less dense materials formed a deep rocky mantle around the core, whose outer surface cooled into a solid crust. Migration of iron-rich material to the core,

known as the iron catastrophe, would have generated large amounts of heat – perhaps enough to melt the entire mantle.

Even after the birth of the Moon and the separation of Earth's layers, it seems that intense heavy bombardment from space continued. High-energy impacts melted Earth's surface, and continued until most of the planetesimals in Earth-crossing orbits had been 'swept up'. The Moon's cratered surface still preserves evidence of this bombardment in the form of the lunar 'seas' formed when lava from later volcanic eruptions filled the low-lying impact basins. Rocks associated with these impacts, brought back to Earth by the Apollo lunar missions, have dates that cluster between 4.1 and 3.8 billion years ago, and provide convincing evidence that this 'Late Heavy Bombardment' peaked around 3.9 billion years ago.

Earth would have suffered just as much as the Moon during this bombardment, and would have been just as heavily cratered. However, our planet's subsequent dynamic history, during which mantle material and most of the crust has been recycled, has wiped out evidence for these early impacts alongside much of the geological evidence for the first 500 million years of Earth's history – but fortunately a few clues have survived.

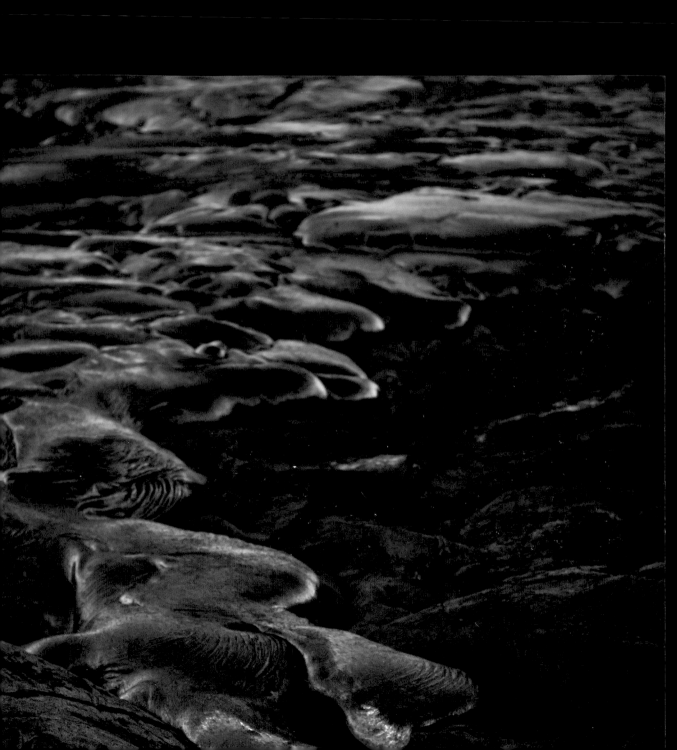

2 Earth's oldest materials

DEFINITION FRAGMENTS OF ROCK AND MINERAL FROM THE EARLIEST
CRUST OF PLANET EARTH

DISCOVERY ZIRCONS IN AUSTRALIA'S JACK HILLS REGION WERE
DISCOVERED IN THE EARLY 1980S AND DATED TO MORE THAN
4 BILLION YEARS OLD

KEY BREAKTHROUGH IN THE EARLY 2000S SCIENTISTS USED THE
ZIRCONS TO PROBE EARTH'S EARLY ENVIRONMENT

IMPORTANCE RESULTS FROM THE ZIRCON STUDIES SHOW THAT WATER
WAS PRESENT ON EARTH'S SURFACE FROM THE EARLIEST TIMES

Earth's first 500 million years are shrouded in mystery because of the lack of any surviving material from the time. As a result, geologists are still arguing about how quickly Earth cooled from a hellish mass of molten lava, and how soon the solid crust and the first oceans formed.

Geological mapping of our planet reveals that, despite billions of years of geological history, there are still a few regions of very old rocks exposed on Earth's surface in the heart of ancient continents such as Australia, South Africa, North America and Greenland. Some of the rocks and minerals that can be identified within these regions date back close to 4 billion years ago, but since our planet is estimated to be approximately 4.54 billion years old, that still leaves some 500 million years of Earth's early history unaccounted for – an eon known as the Hadean. However, recent discoveries are shedding light on this hitherto mysterious period.

Relics of the Hadean

The oldest rocks so far found on Earth are between 3.8 and 4.28 billion years old (see page 21), but some very rare fragments of zircon minerals (zirconium silicate, $ZrSiO_4$) may be as much as 4.36 billion years old. This might seem a trifling difference when dealing with such huge timescales, but the additional 80 million years of history they reveal may offer our best clues to the early stages of our planet's development.

The zircons were discovered embedded in younger rocks of Western Australia's Jack Hills region. With an abundance of less than one part in a million by weight, and typically less than 0.3mm (1/80in) long, the crystals have amazingly survived an entire cycle of erosion that broke down their

OPPOSITE 4.5 billion years ago, before there was any land or water, Earth's surface was a dark crust of lava, erupting in places into a seething, flowing mass of molten rock.

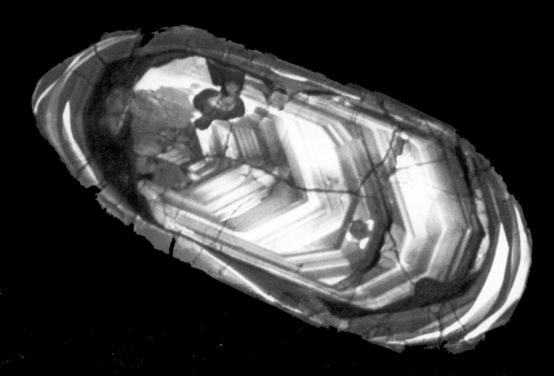

original volcanic parent rock and reworked its debris into much younger sedimentary 'conglomerate' rock. Zircon survives primarily because it is a particularly tough mineral – formed at high pressures and temperatures of up to 1,400°C (2,550°F), it can pass through the erosion cycle without being changed in any significant way, while many other minerals would be degraded and altered by combined physical and chemical processes.

The crystals reveal their age from the ratios of radioactive forms or 'isotopes' of uranium, thorium and lead within them. Over billions of years, uranium in the original sample undergoes radioactive decay to form thorium, and thorium in turn is transformed into lead – by measuring the proportions of each isotope within the present-day sample, geologists can work out how long ago it formed. However, although the age of the zircons was established soon after their discovery, it was not until the early 2000s that a group of scientists led by Simon Wilde of Perth's Curtin University of Technology used them as a probe for Earth's early environment.

Clues to the past

Zircon crystals preserve useful information about the way they formed through their layered or 'zoned' pattern of growth, and the distribution of stray 'rare-earth' elements within the growth zones. These features can reveal the pressure and temperature conditions at the time, while analysis of mineral inclusions (tiny grains sometimes found embedded and preserved within the crystals) can provide even more information about the original host rocks within which they formed.

titanium, whose presence hints at the temperature of formation (the more titanium is present, the higher the temperature is likely to have been). The ratio of two isotopes of hafnium ($^{176}Hf/^{177}Hf$), meanwhile, indicates whether the molten magma in which the crystal formed was composed purely of mantle-like materials from deep beneath Earth's surface (see page 53) or whether it contained crust-like components as well. Similarly, the higher the content of the 'heavy' oxygen isotope ^{18}O, the greater the likelihood that the magma interacted with liquid water on Earth's surface.

The Jack Hills zircons were originally discovered in the early 1980s, and in the decades since then, geologists have recovered enough individual crystals to build up a detailed picture of the early Earth. The Wilde team's analysis, published in 2001, suggested that the zircons originated in an environment of granite-like magmas, with temperatures no higher than around 680°C (1,260°F), between 4.25 and 4 billion years ago. This magma apparently developed from a mix of mantle and crust rocks, the latter of which had already interacted with water at or near the surface. So by the time the zircons formed, it seemed that Earth was already cool enough to have a solid crust and surface water.

'Analysis of mineral inclusions (tiny grains sometimes found embedded and preserved within the crystals) can provide even more information about the original host rocks within which they formed.'

However, the recent discovery and analysis of diamonds among the zircons' various mineral inclusions has thrown this picture of Earth's early history into some doubt. Diamonds are not normally found in relatively low-temperature granite magmas because famously they only form at high pressures – either at considerable depths below the surface, or in the immediate aftermath of meteorite impacts. The zircons themselves, meanwhile, show no other signs of exposure to such pressures, either at depth or during an impact event.

At present, there are two widely touted explanations for this mystery. One is that the diamonds formed up to 4.36 billion years ago in a single high-pressure event, such as a major impact, and were then recycled into younger rocks and incorporated into the zircons (just as the zircons themselves have survived intact into a later generation of rock). The other is that they were introduced to the developing zircons by some unknown and repeated process during Earth's early history. However, neither of these explanations is entirely satisfactory and both sidestep the question of Earth's surface temperatures at the time. Another possibility, however, retains the 'cool Earth' idea, with the carbon that formed the diamonds originally introduced to the zircons as graphite from the primitive, carbon-rich atmosphere (see page 25). The carbon might then have been transformed into diamond by some form of compression during a later 'subduction' of the zircons

3 Geological dating

DEFINITION THE WIDELY USED SYSTEM OF EONS, ERAS, PERIODS AND
STAGES USED FOR SUBDIVIDING AND DATING EARTH'S HISTORY

DISCOVERY ESTABLISHED BY WILLIAM SMITH USING PRINCIPLES OF
STRATIGRAPHY TO FIND RELATIVE AGES OF DIFFERENT ROCK UNITS

KEY BREAKTHROUGH SINCE THE 1950S, RADIOMETRIC TECHNIQUES
HAVE ALLOWED PRECISE DATING OF GEOLOGICAL PERIODS

IMPORTANCE THE DATING SYSTEM HELPS SUBDIVIDE EARTH'S HISTORY
INTO DISTINCT PHASES WITH UNIQUE CHARACTERISTICS

The names of geological periods such as the Jurassic and Cretaceous, and
dates of events stretching back through hundreds of millions of years, form
the backbone of our planet's history. But what do these names mean, and
how do geologists map out such massive stretches of time?

During 250 years or more of discovery and research into geological processes
and the changing nature of past life, the history of our planet has been
subdivided and dated in many different ways. Placing an accurate timescale
onto the geological past is a much more recent endeavour, dependent on
sophisticated technologies that have only been refined since the 1950s.

Historic investigations

Early ideas about Earth's history were heavily influenced by religious, cultural
and philosophical attitudes – the sort of thinking that led 17th-century Irish
cleric James Ussher to calculate the date of creation as 4004 BC, based largely
on biblical chronology. However, the spread of Enlightenment philosophy in
the 18th century, and the establishment of recognizably 'scientific' principles
of investigation and inquiry, led to an inevitable change in attitudes. In 1788,
Scottish geologist James Hutton (1726–97) published his highly influential
Theory of the Earth, a work that outlined the basic principles by which
sedimentary rock strata are laid down, and which argued that Earth had a
much longer history, with 'no vestige of a beginning, no prospect of an end'.

But it was not until the early 19th century that the rock record was first
carved up into a series of 'systems', each representing a distinct period of
time. In a set of pioneering geological maps, English geologist William Smith
(1769–1839) subdivided strata according to the rock types and fossil deposits

OPPOSITE Layer upon
layer, the history
of Earth's surface
deposits has piled
up into rock strata.
Once you know the
geological language
in which it is written,
the mineral and fossil
debris of the remote
past can be read like
the pages of a book.

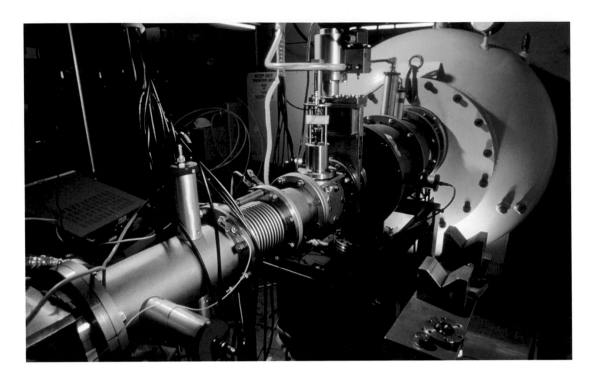

ABOVE Radiometric dating involves measuring the ratio of radioactive 'parent' isotopes and their decay products or 'daughter' isotopes within a sample of organic or inorganic material. In the widely used technique of 'carbon dating', for example, equipment such as this linear accelerator is used to count the ratio of carbon isotopes in a sample, revealing its age provided it is less than around 50,000 years old.

within them. French geologists Georges Cuvier (1769–1832) and Alexandre Brongniart (1770–1847) independently came up with the same technique for mapping the relative age of strata.

By the mid-19th century, all the main subdivisions of the rock record had been named, giving rise to the various geological periods that are the most familiar way of cataloguing Earth's history. The names for these rock systems come from a variety of sources – for example, the Jurassic system and period, first distinguished by Leopold von Buch (1774–1853) in 1839, was based on the marine strata of the European Jura Mountains, while the Cretaceous, named by the Belgian geologist J.J. D'Omalius d'Halloy (1783–1875) in 1822, comes from '*craie*', the French word for chalk, on account of its characteristic chalk limestones in northwestern Europe. Today, countless refinements have been added to the scheme, and the definition of the subdivisions and dates is regulated by an international commission (see page 410).

The Geological record and timescale

Today, Earth's age is generally accepted to be about 4.54 billion years, of which the first 4 billion years or so, known as the Precambrian, is divided into three 'eons' – the Hadean, Archean and Proterozoic. Although fossils appear as far back as 3.2 billion years ago, such remains of marine microbes are rare, and it is only in the Proterozoic (around 635 million years ago) that readily visible fossils start to appear (see page 149). Because of the rarity of Precambrian fossils, further subdivisions within the early eons are largely based on radiometric dating and scattered evidence of ice ages.

The end of the Precambrian, 542 million years ago, marks the beginning of the current Phanerozoic Eon, and the Cambrian Period. The Cambrian is the first of six periods in the Palaeozoic ('ancient life') Era, followed by the Ordovician, Silurian, Devonian, Carboniferous and Permian. Each can be recognized by characteristic fossils, which change over time as a result of evolution. The Palaeozoic was brought to an end by the biggest extinction event in the history of life (see page 209), and followed by the Mesozoic ('middle life') Era, often characterized as an 'age of reptiles'. The Mesozoic began 251 million years ago with the Triassic Period, followed by the Jurassic and Cretaceous, and came to an end 65.5 million years ago with another major extinction event (see page 261). Most recently, the Cenozoic ('recent life') Era, also known as the age of mammals, is subdivided into three periods, the Paleogene, Neogene and Quaternary. The Paleogene and Neogene (formerly grouped together as the 'Tertiary Period') are defined by their sedimentary rocks and fossils, while the base of the Quaternary, 2.58 million years ago, is marked by the onset of the recent Quaternary Ice Age.

'In 1788, Scottish geologist James Hutton published his highly influential *Theory of the Earth*, a work that outlined the basic principles by which sedimentary rock strata are laid down, and which argued that Earth had a much longer history, with "no vestige of a beginning, no prospect of an end". '

Radiometric dating

Thanks to modern scientific advances, there are now numerous methods for dating rocks, most of which depend on physical and chemical processes of change and decay associated with radioactivity. These processes affect various naturally occurring substances and occur at well-defined rates, so that changes in the ratio of 'parent' and 'daughter' atoms, or physical traces created within minerals, can be measured to reveal the time since a rock sample's radiometric 'clock' was initially set.

Radiocarbon dating is perhaps the most familiar of these processes. It relies on the decay of radioactive carbon-14 (^{14}C) – an 'isotope' of carbon that occurs in a small but constant proportion in Earth's atmosphere. This isotope is absorbed by living organisms such as plants and animals and incorporated into their bodies. After the organism dies, however, its ^{14}C decays. Half of its atoms transform into other, non-radioactive substances every 5,730 years (the isotope's 'half-life'). By measuring the proportion of radioactive carbon in a sample of once-living material, archaeologists can calculate the time since the original organism died.

Of course, the relatively rapid decay rate of ^{14}C means that it can only be used to date carbon-based materials back to around 50,000 years ago. The dating of older rocks requires other elements and other radioactive isotopes, such as potassium/argon decay (^{40}K to ^{40}Ar) useful back to 1.2 billion years ago, and uranium/lead (^{238}U to ^{206}Pb), which allows dating across 4.5 billion years. It was this system that was first used in 1953 by American geochemist Clair Cameron (Pat) Patterson (1922–95) to determine Earth's age.

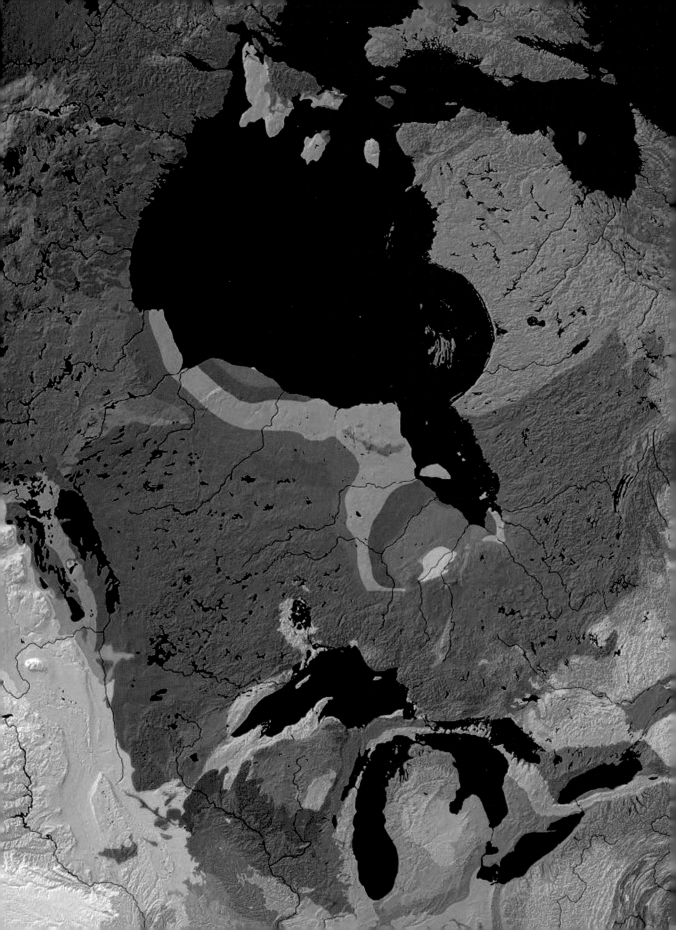

4 The oldest rocks

DEFINITION THE OLDEST SURVIVING ROCK OUTCROPS ON THE
SURFACE OF OUR PLANET, AROUND 4 BILLION YEARS OLD

DISCOVERY THE ACASTA GNEISS OF NORTHWESTERN CANADA WAS
FIRST DATED IN 1989

KEY BREAKTHROUGH MEASUREMENTS FROM WIDESPREAD ANCIENT
ROCKS SHOW THAT THE EARTH'S CRUST SPLIT INTO DISTINCT
CONTINENTAL AND OCEANIC ROCKS EARLY IN ITS HISTORY

IMPORTANCE ANCIENT ROCKS CAN REVEAL A GREAT DEAL ABOUT
CONDITIONS ON THE EARLY EARTH

Earth's earliest known rocks, are dated to around 4 billion years old, and were discovered close to the Arctic Circle in Canada's remote Northwest Territories. They originated in a period of history known as the Hadean Eon on account of the hellish conditions on Earth's surface at the time.

For decades geologists have searched for the oldest preserved rocks, looking for clues about the circumstances and conditions of our planet's origin. The basic challenge they face is that the process of Earth's formation around 4.567 billion years ago was prolonged and intense, involving extreme pressures and temperatures that no longer exist at the surface. Throughout the first few hundred million years of geological history, mineral and rock materials were repeatedly being reworked and recycled almost as soon as they had formed, and as a result none of them have survived.

Fortunately, there are at least some very ancient rocks, well over 2 billion years old, preserved within cratons, the surviving fragments of the earliest continents (see page 65). Here, they can still occasionally be found exposed at the surface. But the search for the very oldest rocks within the cratons is further complicated by the considerable alteration ('metamorphism' in geological terms) that these rocks have suffered throughout their history. For example, the increased temperatures associated with burial and transformation deep within the crust can 'reset' the ratios of different radioactive isotopes within a rock, robbing geologists of their most valuable tool for assessing the age of rock samples (see page 19). Since most really ancient rocks have been through at least one, if not several, phases of metamorphism, the chances of finding relatively unaltered and truly ancient rocks is exceedingly remote.

OPPOSITE Some of Earth's oldest known rocks have been found around Canada's Hudson Bay. Geological maps such as this one reveal the ancient Precambrian rocks (here coloured orange and pink) that make up much of the Canadian Shield.

The Acasta gneisses

Fortunately in the mid-1980s, scientists discovered a veritable motherlode of just such ancient rock within the vastness of the Canadian Shield (the geological 'heart' of an ancient landmass called Laurentia). One of the world's largest areas of Precambrian rock, covering some 8 million sq km (3 million sq miles), this region underlies a huge area of northern North America. What's more thanks to the glacial 'scouring' inflicted by recent ice ages (see page 293), it is also a region where a considerable quantity of ancient rocks are exposed at the surface.

The Slave Craton, some 300,000 sq km (120,000 sq miles) in area, is just one of several cratons that make up the Laurentian Shield. During systematic mapping of this area in the 1980s, the Geological Survey of Canada identified some metamorphosed rocks on an island in the Acasta River, east of the Great Bear Lake. Forming part of an outcrop known as the Acasta gneisses, they contained zircon materials that offered the potential for isotopic analysis and radiometric dating. Scientists Sam Bowring from Washington University and Ian Williams and William Compston of the Australian National University in Canberra analysed uranium and lead isotopes within tiny individual zircon crystals, initially obtaining a minimum age of 3.84 billion years. A further study in 1999 dated the rocks' formation more accurately to between 4.00 and 4.03 billion years. The scientists also argued that these banded gneisses, with their alternating layers of granitic and amphibolite composition, were probably derived from small island 'nuclei' of chemically differentiated, continental-type crust magmas. Other experts have argued that while the dates of the Acasta zircons may be secure, the metamorphic 'host' rocks that contain them could be somewhat younger.

'Throughout the first few hundred million years of geological history, mineral and rock materials were repeatedly being reworked and recycled almost as soon as they had formed.'

Since the discovery of the Acasta gneisses, competing contenders for the oldest rocks have emerged from other Precambrian cratons, such as the Itsaq metamorphic gneisses from the southern part of Western Greenland. Dated at 3.7–3.75 or possibly 3.8 billion years old, these rocks were originally shallow-water marine sediments and volcanic rocks such as pillow lavas (see page 63). Their existence indicates the presence of seas and landmasses by this stage in Earth's development. In 2001 another, more problematic, contender for the title of 'most ancient rock' was found discovered back in the Laurentian Shield of Canada. Known to geologists as the Nuvvuagittuq greenstone belt, it lies along the eastern shore of Hudson Bay, in northern Quebec.

In 2008, geologists Jonathan O'Neill and Don Francis from Montreal's McGill University, together with Richard Carlson of Washington's Carnegie Institution, used a new radiometric dating technique based on isotopes of the rare earth elements samarium (^{146}Sm) and

neodymium (^{142}Nd), found in metamorphic rocks known as an amphibolites, to estimate the age of the rocks with renewed accuracy. Their results pointed to an astonishing age of between 3.8 and 4.28 billion years old, making Nuvvuagittuq the most ancient large rock mass known. However, other experts, such as Stephen J. Mojzsis of the University of Colorado, disagree and argue that the Nuvvuagittuq rocks are no more than 3.8 billion years old, and therefore similar in age to the Greenland rocks.

Dating problems

Unfortunately, like most very ancient rocks, the Nuvvuagittuq greenstones have been metamorphosed by a later phase of burial or 'subduction' deep within the crust, accompanied by heating to around 500°C (930°F) that has erased many signs of their formation. Luckily, however, they were never heated above their melting point, so their 'radiometric clock' has not been reset. Despite this, their age may still be somewhat exaggerated – the magma from which their 'greenstone' lavas formed originated as melting rock many kilometres below the surface, and it is possible that the date obtained from the isotopes is actually that of the magma's subterranean formation, rather than its eruption to the surface as lava, which could have been several million years later.

ABOVE Recognition and dating of Earth's oldest rocks is a difficult and contentious task. The most widely accepted outcrop of such rocks is the Acasta Gneiss Complex of Canada's Northwest Territories. These metamorphic rocks range in age from 3.8 to 4.02 billion years old and represent the remains of igneous activity in Earth's early crust.

5

The early atmosphere

DEFINITION THE EARLIEST ENVELOPE OF GASES AROUND THE EARTH, FORMED LARGELY THROUGH VOLCANIC 'OUTGASSING'

DISCOVERY WIDESPREAD ROCKS AND MINERALS CONFIRM THAT THERE WAS VERY LITTLE FREE OXYGEN IN EARTH'S ATMOSPHERE PRIOR TO 2.5 BILLION YEARS AGO

KEY BREAKTHROUGH MEASUREMENTS OF ISOTOPES IN SOUTH AFRICAN ROCKS REVEAL THE FIRST TRACES OF OXYGEN IN OCEAN WATERS

IMPORTANCE THE TRANSFORMATION OF EARTH'S ATMOSPHERE WAS VITAL IN CREATING A WORLD SUITABLE FOR ANIMAL LIFE

Our planet's atmosphere is a thin shell of gases, concentrated mostly within 8km (5 miles) of the surface. Today it is dominated by nitrogen and oxygen, but Earth's original atmosphere, formed around 3.8 billion years ago, had a radically different composition.

Earth's present-day atmosphere consists primarily of nitrogen (78 percent by volume), oxygen (21 percent), argon (0.93 percent), carbon dioxide (0.39 percent) and numerous other gases in ever diminishing amounts, plus a small but vitally important quantity of water vapour (1 percent). However, when the first stable envelope of gases formed around the planet a little more than half a billion years after its formation, conditions were very different – most notably in the complete lack of 'free' oxygen. Intense volcanic activity, known as outgassing, generated huge volumes of volatile gases, especially nitrogen, carbon dioxide and water vapour, along with ammonia, methane and others. Water vapour condensed and precipitated to form surface water, which flowed under gravity to form the first significant bodies of water, and eventually the earliest seas and oceans.

Oxygen appears

According to widely accepted models of the early atmosphere, production of oxygen only began through the activity of photosynthetic ocean-dwelling microbes, called cyanobacteria, around 2.7 billion years ago. These microbes began to generate energy just like modern plants, using sunlight and water vapour to process carbon dioxide from the atmosphere, creating energy-rich chemical compounds and producing oxygen as a by-product. The process is thought to have been underway for some considerable time before levels of atmospheric oxygen began to rise significantly.

OPPOSITE Shallow tropical waters are home to primitive life forms of the kind that originally pumped the first oxygen into Earth's atmosphere around 2.7 billion years ago. These microbial cyanobacteria build up laminated mounds and clumps known as stromatolites, made from alternating layers of sediment and microbial sheets.

However, it has proved very difficult to find geochemical evidence for the event and the rise of these first photosynthetic microbes. As a result, our picture of the relative abundance and distribution of oxygen in Earth's oceans prior to late Archean times is not well defined.

Nevertheless, there are a number of chemical indicators for the presence of oxygen as far back as 2.7 billion years ago. The discovery of sulphur and nitrogen isotopes in South Africa's Kaapvaal Craton and at Hamersley Basin (within the Pilbara Craton of Western Australia), along with chromium isotopes in banded iron formations (see page 125) have all been taken as signs of atmospheric oxygen dissolved in late Archean oceans. What's more, some scientists have claimed that hydrocarbon 'biomarkers' – a kind of chemical fossil, found in black shales of this date – are direct traces of the hypothetical oxygen-producing marine cyanobacteria (see page 121). Even so, the origin of these biomarker molecules is far from certain.

Fossil evidence?

The earliest known fossils of cyanobacteria themselves, found in the Apex Chert of Western Australia, date back to 3.5 billion years ago, but the organic nature of these 'microfossils' has also been questioned (see page 121). Fortunately, slightly younger, 3.46 billion-year-old structures called stromatolites, related to the presence of ancient cyanobacteria, have been identified from the nearby Strelley Pool Chert with much greater confidence. Of course, the physiological capabilities of the micro-organisms that left these fossil traces are also far from certain, and recent discoveries of 'extremophile' organisms capable of generating energy in strange ways have clouded the issue. For example, some modern microbes that derive energy through photosynthesis (so-called 'photoautotrophs'), can nevertheless oxidize iron in the absence of oxygen, creating biomarkers similar to those that have, until recently, been seen as inarguable signs of an oxygenated environment.

New geochemical measures

Faced with this uncertainty, geologists have renewed their search for other measures of oxygen in late Archean rocks. One promising source are black shales found in the Griqualand West Basin of the Kaapvaal Craton. A new technique uses measurements of the elements rhenium and molybdenum found in these rocks. Weathered and released from continental crust rocks by reactions with atmospheric gases, compounds containing oxides of these elements were carried by rivers into the ocean, where they accumulated just off the margin of an extensive limestone continental shelf, between 2.59 and 2.46 billion years ago. Introduced to the water, these oxides are both vulnerable to being 'reduced' (having their oxygen removed by other elements that have a stronger affinity for it), and crucially molybdenum compounds are more easily reduced than those of rhenium.

'Intense volcanic activity, known as outgassing, generated huge volumes of volatile gases, especially nitrogen, carbon dioxide and water vapour, along with ammonia, methane and others.'

The Griqualand shales turn out to be strongly enriched with rhenium, but contain only small amounts of molybdenum. This suggests that small amounts of dissolved oxygen were present in the bottom waters of this submarine slope. Since water depths here reached several hundred metres and were impenetrably dark, the oxygen was probably being produced in large quantities by abundant photosynthetic microbes in the sunlit 'photic zone' above and transported downwards. However, it's clear that oxygen was still not being produced in large enough quantities to satisfy *all* the oxygen-stealing deep-ocean 'reductants', so below these depths it's likely that ocean waters were still largely oxygen-free.

So it seems that surface waters around continental margins at least were becoming widely oxygenated by late Archean times, though we still cannot know whether the sea's upper levels were widely oxygenated beyond the shelf areas. Nevertheless, there was certainly enough oceanic oxygen to produce a small increase in atmospheric oxygen around 2.5 billion years ago, some 100 million years before a far steeper rise known as the Great Oxidation Event (see page 126). Atmospheric oxygen levels did not build up to any significant extent before this time because the cyanobacteria were not producing enough to counteract the actions of reducing elements in the atmosphere and ocean waters of the time. It was only when a critical threshold was crossed that oxygen finally began to transform the atmosphere, paving the way for more complex forms of life.

6

The evolving hydrosphere

DEFINITION THE HYDROSPHERE IS THE LAYER OF WATER THAT COVERS 71 PERCENT OF EARTH'S SURFACE, AND FORMED SHORTLY AFTER THE EARTH ITSELF

DISCOVERY THE ORIGIN OF THE HYDROSPHERE IS OFTEN ATTRIBUTED TO COMETS RAINING DOWN ON A COOLING EARTH

KEY BREAKTHROUGH NEW STUDIES SHOW THAT MUCH OF EARTH'S WATER COULD HAVE COME FROM WITHIN ITS ROCKS AFTER ALL

IMPORTANCE THE ORIGIN OF EARTH'S WATER HAS IMPLICATIONS FOR THE BEGINNINGS OF LIFE AND SUGGESTIONS THAT COMETS 'SEEDED' EARTH WITH ORGANIC CHEMICALS

Today, more than two-thirds of Earth's surface is covered with water, to an average depth of 3,700m (12,100ft), but where did it all come from? The idea that most was delivered from icy comets is now being challenged by evidence that water could have come from Earth's rocks themselves.

Earth is distinguished from every other world in the solar system by its abundance of surface water. Around 71 percent of the planet's surface is covered with salty seawater, and the total volume of the oceans is immense (1.34×10^{18} cubic m or 4.73×10^{19} cubic ft). According to one estimate, if all the salts were extracted from ocean water and spread out over Earth's entire surface, they would form a layer more than 150m (500ft) thick. Fresh water, meanwhile, amounts to 3.2 percent of the global total, with ice caps and glaciers comprising about 2 percent and ground water contributing around 1.1 percent. These might seem like small percentages, but given the vastness of the total volume, they comprise significant amounts. The presence of fresh water, and its complex cycle of transfer between atmosphere, rocks and oceans, is also vital to the survival of all terrestrial life.

OPPOSITE Life has only been able to colonize land thanks to the recycling of water from Earth's oceans up into the atmosphere where it forms clouds and is precipitated back to the surface as rain and snow. The complex dynamic of this circulation depends upon many factors from solar radiation to cycles in Earth's orbit and patterns of ocean circulation.

Origin of the oceans

When Earth first formed, its surface temperatures exceeded the melting point of silicate rocks, far above the boiling point of water. As a result, there could have been no surface water until the surface had cooled considerably. The origin of Earth's water has been debated for decades, with plenty of speculation but, until recently, very little hard evidence. However, today geochemical analysis of Earth's oldest rocks, minerals and some of the isotopes they contain is revealing indirect traces of the composition of the early oceans and the processes by which they formed.

One long-standing theory suggested that the majority of Earth's water was introduced during the Late Heavy Bombardment (see page 11), as icy asteroids and comets rained down onto our planet around 3.9 billion years ago. However, individual water molecules can also be trapped within the crystalline structure of common minerals such as olivine and orthopyroxene. As a result, magmas generated by melting these minerals could have released considerable volumes of gas and steam into the atmosphere as they escaped to the surface during Earth's intense early phase of volcanic activity – a process known as outgassing. But it is unclear how much water this process alone would have released, and some scientists claim that it could only have formed oceans hundreds rather than thousands of metres deep. But recent analysis of meteorites made from similar materials to Earth supports the idea that a much larger proportion of Earth's water is derived from its rocky material, while computer modelling shows that early melting and volcanism would quickly create a steamy atmosphere that could have cooled and condensed to form oceans as early as 4.4 billion years ago.

'One long-standing theory suggests that the majority of Earth's water was introduced during the Late Heavy Bombardment, as icy asteroids and comets rained down onto our planet around 3.9 billion years ago.'

The first seas

Geologists can only make educated guesses at the temperature and chemistry of this early standing water. One complication is that the Sun was some 25 percent weaker than it is today – incapable on its own of heating any surface water above melting point. However, volcanic outgassing of carbon dioxide and other greenhouse gases could have warmed the planet sufficiently to allow water to run freely across most of the planet, except perhaps at the poles.

Alongside carbon dioxide, outgassing would also have released compounds such as sulphur dioxide, hydrogen sulphide, hydrogen chloride, nitrogen, nitrous oxide and steam. Several of these chemicals dissolve in water to form acids such as carbonic and sulphuric acid, and these would react with minerals in surface rocks and be neutralized. The earliest rocks of the crust were volcanic in origin, derived from the mantle and composed of silicate minerals that were easily broken down by these acidic waters. As a result, the surface waters would soon have become chemically complex and active, enriched with dissolved salts of elements such as iron, aluminium, magnesium and silicon.

Meanwhile, the actions of running water – its ability to erode rock and transport both dissolved and granular material – would soon have led to the creation of the first sedimentary rock deposits. The oldest of these rocks to be preserved in the geological record are 3.75 billion-year-old iron-rich sediments from northern Quebec. These strata have been altered by heat and pressure throughout their subsequent history, but still retain their original layering. Analysis suggests that the iron within them was probably released from volcanic vents on the ocean floor and transported in solution

LEFT The question of how Earth got all the water that fills its oceans, lakes and rivers, ice caps and atmosphere is still a matter of debate. One possibility is that it came from icy comets that bombarded Earth early in its formation some 4.4 billion years ago.

to relatively shallow waters before being deposited in the sediments called banded iron formations (see page 125). Their existence confirms that oceans predate this time, and there is also slightly older geological evidence for oceans from Greenland, in the form of 3.8 billion-year-old 'pillow lavas' formed when molten rock erupting underwater cooled rapidly into globules and cylindrical shapes.

Water temperature

Previous studies based on the proportions of oxygen isotopes in rock sediments had also suggested that seawater temperatures around 3.5 billion years ago were in the range of 55–85°C (130–185°F). But a 2009 analysis of hydrogen and oxygen in 3.4 billion-year-old chert sedimentary rock from South Africa, carried out by researchers at California's Stanford University, suggests that ocean waters could not have been warmer than 40°C (104°F), and that the ancient ocean was much richer in hydrogen than modern oceans. The implication is that the atmosphere was also richer in hydrogen than previously thought.

So it would seem that oceans had possibly formed as far back as 4.4 billion years ago, and certainly by 3.8 billion years ago. And while they were probably not as hot as previously thought, their composition was radically different from that of today. The persistence of the banded iron formations up until 1.8 billion years ago shows that it took a long time for the chemistry of the oceans to change.

Layered Earth

DEFINITION EARTH'S LAYERED STRUCTURE IS COMPOSED AT ITS MOST
BASIC LEVEL OF A CORE, MANTLE AND CRUST

DISCOVERY THE FIRST SIGNS OF INTERNAL LAYERING WERE
DISCOVERED BY ANDRIJA MOHOROVICIC IN 1909

KEY BREAKTHROUGH CHANGES TO SEISMIC WAVES FROM DISTANT
EARTHQUAKES REVEAL THEIR ROUTE THROUGH THE LAYERS

IMPORTANCE HEAT ESCAPING FROM THE INTERIOR OF OUR PLANET
IS THE FUNDAMENTAL DRIVING FORCE BEHIND MOST OF EARTH'S
SURFACE ACTIVITY

Today, we have a good idea of Earth's internal layered structure – a cool rocky outer crust, a hot central core and a mantle in between. This layering developed soon after Earth's formation and has been maintained ever since as part of the planet's internal dynamic.

Ever since miners first penetrated deep below ground in medieval times, it has been clear that temperature increases with depth. Volcanic eruptions reinforced the view that Earth's interior is hot enough to melt rock, producing molten magma that erupts onto the surface as lava. Historically, a range of other measurements of Earth's physical properties, such as its magnetic and gravitational fields, and its overall mass compared to the density of its surface materials, all contributed to the idea that Earth has some kind of layered structure. But it was the study of seismic waves, propagated across Earth's surface and through its interior (see page 37), that provided the key to reveal Earth's layering in detail.

OPPOSITE Earth's metallic core lies far beyond the reach of any direct sampling, but measures of density, fluidity and magnetism suggest that it has an inner core with a metallic composition of nickel and iron and a crystalline structure with 'Widmanstätten patterns' similar to those found in iron meteorites (shown here).

The central core

At Earth's centre lies the dense hot core, roughly 6,800km (4,220 miles) in diameter and with an outer edge about 2,900km (1,800 miles) beneath the crust. Dominated by heavy metals (principally iron, with some nickel) the core is under intense pressure from the weight of the overlying material, and reaches temperatures of 3,570–5,800°C (6,470–10,470°F) – about as hot as the surface of the Sun. The fact that certain seismic waves do not pass into the core from the mantle suggests that the outer core, from 2,890km (1,800 miles) down to 5,140km (3,190 miles) below the surface, is molten liquid. However, there is further seismic evidence that, despite the increasing temperature, the pressure in the very centre of the core is so great that the nickel-iron mix solidifies, forming an inner core around 2,440km (1,520 miles) across.

Swirling motions in the electrically conductive liquid metal of the outer core are thought to be responsible for generating Earth's powerful magnetic field, which emerges close to the poles, and extends for thousands of kilometres out into space, creating a protective cocoon around the planet that is vital to the survival of life (see page 47). Further evidence for the core's composition comes from the nickel-iron meteorites that occasionally fall from the sky – these are thought to be the fragmented cores of ancient asteroids, formed early in the solar system's history by the same process of melting and differentiation that created Earth's core.

The mantle

The intermediate layer of our planet's structure, the mantle, extends from close to the surface, beginning at depths of 1–30km (0.6–19 miles), to the boundary with the core around 2,900km (1,800 miles) down. The mantle is thought to be fairly uniform in composition – it has five times the volume of the core, but only twice its mass because its rocky material, rich in silicate minerals, is so much less dense than the metallic core.

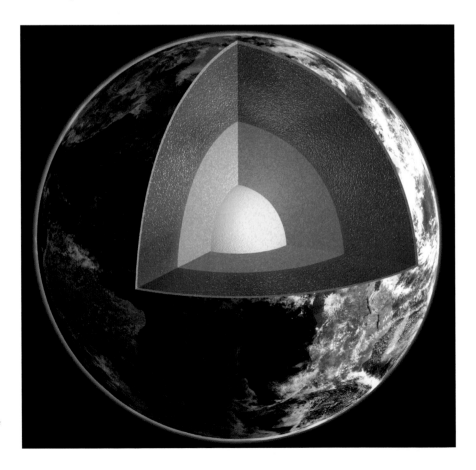

RIGHT After 200 years of investigation, geoscientists can now model the broad outlines of Earth's layered interior with considerable confidence. However, the details are still a matter of ongoing investigation and considerable debate as the scientists seek to understand the processes that transfer heat from the core to the surface and recycle crustal rocks back into the interior.

Across its immense depth, the mantle's temperature more than doubles, climbing from some 1,270°C (2,290°F) near the surface, to around 3,000°C (5,400°F) close to the core. Geophysical evidence, such as changes in the travel times of seismic waves, reveals that its properties change with depth. For instance, the boundary between the base of the crust and the top of the mantle, discovered by Croatian seismologist Andrija Mohorovicic (1857–1936) in 1909, is known as the Mohorovicic discontinuity or Moho. Another boundary, at a depth of 410km (255 miles), separates the upper mantle from a transition zone, while a further boundary at 660km (410 miles) marks the top of the lower mantle. The various boundaries are thought to coincide with changes to the mineralogy and structure of the mantle's silicate material. Changes to rocks such as eclogite and peridotite influence the convective motions that transport hotter materials towards the surface, driving plate tectonics (see page 61).

The crust and beyond

Inevitably, Earth's outermost layer is the most familiar to us, but it is also the most complex and variable in composition, with a long history of differentiation and processing since it first solidified more than 4 billion years ago. The crust is the coolest, most brittle and thinnest layer, less than 1km (0.6 miles) deep in some places but extending to around 30km (19 miles) deep in others. It is fundamentally divided into two types of material – the rocks that form the ocean floors, and those that form the continents.

'Evidence for the core's composition comes from the nickel-iron meteorites that occasionally fall from the sky – these are thought to be the fragmented cores of ancient asteroids, formed early in the solar system's history by the same process of melting and differentiation that created Earth's core.'

The relatively thin ocean crust, from a few hundred metres to about 6km (3.7 miles) deep, is predominantly volcanic or 'igneous' in origin, while the generally thicker continental crust combines a huge variety of igneous, metamorphic and sedimentary rocks made from thousands of different minerals. Further adding to its complexity, the crust, along with the brittle upper mantle (together known as the lithosphere) is broken into seven continent-sized plates and another dozen or more smaller plates. The formation, destruction and rearrangement of these 'lithospheric plates' over billions of years has profoundly affected both the outward appearance of our planet, its geology, and the evolution of its life.

But Earth's layered structure does not stop with its crust – above this lie the closely linked layers of the hydrosphere and atmosphere. The hydrosphere is formed of water, ice and water vapour, mostly lying on top of the crust but also penetrating the surface to some extent and reaching up into the atmosphere as fog, mist and clouds. The atmosphere itself is a layer of gases that extends for more than 200km (125 miles) above the crust. Today, it is predominantly made up of nitrogen and oxygen, with important traces of other gases such as hydrogen and ozone along with greenhouse gases such as carbon dioxide and methane. Without this outermost atmosphere and ocean, life as we know it would not be possible.

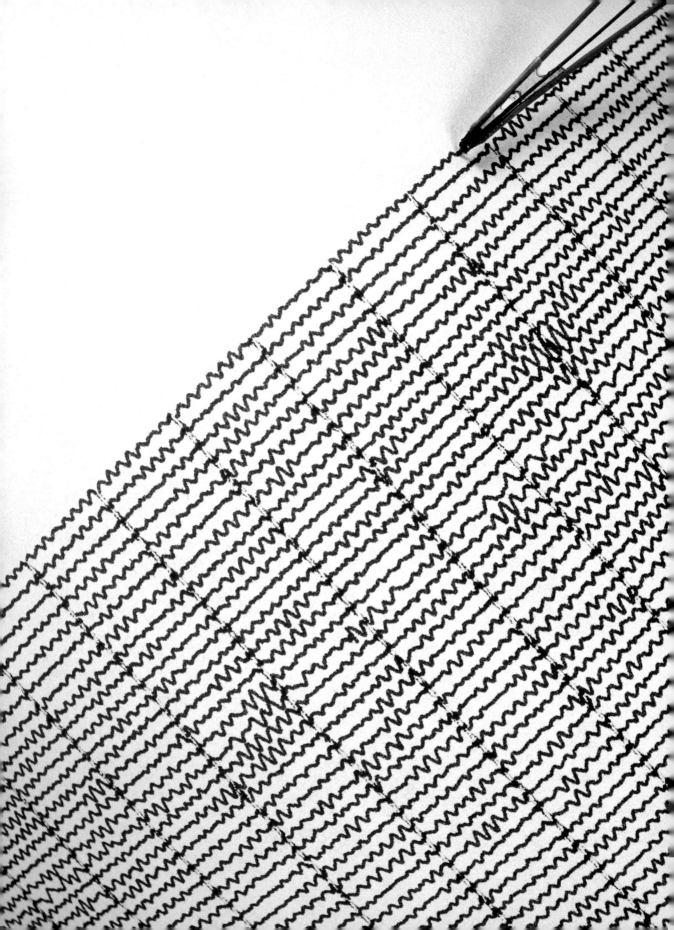

8 Probing Earth's interior

DEFINITION SEISMOGRAPHY ENABLES THE MAPPING OF LAYERS WITHIN THE EARTH, AND CONCENTRATIONS OF ACTIVITY ON ITS SURFACE

DISCOVERY 19TH-CENTURY SCIENTISTS WORKING IN JAPAN INVENTED THE SEISMOMETER TO MEASURE THE SPEED AND DIRECTION OF VIBRATIONS GENERATED BY EARTHQUAKES

KEY BREAKTHROUGH ANDRIJA MOHOROVICIC REALIZED THAT EARTH'S INTERNAL STRUCTURE AFFECTS THE MOVEMENT OF WAVES

IMPORTANCE SEISMOGRAPHY HELPS GEOLOGISTS TO MAP EARTH'S INTERNAL STRUCTURE AND TECTONIC PLATES

Earth's deep interior lies far beyond the reach of any mechanical sampling, but by 'listening in' to earthquake shock waves passing through the inner layers, geologists can discover a surprising amount about our planet's internal properties and structure.

If we were restricted to studying the structure, dynamics and history of the Earth purely through the appearance of its surface rocks, our understanding of the planet would be severely limited. Fortunately, however, there are other physical phenomena that can inform us about what is going on deep within Earth. Some of these – such as earthquake vibrations, gravity, heat flow, electricity, radioactivity and magnetism – have an obvious influence on our everyday lives, while others, such as electromagnetism and fluid dynamics, are less obvious. The study of the way these phenomena affect Earth is known as geophysics, and it has been responsible for revolutionizing our understanding of our world, and other planetary bodies. When it comes to investigating Earth's deep structure, seismography – the study of earthquake wave propagation – is perhaps our single most powerful tool.

Shaking the Earth

People have been aware of earthquake waves' ability to travel through the Earth from origins deep underground for hundreds of years, but a proper scientific understanding of the phenomenon had to await technological developments at the end of the 19th century. One of the first seismograph instruments was invented in 1880 by John Milne (1849–1913), James Ewing (1855–1935) and Thomas Gray (1850–1908), three British scientists working in Japan. It allowed the detection and identification of earthquake waves, as well as measurement of their velocity. Milne went on to establish

OPPOSITE Shock waves generated by the release of pent-up energy from the movement of Earth's interior are transmitted through the rocks that make up Earth's solid crust and semi-solid mantle. Seismographs detect and record these vibrations, helping to pinpoint their sources and causes.

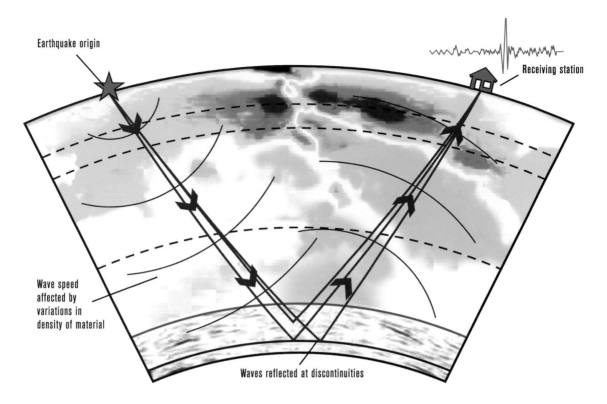

Earthquake origin

Receiving station

Wave speed affected by variations in density of material

Waves reflected at discontinuities

ABOVE Recent advances in the computerized processing of seismic data has allowed the deep structure of Earth's mantle to be imaged using a process analogous to ultrasound. Known as seismic tomography, it builds up an image of density differences, which can then be combined to produce 3-D models of deep structures such as subducted slabs of crustal material.

the first worldwide network of earthquake observatories at the end of the century, which recorded the Great Assam earthquake of 1897 and allowed Richard Dixon Oldham (1858–1936) to further distinguish the different types of pressure waves produced by earthquakes.

In essence, earthquakes produce two main types of seismic wave – surface and body waves. As their name implies, surface waves are like waves on water – they travel along Earth's surface with relatively long duration, low frequency and large amplitude, and as a result are the most destructive type of wave. In contrast, body waves travel through Earth's interior, following paths that depend on the relative density and stiffness of the rocks (which in turn are governed by factors such as composition, temperature and mineral structure). Body waves can be subdivided into P (primary) waves and S (secondary) waves. P-waves are compressional (similar to sound waves or a pulse sent down a stretched spring) and can travel rapidly through any material, liquid or solid. S-waves, meanwhile, are transverse or shear waves (similar to water waves, they involve movements of material perpendicular to the movement of the wave itself) that only travel through solids. S-waves are significantly slower than P-waves and consequently arrive later at seismic stations. It was Oldham's 1906 analysis of the time taken for these different waves to pass through Earth's interior that first indicated our planet had a liquid core. In 1936, Danish seismologist Inge Lehmann (1888–1993) made a closer study of the arrival times of P-waves to show that there was a solid interior at the centre of the liquid outer core.

Meanwhile, in 1909, Croatian seismologist Andrija Mohorovicic saw that seismograms of shallow earthquakes recorded *two* sets of P- and S-waves. One set followed a shallow path near Earth's surface, while the others followed a deeper path with greater speeds, showing that they were refracted or bent by interaction with a higher-velocity medium. The seismic change from the 'slow' surface material to the deeper high-velocity material marks the boundary between the crust and mantle, and is now known as the Mohorovicic discontinuity or 'Moho'. It typically lies 5–10km (3–6 miles) below the ocean floor, and 20–90km (12.5–55 miles) below the continents.

Lessons from the waves

Understanding how Earth's materials behave in the very different conditions of the surface and the deep interior has allowed us to begin to appreciate the way our planet works. Earth's surface 'lithospheric' rocks are cool and brittle, with a tendency to break catastrophically (fault) under pressure. In contrast, the hotter (though still essentially solid) rocks of the asthenosphere are mechanically weaker and can deform or 'creep' over geological timescales. The hotter the rocks, the more easily they creep, especially as they approach their melting point. Consequently, as temperature increases with depth, the behaviour of rocks changes from brittle fracture to ductile flow, depending upon their composition. Below around 100 km (62 miles), however, most rocks have lost their strength and are so weak that they will all creep.

'When it comes to investigating Earth's deep structure, seismography – the study of earthquake wave propagation – is perhaps our single most powerful tool.'

By the 1960s, seismic studies had revealed that Earth's major earthquake zones lay along the boundaries between lithospheric plates, and that deep earthquakes mostly have their origins along the boundaries between oceanic and continental plates, almost always above 650km (400 miles) in depth. However, it seemed clear that classic techniques of seismic wave analysis were not going to provide any more details of the mantle.

Then, in the 1970s, Polish-American geophysicist Adam Dziewonski and colleagues at Harvard University digitized seismograms from the great Alaskan earthquake of 1964 and used computer analysis to discover many previously unidentified 'modes' of free vibration. By comparing the travel times of the waves with those predicted by the simple 'layered Earth' model, they were able to generate computerized images of Earth's interior, showing variations in the speed of seismic waves passing through different areas. This technique, now known as seismic tomography, produces a 'sliced' image analogous to an ultrasound picture of a baby in the womb. Multiple slices can be combined into three-dimensional models, revealing density differences that extend deep into the lower mantle. Today, seismic tomography is offering geologists unprecedented insights into the deep structure of the Earth, and revealing hitherto unexpected features.

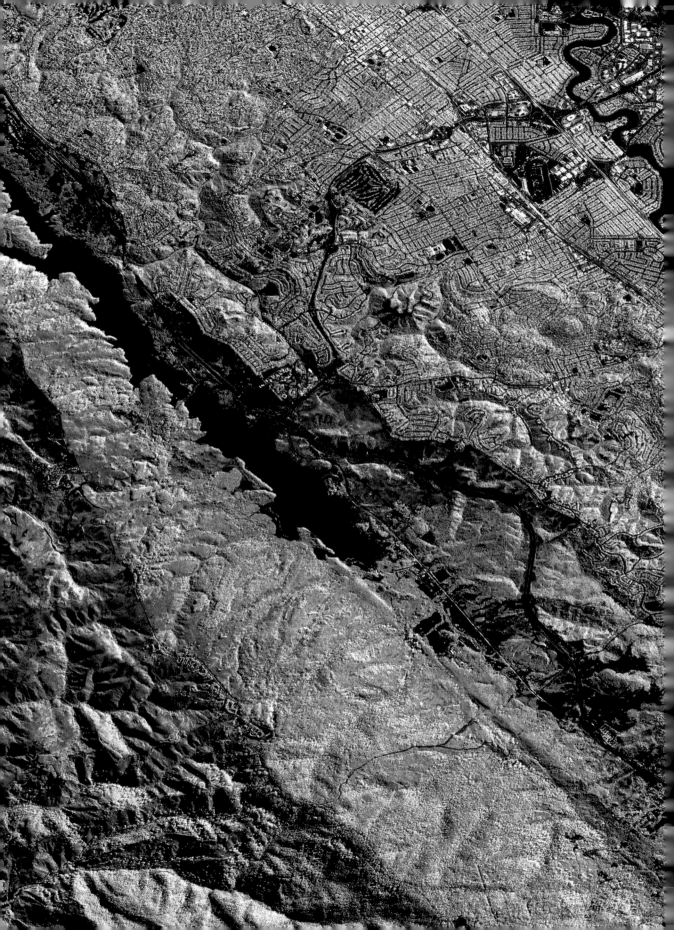

Tectonic Earth

9

DEFINITION PLATE TECTONICS IS THE THEORY THAT DESCRIBES LONG-TERM MOTION OF CONTINENTS ACROSS EARTH'S SURFACE AND ALSO EXPLAINS MANY LARGE-SCALE GEOLOGICAL PHENOMENA

DISCOVERY THE FIRST DETAILED MODEL OF 'CONTINENTAL DRIFT' WAS PROPOSED BY ALFRED WEGENER IN 1915

KEY BREAKTHROUGH DISCOVERY OF MID-OCEAN RIDGES AND SEA-FLOOR SPREADING IN THE 1950s CONFIRMED TECTONIC THEORY

IMPORTANCE TECTONICS HAVE BEEN A DRIVING FORCE IN THE STORY OF THE EARTH, AFFECTING GEOGRAPHY, CLIMATE AND LIFE ITSELF

The idea that continents can move across Earth's surface might seem improbable, but today there is incontrovertible evidence not only that this has happened throughout geological history, but also that it is the driving force behind the creation of many of Earth's unique geological features.

As early as the 16th century, geographers noted the remarkably well-matched shapes of widely separated coastlines such as those of West Africa and eastern South America. At the time, this was little more than a curiosity, but in the 19th century some scientists began to see it as evidence for a controversial theory of 'continental drift' – the idea that landmasses moved around on the surface of the Earth on a timescale of tens of millions of years.

The best-known proponent of the theory was German meteorologist Alfred Wegener (1880–1930). As well as the apparent matching-up of continents, he marshalled evidence including the presence of similar fossils and geological features on widely separated landmasses. However his hypothesis, published in 1915, foundered on the question of what force could power such motions. As a result, continental drift was widely dismissed for several decades, until British geologist Arthur Holmes (1890–1965) suggested that convection of hot rock in the mantle could provide the missing driving force. In the 1950s, the discovery of sea-floor spreading (see page 61) finally clinched the case.

Since the time of its formation, Earth's outer layers have cooled sufficiently to form a brittle and rocky outer layer, the lithosphere, with a hotter and more 'plastic' asthenosphere below. The asthenosphere is capable of moving at slow rates, transferring heat from Earth's deep interior to its surface. Between 6km (3.7 miles) and 180km (110 miles) thick, the lithospheric

OPPOSITE A false-colour satellite radar image clearly shows the line of California's San Andreas Fault running through the Crystal Springs Reservoir (black) to the west of San Mateo (upper right). San Andreas is a complex and active fault zone that forms the boundary between the Pacific and North American plates.

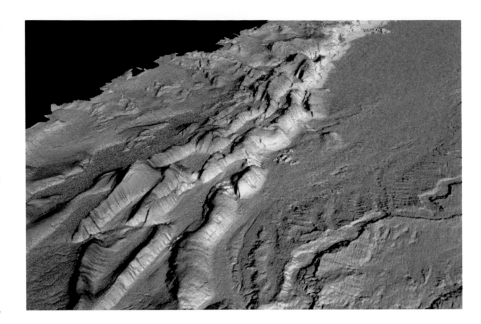

RIGHT New techniques for high-resolution imaging of seabed topography are revealing astonishing new details of the effects of major geological processes. Here, a false-colour image from satellite measurements shows a series of landward-tilted fault blocks (orange) lying in deep water off the Pacific coast of Oregon (the black area at top left), where ocean crust is being subducted below the North American Plate. The ocean floor (green) also shows a submarine channel cut into it by turbidity currents.

crust responds to the motion of the interior by breaking up into a handful of continent-sized tectonic plates and a number of smaller fragments. The boundaries between plates are marked by frequent earthquakes, faulting and volcanic activity.

There are some important variations between different plates – most significantly between plates composed entirely of oceanic crust and those that combine oceanic and continental crust. At present, most tectonic plates contain a mix of crust types, with oceanic crust forming a margin around the edges of continental landmasses. A few plates, such as the Pacific, Nazca and Philippine Sea plates, are composed of ocean crust alone.

Plates on the move

Heat flow within Earth's interior causes the plates to jostle one another. Where heat rises, plates are stretched out, thinned and ultimately fractured, allowing hot molten rock to well up to Earth's surface in the form of magma and volcanic lava. This plate-spreading and rifting process creates new crust (typically along mid-ocean volcanic ridges) but, since the planet as a whole is not expanding, the creation of new crust requires old crust to be destroyed. For example, the floor of the Indian Ocean midway between India and Africa is spreading apart at a rate of 10–20mm (0.4–0.8in) per year, and as a result, the Indian Plate is being driven northwards into southern Asia. Unsurprisingly, the plate boundary between the two colliding landmasses is a massive and continuing collision zone, giving rise to the Himalayas, currently Earth's largest mountain range (see page 241).

A similar spreading ridge exists in the Atlantic Ocean, where the South American and African plates are moving apart. The Atlantic margins of both continents are known as passive margins, since continental and oceanic crusts fuse firmly together. As a result, the opening of the Atlantic pushes South America towards the Pacific at a speed of some 25mm (1in) per year.

The Eastern Pacific, meanwhile, has its own spreading ridge, pushing the Nazca Plate east into South America at a rate of 37mm (1.5in) per year. Because this collision involves two different types of crust, the result is a 'dislocated boundary' in which the deep but comparatively light continental crust of South America overrides the thin but dense oceanic crust of the Nazca Plate. As a result, the continental crust of South America has been thickened and deformed by the creation of the Andean mountain belt, so that it is now some 70km (45 miles) thick in places.

The thin but dense ocean floor of the Nazca Plate, meanwhile, has been pushed down and overridden by the oncoming might of the South American continent. The boundary is marked by the Peru-Chile Trench, more than 8km (5 miles) below sea level, where the descending Nazca slab slopes down at an angle of around 30 degrees, beginning its descent along a 'subduction zone' that plunges some 100km (62 miles) into the mantle beneath the Andes.

Subduction is one of the most important geological processes, with effects that have a huge impact on regional environments and life. Friction generated in a subduction zone can produce frequent large earthquakes that are sometimes catastrophic (see page 89). The origins of these quakes grow deeper as the subducted plate plunges beneath the thickening margin of the continental crust, and geologists can plot the structure of the subduction zone by mapping out the locations of these deep disturbances.

'Continental drift was widely dismissed for several decades, until British geologist Arthur Holmes suggested that convection of hot rock in the mantle could provide the missing driving force.'

Another important effect of subduction is to trigger volcanic activity – as the descending slab encounters increasing temperatures and pressures, water is released, lowering the melting temperature and partially melting minerals in the overlying continental crust. The rising magma changes in composition as it ascends through surrounding rocks, increasing its silicate content, becoming more viscous and trapping gases within it under pressure, so that when it finally escapes to the surface, the result is a violent and explosive volcanic eruption (contrasting with the more measured eruptions of oceanic-crust volcanoes – see page 25).

The present configuration of interlocking plates on the surface of the Earth is radically different from that of the deep past, but geologists can reconstruct the history of plate motion over the last 600 million years with some confidence. Further back in time, the distribution and motion of the plates becomes increasingly speculative (see page 133), but this only serves to make it a more exciting area for research.

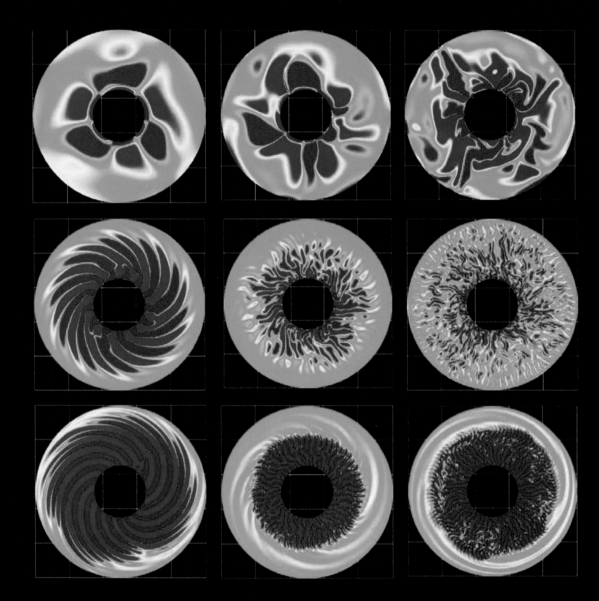

10 The molten core

DEFINITION THE CENTRAL REGION OF THE EARTH, CONSISTING OF SOLID AND MOLTEN LAYERS DOMINATED BY IRON AND NICKEL

DISCOVERY IDENTIFIED THROUGH THE STUDY OF SEISMIC WAVES BY RICHARD DIXON OLDHAM IN 1909

KEY BREAKTHROUGH RECENT STUDIES HAVE SHOWN THAT THE STRUCTURE OF THE CORE IS FAR MORE COMPLEX THAN EXPECTED

IMPORTANCE HEAT FROM THE CORE DRIVES TECTONIC PROCESSES ON EARTH'S SURFACE, WHILE THE MAGNETIC FIELD IT GENERATES PROTECTS LIFE ON EARTH

The most inaccessible region of Earth is the core, whose outer boundary lies some 2,890km (1,800 miles) beneath the surface and has an estimated temperature of more than 3,500°C (6,870°F). However, seismic studies have revealed that it has a layered structure of outer liquid, and inner solid cores.

Our basic model for the structure of Earth's core has remained essentially the same since the 1930s, when the existence of the inner core was identified by Danish seismologist Inge Lehmann. The inner and outer core have generally been treated as more or less uniform masses of solid and molten metal, with flows of electrically conductive material in the liquid outer core creating massive, shifting electric currents that act as a 'geodynamo' to generate Earth's magnetic field. However, in the past few years new investigations have upset this relatively simple model, revealing some surprising and puzzling features of Earth's deep interior.

Unexpected features

Most importantly, new studies using seismic waves (see page 37), have detected a dense layer of stratified fluids at the boundary between the outer and inner cores, some 200–250km (125–155 miles) thick. Perhaps more puzzling, though, is the recent discovery that seismic waves passing through the inner core travel faster from north to south than they do from east to west. This effect, called elastic anisotropy, may be caused by some kind of alignment of iron crystals within the inner core. On top of this, the inner core appears to have some strange asymmetrical features that vary with longitude. Far from being homogenous structures, it is starting to look as if Earth's deep inner layers could be almost as varied as its surface.

OPPOSITE Flows within the molten metal of Earth's rotating outer core generate Earth's magnetic field. Computerized simulation reveals changing patterns of flow, with increasing vorticity and angular momentum producing sheet-like inner radial plumes and an outer, cylindrical, zonal flow (lower right).

ABOVE Polar lights or
aurorae occur some
80km (50 miles) up
in Earth's ionosphere.
This spectacular
display of northern
lights (aurora borealis)
seen above Iceland is
produced by photons
generated from the
collision of ionized
nitrogen and oxygen
atoms with solar wind
particles swept up in
Earth's magnetic field.

Birth of the core

According to our best understanding of Earth's origins, the core formed
around a billion years after the planet's initial creation. Around 3.5 billion
years ago, the long, slow process of cooling had progressed sufficiently for
dense, iron-rich materials, which had already descended to the centre of the
planet during the early stages of differentiation (see page 10), to begin to
solidify. Despite the higher temperatures, the process of solidification began
at the centre and spread slowly outwards – an apparent paradox resulting
from the high pressures of the centre increasing the melting point of iron.

As the inner core solidified and grew increasingly dense, lighter elements
were driven outwards into the upper core. This 'compositional convection'
process (the driving out of lighter elements by denser ones) has long
been thought of as the driving force that keeps the outer core in motion
and generates Earth's magnetic field. Most geologists predicted that the
intermediate fluid layer between the inner and outer cores would have a
higher concentration of light elements, and therefore a lower density, than
the surrounding outer core. But surprisingly, the seismic evidence suggests
exactly the opposite – and what's more, the very presence of such a stable

layer in the first place seems to argue against the idea that compositional convection plays a role in establishing the geodynamo and magnetic field.

Core dynamics

However, a new theory suggests that the problematic region of reduced seismic velocity at the base of the outer core might not be all that it seems. A team of French geoscientists from the Grenoble Observatory have shown that, far from being a stable layer of molten material in which heavier elements are concentrated, the same effect could be produced by the simultaneous crystallization and remelting of material on the surface of the inner core. The theory is that, driven by convection, the inner core is effectively 'migrating' eastwards. As it cools slightly and solidifies by crystallization in the west, it becomes hotter and melts in the east, releasing plumes of lighter elements that escape into the outer core.

'Flows of electrically conductive material in the liquid outer core create massive, shifting electric currents that act as a "geodynamo" to generate Earth's magnetic field.'

The result is an iron-enriched layer some 200–250km (125–155 miles) thick, which is denser than the outer core and can remain stable even when penetrated by the convective plumes of lighter elements. The French theory also has the benefit of explaining the apparent features detected on the surface of the solid inner core. At the current rate of migration of about 1.5cm (0.6 in) per year, it would take about 100 million years for the inner core to be completely renewed by the solidification and melting process, which is significantly faster than its rate of growth and expansion. This inner core dynamic is also thought to have a considerable impact upon the motion of the outer core and Earth's 'geodynamo'.

The complex geodynamo

Other studies are also adding layers of complexity to our hitherto simple model of the generator behind Earth's magnetic field, and in particular the way in which the molten electrified fluid churns and flows in the outer core. Previous computer simulations have successfully reproduced many of the features of Earth's magnetism field (including the periodic reversals that seem to occur every few million years) based on a model in which convection takes place in column-like cells parallel to the axis of Earth's rotation.

However, the latest models have suggested the presence of stable regions, called zonal flows, within the generally turbulent liquid iron – a series of 'jet streams' that encircle the core, rather similar to those found in Earth's atmosphere and oceans. Similar streams can form spontaneously in other turbulent fluids, such as the liquid plasmas of nuclear fusion devices, or the atmospheres of gas giant planets. If the models are proved correct, then it seems that the outer core must have a 'dual-convection' system, with inner, sheet-like radial plumes passing outwards into a westward-trending jet stream.

11   'D double-prime'

DEFINITION A HIGHLY MOBILE LAYER OF EARTH'S INTERIOR AT THE BOUNDARY BETWEEN THE LOWER MANTLE AND OUTER CORE

DISCOVERY FIRST IDENTIFIED IN SEISMIC SIGNALS BY GEOPHYSICIST KEITH BULLEN IN 1959

KEY BREAKTHROUGH ANALYSIS OF MANTLE-LIKE PEROVSKITE MINERALS IN 2004 REVEALED THE LAYER IS PROBABLY FORMED BY A CHANGE IN MINERAL STRUCTURE AT HIGH TEMPERATURES AND PRESSURES

IMPORTANCE THE LAYER PLAYS A VITAL ROLE IN TRANSFERRING HEAT FROM THE CORE TO THE CHURNING ROCKS OF THE MANTLE

Heat escaping from Earth's metallic core keeps our planet alive and active, but the question of how that heat is transferred from the core through almost 3,000km (1,860 miles) of mantle to the overlying crust has puzzled scientists for decades.

There is now plenty of evidence to show that the main process of heat transfer through much of the upper mantle is convection. Plumes of hot rock rise in some places while cooler material sinks downwards in others. Close to the core itself, it seems that heat is transferred outwards by conduction, a process in which the hot material itself does not move, but heat is nevertheless transferred to cooler areas through the vigorous vibration of its atoms and molecules. The most problematic area for our understanding of the Earth's internal heat transfer, however, has always been the boundary between the rocky lower mantle and the fluid metallic outer core, some 2,900km (1,800 miles) below the surface. There is plenty of evidence to show a sudden and remarkable change in density, chemistry and temperature across this boundary, with the iron-rich core more than 1,000°C (1,800°F) hotter than the mantle.

The marked temperature change from core to lower mantle creates a unique boundary environment within the lowest 200–400km (125–250 miles) of the mantle, with its own mineralogical and seismic properties. Little is known about this enigmatic region, first identified by New Zealand geophysicist Keith Bullen (1906–76) in 1950 and designated as D" (pronounced 'D double-prime') in his alphabetical scheme for naming Earth's layers. Seismic signals revealing its presence (see page 37) are hard to detect, but when they are picked up by surface stations, they reveal abrupt changes to the speeds at

OPPOSITE Although it is rarely found at Earth's surface, perovskite (magnesium silicate) is the most common mineral in the mantle and therefore in our entire planet. Laboratory studies show that changes to its structure and properties produce distinctive layers in the lower mantle.

which seismic waves propagate, suggesting a 30km (19-mile) 'discontinuity' zone (known as the Gutenberg discontinuity) at the top of the D" layer, and a low-density layer at the base. Curiously, the signals also point to some kind of orientation or directionality in the mineral content.

However, much of our knowledge of this transitional region comes not from seismology, but from high-tech laboratory experiments studying the behaviour of the minerals that are thought to exist there under extreme conditions.

Earth's commonest mineral

Although most people have never heard of it, the magnesium silicate mineral perovskite ($MgSiO_3$) is the most common mineral on Earth, simply because it comprises so much of the planet's interior – an estimated 70–80 percent of the lower mantle between 670 and 2,900km (420 and 1,800 miles) down. Since there is no reason to believe the mantle's mineral composition suddenly changes at D", the abrupt changes found there are likely to be linked to physical changes in this mineral's structure or behaviour. No samples of perovskite have ever been recovered from such great depths, and even if they could be, it's likely that they would be altered by the drop in temperature and pressure on their journey to the surface, so laboratory studies of perovskite seem the obvious path to take. However, generating both the high temperatures and high pressures required to replicate the D" layer presents an enormous technical challenge.

'Although most people have never heard of it, the magnesium silicate mineral perovskite ($MgSiO_3$) is the most common mineral on Earth, simply because it comprises so much of the planet's interior.'

Since the late 1950s, pressures higher than one megabar (around a million times the normal pressure exerted by Earth's atmosphere) have been routinely generated using a device known as a diamond anvil, in which two small, specially cut diamonds with flattened bottom facets, known as culets are forced together to create an enormous pressure in the very small space between them. Studying the way that a small sample of material affects light or X-rays fired through the diamonds and the trapped material can reveal details of its physical and chemical behaviour under pressure.

In order to replicate conditions in the lower mantle, however, pressure is not enough – it's also necessary to heat the experimental sample to a temperature of more than 1,700°C (3,100°F). Normally this kind of temperature can be produced in laboratories by heating with a laser beam, but diamond is such a good conductor that it rapidly dissipates the heat generated by the laser. As a result, the microscopically small sample of perovskite also has to be surrounded by insulating materials.

All this effort has yielded success, and some intriguing results. In 2004, samples produced in a diamond anvil at the Tokyo Institute of Technology

were analysed by a University of Minnesota team led by Jun and Taku Tsuchiya. Their studies revealed that perovskite does indeed change its structure under lower mantle-like conditions, creating a new mineral called post-perovskite. The experiments also proved that at higher pressures (equivalent to greater depths) higher temperatures were needed before the transformation took place. It seems that perovskite and post-perovskite probably co-exist throughout a zone 400–600km (250–375 miles) deep – far greater than expected from the seismic data.

The D" enigma

The discrepancy between the two estimates may be explained by changes in the amount of other mantle materials, such as an increase in the magnesium-iron oxide mineral ferropericlase, which is the next most common of Earth's minerals after perovskite. Rare at the surface, ferropericlase has occasionally been found as tiny 'inclusions' within natural diamonds that may have come from the lowermost mantle. Another possibility is that lightweight surface minerals such as alumina (Al_2O_3) and silica (SiO_2) have been introduced to this region through the subduction of oceanic crust right down to the base of the mantle (see page 55). Any of these minerals could help to 'thicken' the mantle material and disguise the effects of increased post-perovskite. If so then the D" layer may be even thicker than predicted by the experimental results.

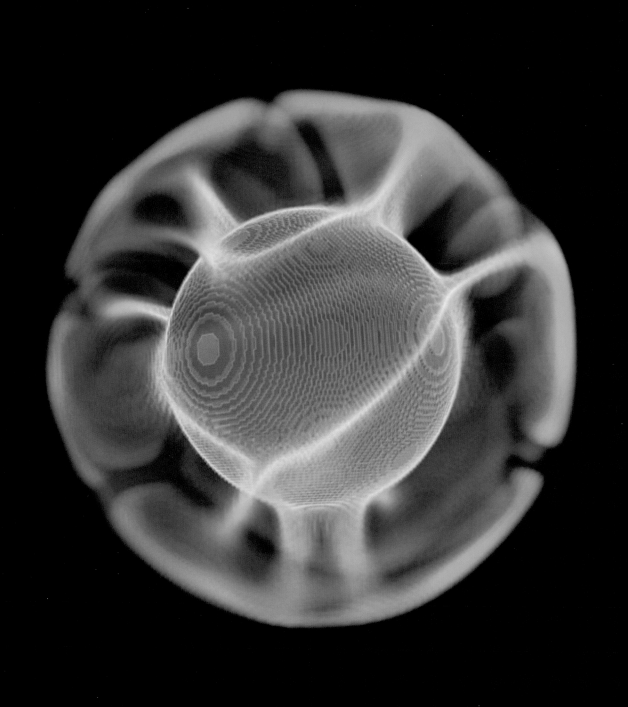

Exploring the mantle

12

DEFINITION THE MANTLE IS THE DEEP LAYER OF ROCK INTERMEDIATE BETWEEN EARTH'S CORE AND CRUST

DISCOVERY FIRST IDENTIFIED AS A UNIQUE LAYER BY ANDRIJA MOHOROVICIC FROM SEISMIC WAVE STUDIES

KEY BREAKTHROUGH SEISMIC STUDIES HAVE REVEALED THAT THE MANTLE IS A REMARKABLY COMPLEX REGION OF UPWELLING HOT PLUMES AND DESCENDING SLABS OF CRUST MATERIAL

IMPORTANCE MANTLE FEATURES INFLUENCE THE DRIFT OF TECTONIC PLATES AND THE FORMATION OF VOLCANOES ON EARTH'S SURFACE

Understanding the structure of Earth's mantle and the processes that shape it relies on the study of a few rare samples found on the surface, seismic data from earthquake shock waves and new techniques of imaging, modelling and experimentation.

The vast bulk of planet Earth is made up from the silicate minerals of the mantle, accounting for some 82 percent of its volume and 65 percent of its overall mass (the discrepancy between these two values is largely due to the much greater density of the metal-rich core). The initial differentiation of the planet concentrated so-called 'siderophile' elements such as nickel, which are soluble in molten iron, at the Earth's centre, leaving the remaining 'lithophile' elements (such as aluminium, potassium, sodium, calcium and magnesium) in the surrounding silicate mantle.

As a result, the mantle is depleted in siderophile elements – particularly iron itself, nickel, sulphur, platinum, gold and lead. The separation of core and mantle can be dated using radiometric measurements to around 4.5 billion years ago – just 100–150 million years after the formation of the Earth itself.

Sampling the mantle

Penetrating beneath Earth's crust to reach the mantle has been a constant challenge to geologists ever since the distinction between the major layers was discovered in the early 20th century. Perhaps the most famous attempt to drill down to the mantle was 'Project Mohole' (named after the Mohorovicic discontinuity it hoped to reach – see page 35), which involved drilling into the ocean floor off the Pacific island of Guadalupe. Although the deepest

OPPOSITE Heat flow through Earth's mantle can be concentrated in rising plumes that convect heat from the core. Computer modelling shows how this can happen on a global scale – hot plumes (light orange) rise from the core (dark orange) through the mantle and spread out when they reach the upper mantle before sinking back (translucent orange).

hole drilled penetrated only to 183m (601ft) below the sea floor, this was still a major achievement in ocean depths of 3,500m (11,500ft). Drilling into the thin oceanic crust presents major challenges but is, on balance, still easier than attempting to drill through tens of kilometres of continental crust – in 1992 a Russian team drilling in the Kola Peninsula reached a depth of 12km (7.5 miles), but like all other attempts this fell far short of their mantle goal.

Fortunately, though, samples of mantle material are occasionally brought to the surface by geological processes. Rare fragments of mantle rocks and minerals, known as xenoliths (literally 'foreign stones'), have been discovered within rocks produced by explosive volcanic eruptions such as kimberlite pipes (see page 66), whose lava channels tap down to the underside of the crust at depths of up to 150km (93 miles). More recently, geologists have realized that tectonic processes such as faulting and uplift (see pages 93 and 81) can bring much larger, kilometre-scale outcrops of mantle material to the surface from great depths. In 2007, a team of British scientists from Cardiff University investigated one such outcrop – an enormous 'gap' in Earth's crust covering thousands of square kilometres of Atlantic sea floor between the Cape Verde Islands and the Caribbean.

Analysis of samples from these various sources suggests that the upper mantle at least is dominated by a silicate rock called peridotite. However, other geochemical evidence suggests this composition may not be typical of the mantle as a whole.

Seismic sources

When seismic data about Earth's inner structure first became available, geologists understandably interpreted it in the light of what they knew about the composition of meteorites – largely unaltered samples of the raw material from which Earth was created. This led to a model in which the overall composition of the interior changed gradually with increasing depth, from silicate-rich to iron-rich. The layered nature of our planet's interior only became clear with the recognition of the distinct boundary between the base of the mantle and the core, where seismic waves suddenly slow down due to the increased density of material.

'In 2007, a team of British scientists from Cardiff University investigated one such outcrop – an enormous "gap" in Earth's crust, covering thousands of square kilometres of Atlantic sea floor between the Cape Verde Islands and the Caribbean.'

Since then, geologists have come to recognize two further seismic discontinuities within the mantle, at depths of 410km (255 miles) and 660km (410 miles). At first, these were also assumed to mark distinct changes in composition from the peridotite-rich upper layers, but newer imaging techniques have led to a revision of this idea. Today, geologists believe that the mantle is dominated by peridotite minerals to great depths, and that phase transformations *within* these minerals (similar to the melting and freezing of water and caused by increasing pressure and temperature with depth) are responsible for the discontinuities. Below 660km (410 miles) there is a steady increase in the speed at which seismic waves travel through the mantle, which does not stop until the D″ layer (see page 49).

At first, geologists were puzzled by the discovery that the mantle's structure was so complex, but by the 1980s, they realized that the subduction of tectonic plates could be playing a key role in the process. The new technique of seismic tomographic imaging (see page 39) has since confirmed this, producing remarkable images of sections through the mantle that confirm the presence of cold slabs of descending crust material. The relatively high density of these slabs compared to the surrounding hot mantle causes them to sink downwards at rates of around 30mm (1.2in) per year, with some slabs descending all the way to the boundary with the core. Over a long period, material in these plates is assimilated back into the mantle, and may eventually be recycled to produce new oceanic crust.

The base of the crust

DEFINITION A BOUNDARY ZONE IN WHICH UPWELLING MOVEMENTS IN THE MANTLE TRANSLATE INTO SIDEWAYS MOTION OF THE CRUST

DISCOVERY THE SUBDIVISION OF EARTH'S OUTER LAYERS INTO LITHOSPHERE AND ASTHENOSPHERE WAS DEVELOPED AS AN INTEGRAL PART OF PLATE TECTONIC THEORY IN THE 1960s

KEY BREAKTHROUGH RECENT STUDIES SHOW THAT WATER RELEASED BY MINERALS AT GREAT DEPTHS LUBRICATES THE ASTHENOSPHERE

IMPORTANCE HEAT-DRIVEN CREEP IN THE ASTHENOSPHERE DRIVES THE MOVEMENT OF EARTH'S UPPER SOLID CRUST

While upward convection in the mantle provides the driving force behind plate tectonics, its outward manifestation takes the form of lateral or sideways motions in the cool, brittle crust. So how do the materials of the mantle and crust interact with each other?

Most people are familiar with the geological division of Earth's interior into the crust, mantle and core (each with various subdivisions and discontinuities) according to their chemical composition. But referring to tectonic processes, geologists also frequently use the terms 'lithosphere' and 'asthenosphere'. Unfortunately, despite first impressions, these terms are not interchangeable with the more familiar crust and mantle – instead, they define mechanical properties of Earth's interior.

In essence, the lithosphere comprises Earth's rocky outer crust together with a region of the uppermost mantle that is relatively cool and brittle. It has slightly different properties depending on whether its upper element is oceanic or continental crust, and generally extends down to depths of 100–200km (62–125 miles) depending on the thickness of the crust involved.

The asthenosphere, meanwhile, is a hotter and far more mobile region that lies directly beneath the lithosphere. Comprising the rest of the upper mantle, it reaches down to roughly 660km (410 miles), where a distinct discontinuity (revealed in seismographic surveys separates it from the lower mantle. The nature of the region where the solid plates of the lithosphere meet the churning rocks of the asthenosphere has been a long-standing source of fascination for geologists.

OPPOSITE This colourful image shows a thin section of eclogite rock viewed through a high-powered microscope in polarized light. Eclogite is formed under intense pressures and temperatures at depths of more than 45km (28 miles). Thin transparent rock sections help reveal the various minerals present. Here, garnet, one of the main minerals in eclogite, is coloured black, while other coloured grains are mostly a sodium-rich pyroxene mineral.

Mineral composition

If scientists are to understand the 'decoupling' between lithosphere and asthenosphere, it's important to have an idea of their overall mineral compositions. Unfortunately, our knowledge of mantle materials is limited to magmas and 'xenoliths' brought up from the depths within certain types of volcanic eruption, and a few exposed regions of mantle-type rocks (see page 54).

From these rare samples, it seems that the crust and upper mantle are most intimately connected beneath the oceanic plates, where basaltic crustal rocks bond to an upper mantle dominated by a coarse-grained rock known as peridotite, rich in the minerals olivine and pyroxene, but relatively poor in the common surface mineral silica (SiO_2). This rock is sometimes referred to as 'lherzolitic', on account of its similarity to lherzolite, a rare rock found at the Etang de Lers in the French Pyrenees.

The importance of water

Detailed analysis of these rare mantle rocks also shows that they contain small amounts of important volatiles such as water and carbon dioxide. Water in particular is thought to play a critical role in determining the strength of mantle rocks, either by weakening the crystal structure of olivine (one of the dominant minerals within these rocks), or by lowering their melting point and thus promoting a partial melting of the mantle rocks, which in turn lowers their strength.

Water seems to be introduced to the mantle via plate tectonic processes, mostly along the mid-ocean ridges where oceanic plates are spreading apart and new ocean-floor rocks are being formed. But the diverging plates also release pressure on the underlying hot mantle rocks, which in turn leads to partial melting through decompression and the eruption of magma onto the ocean floor through long fissures and numerous small volcanic vents. This fissuring of mid-ocean ridges allows ocean-floor water (which is under considerable pressure from kilometres of overlying water) to penetrate into the magma and its mineral components.

'Basaltic crustal rocks bond to an upper mantle dominated by a coarse-grained rock known as peridotite, rich in the minerals olivine and pyroxene, but relatively poor in the common surface mineral silica (SiO_2).'

When ocean-floor lavas and sediments are recycled back into Earth's interior through subduction, they can also carry trace amounts of water into the mantle. At great depths, the pressures become high enough to force this water out of the subducting slab, and the released water fuels a partial melting of the surrounding mantle rocks, creating magmas with a significant water content.

Experimental studies

Geologists believe that the melting conditions of lherzolite on Earth's surface are similar to those of lherzolitic rocks in the upper mantle, and

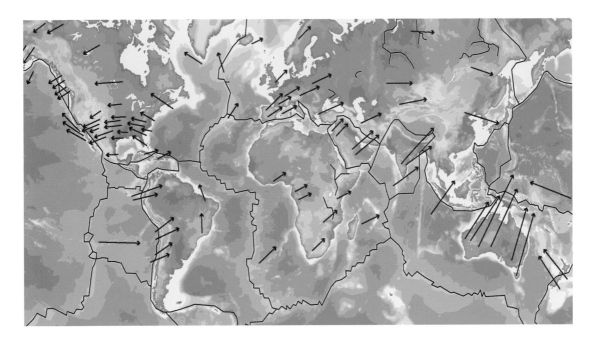

ABOVE Earth's outer rocky crust is broken into eight continent-sized plates and a large number of smaller ones, called microplates, all of which move relative to one another. As a result, their boundaries are the sites of most of Earth's seismic and volcanic activity.

have used laboratory studies to reveal important clues to the way that basalts develop at mid-ocean ridges and hotspots within volcanic island arcs (see page 101).

Other experiments, meanwhile, can model the behaviour and interaction of important mantle minerals to explain the behaviour of the lithosphere–asthenosphere boundary. The major challenge for scientists, however, is in replicating the pressures and temperatures (higher than 900°C or 1,650°F) at the boundary.

According to recent research by geologists at the University of Tasmania, it seems that as pressures increase and temperatures reach up to 950°C (1,740°F) at depths of more than 90km (56 miles), water is released by some minerals and absorbed by others. The presence of water vapour increases the temperature at which silicate rocks melt from 970°C (1,780°F) at shallow depths and low pressures to 1,350°C (2,460°F) at pressures equivalent to depths of more than 190km (118 miles). A mineral called pargasite retains most of the water in upper mantle lherzolitic rock material, but at depths of more than 90km (55 miles), pargasite becomes unstable and releases its water, thus lowering the melting point of the surrounding rock material, leading to partial melting. The appearance of this melt, containing more than 180 parts per million of water weakens the mantle significantly, altering its physical properties, and this marks the lithosphere–asthenosphere boundary. The composition of the melt matches modern mid-ocean ridge basalts, created from mantle asthenosphere material, which have a similar water content of between 50 and 200 parts per million.

14

Oceanic crust

DEFINITION PLATES OF RELATIVELY THIN VOLCANIC ROCK FORMING THE CRUST BENEATH EARTH'S OCEANS

DISCOVERY FIRST MAPPED BY BRUCE HEEZEN AND MARIE THARP IN THE 1970S

KEY BREAKTHROUGH DISCOVERY OF PALAEOMAGNETIC 'STRIPES' ON EITHER SIDE OF MID-OCEAN RIDGES

IMPORTANCE PALAEOMAGNETISM CONFIRMS THAT THE OCEANIC CRUST IS GENERATED AND EXPANDS FROM THE MID-OCEAN RIDGES, PROVING THAT EARTH'S CONTINENTS ARE ON THE MOVE

In geological terms, Earth has two fundamentally different kinds of crust – the type that forms the ocean floors and that which makes up the mass of the continents. They have very different compositions, structures and histories, and until recently very little was known about oceanic crust.

While the geography of Earth's land surface has been well known since 19th-century surveyors mapped out pretty much every bump and hollow from Mount Everest to the Grand Canyon, the shape of the ocean floor remained largely unknown. It is only since the abundant technological innovations of the Second World War that it's been possible to explore the deep seabed in any detail. With the Cold War and the development of nuclear submarines capable of remaining submerged and undetectable for lengthy periods, mapping potential undersea hazards became a matter of urgency, although at first this was done in secrecy, and such maps were for military eyes only.

It was not until 1977 that the first public map of the ocean floor was published, by Bruce Heezen (1924–77) and Marie Tharp (1920–2006) of Columbia University. Their map was still highly interpretative, with artistic licence helping to fill gaps between the seismically surveyed strips of ocean-floor topography. But despite this, it still clearly revealed some very striking features – especially the mid-ocean mountain range that extends from the Arctic Ocean down through Iceland and the Atlantic Ocean, parallel to the complementary curves of the North and South American coastline to the west and those of Europe and Africa to the east. Today we know that this ridge wraps its way around the entire planet, branching in some places and merging together again in others.

OPPOSITE New ocean crust forms where tectonic plates are pulled apart, creating faulted rifts on either side of a central valley in which basalt lavas erupt along fissures and through small conelets. Most of this activity takes place deep on the ocean floor and can only be revealed by sonar techniques.

Rising up to a few thousand metres from the ocean floor, this range is distinguished by its symmetry and the presence of a rift valley along the length of its summit – features that show it is fundamentally different from mountains on land. While land mountains are often the result of compression and shortening of the crust, this mid-ocean range is evidently the result of increased heat flow from the mantle below, leading to tension, stretching and partial melting of the crust, coupled with the formation of magma that erupts to build up the ridge. Another striking feature is the presence of numerous cross-cutting fractures, known as transform faults, which displace the crest of the ridge sideways, especially in places where its main line curves.

'Volcanic rocks typically contain magnetic iron minerals that can become aligned with Earth's overall magnetic field, so that when the rocks cool and harden they preserve a record of planetary magnetism at the time of their formation.'

The flanks of the range are heavily ridged near the crest, but at lower depths are smoothed over by a blanket of sediment, which deepens with distance from the ridge, merging imperceptibly into remarkably flat ocean-floor basins that have an average depth of 3,795m (12,451ft) and stretch out across hundreds or even thousands of kilometres to meet the edge of the continental rise. In places (most notably in the Pacific), chains of individual mountains rise from the ocean floor, forming volcanic islands such as Hawaii, while near the continental margins of some landmasses, such as the west coast of South America and the east coast of Japan and the Kamchatka Peninsula, the ocean floor suddenly plunges into exceedingly deep narrow trenches, the deepest of which lie some 11,040m (36,220ft) below sea level.

It is no accident that the land close to these deep trenches is highly mountainous and rich in volcanic activity. For instance, the numerous volcanoes of the Peruvian Andes are adjacent to the Peru-Chile Trench and the iconic Mount Fuji lies close to the Japanese Trench. The connection between the two lies in the plate tectonic process of subduction (see page 43).

A magnetic record

Today, we know that mid-ocean ridges are essentially volcanic – they mark tectonic rifts where new ocean floor is created and spreads apart as volcanic basalt rocks are erupted from within the ridge's central rift. The most important evidence for this comes from maps of the 'palaeomagnetism' compiled by ships towing magnetometers back and forth above the ocean floor. Volcanic rocks typically contain magnetic iron minerals that can become aligned with Earth's overall magnetic field, so that when the rocks cool and harden they preserve a record of planetary magnetism at the time of their formation. In the early 1960s, undersea surveys discovered that the sea floor around the spreading ridge in the Eastern Pacific displayed remarkably symmetrical 'stripes' of rock, with the palaeomagnetic field flowing first in one direction, then abruptly reversing and pointing in the other. In 1963, Canadian geologist Lawrence Morley and the British geologists Frederick

Vine and Drummond Matthews (1931–97) independently realized that this barcode-like pattern of magnetic stripes was produced by sequential reversals of Earth's entire magnetic field over time. These reversals are themselves caused by changing flows of liquid metal in the outer core (see page 47).

Three years later, the same geologists were able to match the pattern of ocean-floor magnetic stripes with dates for reversals over the last 4 million years obtained by US geophysicists Allan Cox (1927–87) and Richard Doell (1923–2008) from land-based radiometric measurements. The correlation relied on the assumption that ocean-floor basalts on either side of the ridge were moving away from it at a rate of a few centimetres a year. Similar correlations were soon achieved elsewhere and confirmed similar spreading rates around the world. It soon became clear that even the oldest ocean-floor rocks are no more than 180 million years old – remarkably young in geological processes. So what happened to the older seabed rocks? The answer would prove to lie in the darkness of the deep ocean trenches and the processes of tectonic subduction (see page 43).

ABOVE Where lavas erupt deep on the sea floor, their surfaces cool rapidly while the hot interior continues to flow, forming characteristic globular shapes called pillow lavas. Here on the deep-ocean floor, pillow lavas have been colonized by a variety of organisms including tube worms and clams, forming unique communities that live independently of sunlight (see page 337).

Long-lived continents

15

DEFINITION ANCIENT, STABLE BLOCKS OF EARTH'S CRUST THAT
RESIST RECYCLING INTO THE MANTLE

DISCOVERY THOMAS H. JORDAN IDENTIFIED DEEP 'KEELS', WHICH
KEEP CRATONS BUOYANT IN THE MANTLE, IN THE 1970s

KEY BREAKTHROUGH RESEARCH BY NASA SCIENTISTS SUGGESTS THE
KEELS RESIST EROSION BY THE SURROUNDING MANTLE BECAUSE
THEIR MINERALS LACK WATER

IMPORTANCE CRATONS PRESERVE THE OLDEST ROCKS ON EARTH,
ALONG WITH TRACES OF THE MOST ANCIENT LIFE

At the heart of the major continents lie the rocky remains of Earth's early history – cratons that have remained largely undeformed since Archean times. Why have they survived so long, when the rest of Earth's rocks have been recycled by plate tectonic processes?

Earth's continents are built around very ancient rocks, more than 2.5 billion years old, called cratons. They were originally discovered and named as kratogens by Austrian geologist Leopold Kober (1883–1970) in the 1920s, to differentiate these ancient stable blocks of continental crust from younger 'orogens', or mountain belts. However, the term is nearly synonymous with the older term 'shield', which has been applied in a more general sense to large regions of Precambrian rocks such as the Canadian Shield since the end of the 19th century. The fact that these cratons have survived for such a long time has presented something of a mystery to geologists studying Earth's large-scale plate tectonics. Their longevity contrasts markedly with the rocks forming the semi-rigid ocean plates, none of which are more than 250 million years old. Oceanic plates are constantly being recycled into Earth's interior by the processes of subduction (see page 43), whereas cratons are apparently able to resist subduction altogether. It is only recently that geologists have discovered exactly how this happens.

Cratonic heartlands

Although these ancient rock masses are often covered with younger rocks, subsurface mapping has shown that they extend across hundreds of kilometres and have deep rocky roots that descend up to 250km (155 miles) into Earth's hot and semi-plastic mantle. A number of minerals, including diamonds, are only found within cratons or their boundary regions.

OPPOSITE Many of the rocks comprising the ancient cratonic cores of today's continents have been greatly deformed during ancient episodes of mountain building. In the process, the original fabrics have been altered almost beyond recognition – here a deformed sedimentary layer of iron oxide (haematite) from a banded iron formation has been cut and polished as an ornamental stone.

In the 1970s, US geophysicist Thomas H. Jordan discovered that the deep cratonic roots of the continents have chemical and physical characteristics that distinguish them from the surrounding mantle rocks. These cratonic 'islands' comprise the thickest parts of Earth's cool, brittle outer lithosphere, and have somehow remained buoyant and more rigid than their surroundings. Their persistence over billions of years seems to be due to the protective nature of their roots, which act as toughened rigid 'keels' around which the hotter and more mobile rocks of the surrounding asthenosphere flow.

Protecting the roots

Two essential properties have generally been thought to protect the cratonic keels from being subsumed into the surrounding mantle. Firstly, they appear to be more buoyant than the rest of the lithosphere, most likely due to the chemical composition of the minerals from which they are formed. Secondly, they are physically stiffer, due to the relatively low temperature of their rocks compared to the surrounding mantle material. However, calculations made since the 1990s suggest these differences alone would be insufficient to maintain the integrity of the craton through 2 billion years or more, and that there must be some other factor at work.

'Fragments of these mantle peridotites, drawn up from great depths and known as xenoliths are preserved in spectacular cone-shaped structures called kimberlite pipes.'

Unfortunately, the composition of the deep cratonic rocks remains speculative, since physical samples are impossible to obtain directly – no borehole can be drilled to anywhere near this kind of depth, where pressures are intense and temperatures reach more than 800°C (1,470°F). However, we do know that the most abundant rock of the upper mantle is the magnesium iron silicate called peridotite (see page 55), and the main mineral constituent within this, comprising 50–80 percent, is olivine (perhaps more commonly known in its olive-green gem form as peridot). Experiments have revealed that tiny increases in the water content of olivine, in the order of just a few tens or hundreds of parts per million, weaken its structure considerably, making it less viscous and therefore more likely to 'creep' under high pressure and temperature. With this evidence to back them up, some geologists have suggested that similar differences in water content may explain the essential difference between the rigid keel rocks at the base of the cratons and the surrounding mantle rocks of the asthenosphere.

Clues from the depths

While it's impossible to drill down to the base of the cratons, geologists can occasionally get a glimpse of their materials brought to the surface. The Kaapvaal Craton of South Africa, along with other similarly ancient regions, is punctured by a series of very deep-seated extinct volcanoes. The roots of these volcanoes once penetrated through the crust, allowing their explosive eruptions to draw on peridotite rock material from the mantle. Fragments of these mantle peridotites, drawn up from depths of 250km (155 miles) or so and known as xenoliths (from the Greek for 'foreign stones') are preserved in spectacular cone-shaped structures called kimberlite pipes, which are frequently mined for their diamond content.

New analyses of olivine from these xenoliths, carried out by an international team led by NASA scientist Anne Peslier, reveal that, at high pressures and temperatures above 1,100°C (2,010°F), their water content starts to decrease with further increases in pressure, and perhaps temperature. This effectively confirms that peridotites from the base of the cratons should indeed be deficient in water. What's more, it seems that 'dry' olivine from the deep cratonic keels, which has less than 10 parts per million of water, has an effective viscosity or stiffness that is between 20 and 3,000 times greater than typical asthenospheric rocks.

Such differences are more than enough to make craton roots resistant to erosion or deformation by the surrounding flow of mantle material. However, with further increases in pressure and depth, it seems that the effect of the water difference is lost, so that even water-poor olivine loses its viscosity and becomes similar to the surrounding asthenosphere once again. This prevents the cratonic roots from persisting even further into the mantle, and limits them to a maximum depth of around 250km (155 miles).

BELOW This flooded crater-like hole is an abandoned diamond mine excavated in an ancient kimberlite pipe in South Africa. The pipes were formed by ancient volcanic eruptions that originated from the base of the crust where pressures and temperatures are very high. The explosive debris that fills the pipes contains a variety of high-pressure rocks and minerals including diamonds.

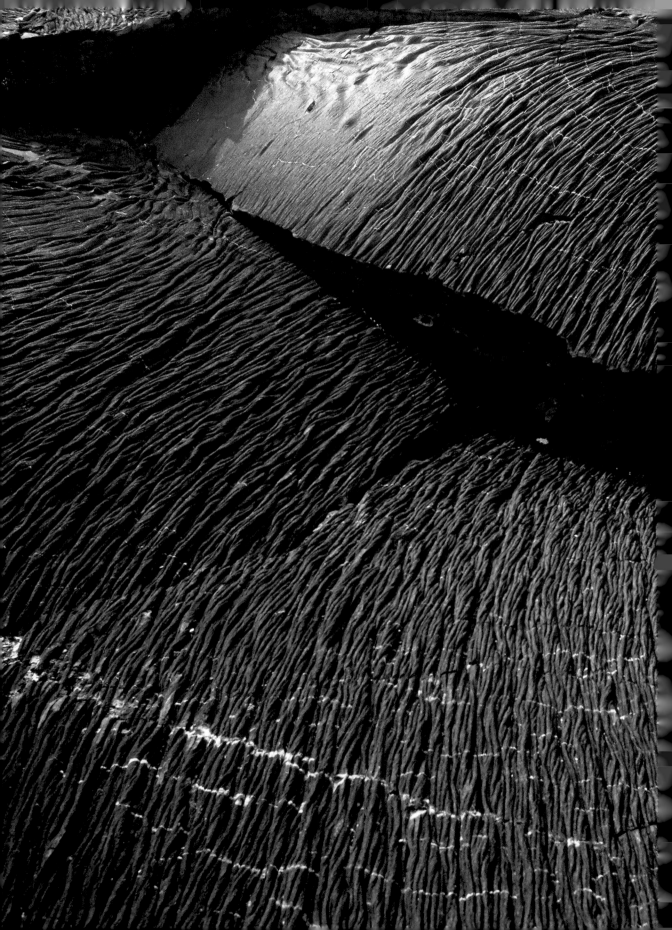

16 Earth's continental rocks

DEFINITION THE VARIETY OF ROCKS ON EARTH'S LAND SURFACE,
CATEGORIZED AS IGNEOUS, SEDIMENTARY AND METAMORPHIC

DISCOVERY JAMES HUTTON ESTABLISHED THAT EARTH'S CRUST
MATERIALS WERE FORMED BY CYCLES OF HEAT-DRIVEN IGNEOUS
ACTIVITY AND SUBSEQUENT EROSION

KEY BREAKTHROUGH HUTTON'S DISCOVERY OF MAGMATIC 'DYKES'
PROVED THAT SOME ROCKS ORIGINATED IN MOLTEN FORM

IMPORTANCE MUCH OF HUMAN CIVILIZATION IS DEPENDENT ON
MINERAL AND ROCK-DERIVED MATERIALS FROM THE CRUST

Earth's continental crust consists of three major rock types – igneous, volcanic and sedimentary. Our modern way of life depends upon this variety of rock and mineral matter, but what are these materials and how do they form?

The first archaeological evidence for human mining of metal ores dates back to the Middle East at least 8,500 years ago, and by 5,300 years ago people were already producing technically sophisticated weapons (the oldest known copper axe head is that belonging to Otzi, the so-called 'Iceman' whose mummified body was found in the Tyrolean Alps in 1991). Greek philosopher Theophrastus (371–287 BC) wrote the first treatises about mineralogy, *On Stones* and *On Mining*, but it was in Renaissance Germany that the scientific study of rocks and minerals really began, with works such as Agricola's *De Re Metallica* (1556). A century later, in 1669, the Danish scholar Nicolas Steno (1638–86) proved that the natural crystal faces of minerals such as quartz always had the same angular relationships, initiating the technical study and analysis of crystal structure (today called crystallography).

Conflicting theories

It was another German, Abraham Gottlob Werner (1750–1817), who published the first modern textbook on mineralogy and its chemical links in 1774. One of the greatest teachers of the time, he attracted students from all over the world, who subsequently spread his ideas. Werner's 'Neptunian' theory argued that Earth formed from an universal ocean that deposited five major rock formations. Werner's Scottish contemporary James Hutton took the opposing view, the so-called 'Plutonist' theory, proposing that Earth's internal heat engine drives the formation of new

OPPOSITE The basalt lavas of the Galapagos Islands were originally formed at an ocean-floor ridge where the Cocos and Nazca plates are separating. Rapid cooling underwater created these distinctive grooved lava plates.

Continental rifting

DEFINITION THE PROCESS BY WHICH EARTH'S LARGE LANDMASSES SPLIT APART, WITH OCEAN BASINS FORMING BETWEEN THEM

DISCOVERY THE IDEA OF RIFTING WAS AN INTEGRAL PART OF ALFRED WEGENER'S 'CONTINENTAL DRIFT' THEORY, THE PRECURSOR OF MODERN PLATE TECTONICS

KEY BREAKTHROUGH RECENT SURVEYS OF ATLANTIC COASTS HAVE REVEALED NEW DETAILS OF THE RIFTING PROCESS

IMPORTANCE CONTINENTAL RIFTING CREATES NEW OCEAN BASINS AND LAYS DOWN HYDROCARBON DEPOSITS OF OIL AND GAS

When plate tectonic forces pull cold slabs of continental crust apart, they thin and subside to form linear depressions called rift basins. Further stretching and sagging eventually splits the crust in two, with the depression between them flooded by an influx of seawater.

Rifting of the continental lithosphere is a fundamentally important geological process that has operated ever since the continents first formed in Earth's early history. Where continents split in two and their sides separate, the opposing edges, known as rifted or passive margins, become sites of thick sediment accumulation as flowing waters on the land find new routes down to sea level. The material dumped along these offshore margins often includes significant amounts of organic matter that, when buried and compressed over time, can be transformed by pressure and temperature to create hydrocarbon reservoirs. Understanding exactly how the rift margins and basins form is important not only for a better understanding of our planet's history, but also for the practical location of oil and gas reserves.

Stretching the crust

Since the 1980s, the development of seismic mapping (see page 39) and improved knowledge of plate tectonics have led to a better understanding of how continents break apart and form new ocean basins. A key element in this understanding is the study and interpretation of the rifted margins themselves – the remnants of the sites where the separation first begins. For instance, the eastern margins of the Americas and western margins of Europe and Africa form rifted margins flanking the Atlantic Ocean, while the southern rifted margin of Australia is mirrored across the Southern Ocean by a rifted margin along the coast of Antarctica.

OPPOSITE Lake Baikal in southern Siberia is one of the most impressive geological structures on Earth, lying in the central part of Baikal-Vitim Rift Zone, some 1,800km (1,100 miles) long. The rift is on the southeastern edge of the Siberian craton, and lies within a fold and thrust belt (mountainous foothills) that originated during Precambrian and Palaeozoic collisions between separate elements of the craton.

RIGHT This false-colour satellite radar image shows the Crater Highlands of Tanzania, with its numerous spectacular volcanoes, including the gigantic Ngorongoro Crater and lake (centre). These volcanoes are created by plate tensions associated with the formation of the Great Rift Valley (see page 277).

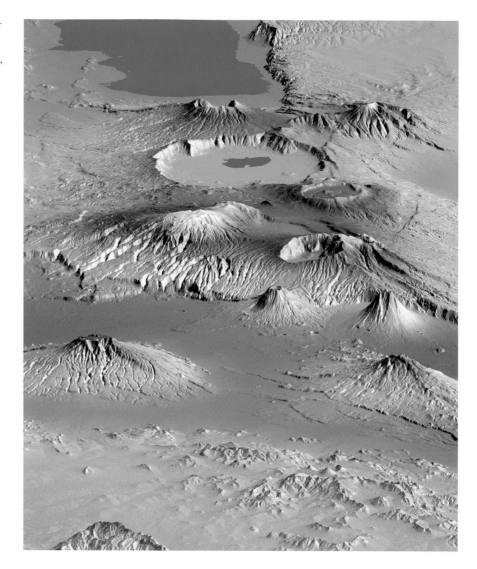

The traditional model for the formation of such rifts envisioned it as a process of stretching, thinning and subsidence in the continental lithosphere, accompanied more or less simultaneously by faulting to form rift basins. At first, brittle faulting of the rocks, involving downward and sideways slippage to fill the opening gaps, was the most important means of extension, but as the stretching continues crustal thinning (in which the lithosphere grows thinner simply because its material is being stretched across a wider area) becomes more important. Geologists had assumed that differences in behaviour between the upper and lower lithosphere explained these discrepancies – the cooler and more brittle rocks of the upper layer, approximately 30km (19 miles) thick, responded first to the stretching forces through brittle faulting, while as the process continued, the warmer and more ductile lower layer, between 80 and 110km (50 and 70 miles) thick, began to stretch.

Asymmetry in rifted margins

However, this explanation does not explain a curious asymmetry that is typically found between opposing pairs of rifted margins. At first glance, we might expect the fault structures on either side of the margin to be pretty much symmetrical, but seismic profiling of margin rocks over recent decades has revealed characteristic patterns in which one margin is greatly thinned over a short distance with relatively small-scale faulting, while the opposite margin shows much more gradual thinning over a greater distance, with much larger-scale faults. Until recently, the definition of images generated by seismic technology has not been good enough to clarify the nature of the extension process, particularly at depths greater than about 10km (6 miles).

However, new detailed geophysical surveys, combined with information from deep-ocean drilling and dredge sites and direct visual observations from submersibles, now allow profiling of opposing rifted margins in Newfoundland on the North American Plate and the Iberian Peninsula on the Eurasian Plate. New techniques for processing seismic data can also extract more detail to reveal the structure of these margins. The net result is the production of a much more accurate model for the process of rifting as these two once-united landmasses separated around 130 million years ago.

Opening the North Atlantic

The North Atlantic Basin began to open up in late Jurassic times from the Central Atlantic, and spread progressively northwards as Europe and North America rifted apart. The oldest sediments preserved in the downfaulted rift basins off the Iberian coast are of late Jurassic age and help to date the flooding of the continental margin as it subsided. Successive strata in deeper basins show an accumulation of early Cretaceous deposits, which are around 4km (2.5 miles) thick and provide an ideal setting for the deep burial and transformation of organic remains into oil and gas.

'Understanding exactly how the rift margins and basins form is important not only for a better understanding of our planet's history, but also for the practical location of oil and gas reserves.'

The new studies also help to resolve the discrepancy between crustal thinning and faulting on either side of the margin. They show that the faults develop in sequence, with each successive fault cutting the wall of the previous one to increase the overall extension. The asymmetry of the two opposite margins can also be explained – it seems that when rifting begins, the entire lithospheric crust, with its brittle upper layer and more ductile lower layer, responds to the extension by producing a number of small faults across a large area. The stretching also thins the brittle crust from 30km (19 miles) to around 5 km (3 miles) thick, exposing the lower crust, which also now cools and grows more brittle. It too fractures, but with fewer major faults, each slipping by a larger displacement of some 4–6km (2.5–3.7 miles).

18

Making mountains

DEFINITION MOUNTAIN BUILDING INVOLVES SHORTENING OF
EARTH'S CRUST DURING COLLISIONS BETWEEN TECTONIC PLATES

DISCOVERY ELIE DE BEAUMONT WAS THE FIRST TO SUGGEST THAT
MOUNTAINS WERE FORMED THROUGH THE COMPRESSION OF ROCKS

KEY BREAKTHROUGH THE ARRIVAL OF PLATE TECTONIC THEORY
EXPLAINED WHERE THESE COMPRESSION FORCES CAME FROM

IMPORTANCE TERRESTRIAL MOUNTAIN FORMATION INVOLVES
SHORTENING AND THICKENING OF THE CRUST, ELEVATING THE
ROCKS AND INCREASING EXPOSURE TO WEATHERING AND EROSION

The process of mountain building, or 'orogeny', is one of the most impressive manifestations of plate tectonics. A result of collisions between plates, it involves thickening, uplift and displacement of the crust, with igneous activity, folding and faulting creating long scars over Earth's surface.

Mountains have always fascinated humans, and for many thousands of years they were associated with dangerously powerful forces beyond our control. These were the homes of gods who could invoke earthquakes, volcanic eruptions, avalanches, storms and floods. It is only since the late 18th century that mountains and their scenery have come to be seen as attractive and romantic, if still tinged with danger.

For geologists, however, the formation, structure, and even the very existence of high mountain ranges on land has been a considerable problem until quite recently. The significance of features such as faults and folds came to light in Europe, North America and elsewhere during the Industrial Revolution of the early 19th century, largely because of their effects on the mining industry. In 1852, French geologist Elie de Beaumont (1798–1874), who supervised production of the first geological map of France, was one of the first to link these features to the formation of mountain ranges, and to suggest that mountains were created when rocks were somehow compressed. Over the following decades, geologists James Hall (1761–1832) and James Dwight Dana developed a 'geosyncline theory' that described the vertical uplift of mountain ranges through compression and volcanism from beneath. Dana proposed that the compression was caused by slow shrinking of the entire planet as Earth gradually cooled – a theory that successfully explained some significant geological features, but failed when it came to others.

OPPOSITE The folding and crumpling of hard and normally brittle rocks is a measure of the immensity of tectonic forces within Earth's crust. When deeply buried and confined by high pressure over a long time, rock responds by internal changes to its component mineral grains and plastic deformation rather than brittle fracture.

Fortunately, thanks to the development of plate tectonic theory (see page 41) the processes behind the formation of major mountain ranges can now be explained as a result of the collision (technically 'convergence') of lithospheric plates, resulting in subduction, shortening and thickening of the crust.

The exact process of mountain formation depends on the nature of the plates involved – convergences can involve two oceanic plates (giving rise to volcanic island arcs such as Japan – see page 101), two continental plates (producing crustal thickening and high mountain ranges such as the Himalayas) or one oceanic and one continental plate (producing subduction and volcanism along a continental margin, as seen in the Andes of South America).

Volcanic mountains

Subduction occurs where one of the converging plates is denser than the other. Where oceanic and continental plates converge, it is the oceanic plate that is subducted, while in the case of two converging oceanic plates, it is the older, colder and denser of the two. The subducted plate slides down, at a slope of between 25–45 degrees, to depths of 100km (62 miles) or more, driving its way into the mantle beneath the leading edge of the less dense plate. In doing so, it releases huge amounts of energy in the form of earthquakes, which are generated at all depths down to more than 600km (375 miles). As the subducted slab descends, pressure squeezes water out of its rocks and heats it up. The rising hot water lowers the melting point of the rocks above, partially melting them. This 'rock melt' then continues the upward migration and can eventually erupt through surface volcanoes.

'Once high mountain ranges have been elevated to a certain height, their summit rocks are worn down at more or less the same rate as they are being elevated.'

Continental collisions

Crustal shortening and thickening occurs when two continental plates converge and collide. Since both plates are likely to be tens of kilometres thick and have a roughly equal density, neither will easily give way and be subducted. Instead, the collision triggers intense folding and faulting of the rocks, and pushes the underside of the plates many kilometres down into the mantle. At depth, these rocky roots are 'cooked' by high temperatures and squeezed with high pressure, resulting in a partial melting and the formation of magma. Hot magma then rises up through any structural weaknesses in the crust, erupting at the surface through volcanoes or accumulating in deep magma chambers to form large bodies of igneous rock such as granite.

The crustal thickening also pushes mountain ranges upwards, rising several kilometres above sea level and exposing the rocks to increased gravitational forces, weathering and erosion. Once mountain ranges have been elevated to a certain height, their summit rocks are worn down at more or less the same rate that they are being elevated (see page 353). For instance, it is estimated

that the surface of the Himalayas is currently being eroded at rates of 2–5mm (0.1–0.2in) per year, roughly in balance with the rate of their uplift as the Indian Plate collides with Asia (see page 241).

Exposing mountain roots

Over millions of years, the processes of convergence grind to a halt and mountain-building ceases. Weathering and erosion now strip away the elevated peaks, reducing and 'unloading' the amount of rock raised above sea level. The deep rocky roots of the mountains now rise back to the surface through a process known as isostatic readjustment. These roots are less dense than the mantle, which has been displaced to accommodate them, and so they are pushed back up by tens of kilometres as the mantle returns to equilibrium. As a result, the igneous granites and highly metamorphosed rocks of these ancient mountain roots eventually appear at the surface, where they can be studied in regions such as the Scottish Highlands and Scandinavia (both remnants of the great Caledonian mountain belt of mid-Palaeozoic times – see page 173). In detail, the structure and formation of major mountain ranges is often very complicated. The forces involved are powerful enough to bend great thicknesses of rocks, as if they were made of dough or modelling clay, creating huge folds that stretch over tens of kilometres, and faults that may be hundreds of kilometres long.

ABOVE The Appalachian Mountains of northeast America are a southwestern extension of the Caledonian Mountains of Scandinavia and the British Isles, which were some of the highest mountains on Earth during the Palaeozoic. Eroded almost flat by the end of Mesozoic times, the deep roots of these ancient mountains have since been elevated upwards to form the present-day range.

19

Tectonic uplift

DEFINITION A PROCESS THAT RAISES LARGE AREAS OF EARTH'S CRUST FROM BENEATH WITHOUT THE TECTONIC DISTORTION ASSOCIATED WITH MOUNTAIN BUILDING

DISCOVERY THE TIBETAN PLATEAU, THE LARGEST UPLIFTED AREA ON EARTH, WAS FIRST EXPLORED BY PRZEWALSKI IN THE 1870s

KEY BREAKTHROUGH A RECENT SURVEY HAS SHOWN THAT PLATEAUS SUCH AS TIBET ARE ACTUALLY RAISED IN SEVERAL PHASES

IMPORTANCE UPLIFTED REGIONS CAN HAVE A HUGE EFFECT ON REGIONAL AND EVEN GLOBAL CLIMATE

The Tibetan Plateau is the largest and highest plateau in the world, and scientists have been intrigued by the forces that support it for decades. In contrast, the Colorado Plateau is neither so high nor so extensive, but scientists have been equally puzzled by how and when it was uplifted.

With an area of some 2 million square km (770,000 square miles) and an average elevation of more than 4,500m (14,850ft), the Tibetan Plateau is by far the most extensive and highest uplifted region on Earth today. To its north, the Kunlun Mountains, one of the longest ranges in Asia, extend across 3,000km (1,900 miles), with peaks more than 7,000m (23,000ft) high. The Karakoram Range to its west has peaks more than 8,000m (26,200ft) high and to the south the Himalayas have more than 100 peaks that rise above 7,000m (23,000ft), including the tallest mountains in the world.

The region remained a mystery to outsiders until the late 19th century, however, because of its inaccessibility and the isolationist policies of its rulers. The first scientist to reach the area was the Russian explorer Nikolai Przewalski (1839–88), who approached from the north in the 1870s and brought the region to the attention of European geographers.

A geological puzzle
How can such a large plateau be elevated so high? Its crustal structure is particularly puzzling as it has clearly not been thickened by the kind of mountain-building processes, such as folding and faulting, normally associated with the formation of ranges like the Himalayas. However, the underlying structure of this massive elevated region remained a geological puzzle for many years, until it was finally mapped in the 1980s.

OPPOSITE The uplift of the Colorado Plateau to some 2km (1.2 miles) above sea level has promoted 'downcutting' erosion by the Colorado River to form the world famous Grand Canyon. The gorge cuts down some 1.5km (0.9 miles) through a layercake of rock strata that range in age from late Cretaceous at the top, all the way down to Precambrian.

20 Glacial puzzles

DEFINITION MOUNTAIN GLACIERS ARE SLOW-MOVING MASSES OF SEMI-PERMANENT ICE THAT FORM AT HIGH ALTITUDES

DISCOVERY GLACIERS HAVE BEEN FAMILIAR THROUGHOUT HUMAN HISTORY, AND THEIR EROSIVE POWER WAS THOUGHT TO PREVENT MOUNTAINS GROWING ABOVE A CERTAIN HEIGHT

KEY BREAKTHROUGH RECENT WORK SHOWS THAT IN EXTREME COLD, GLACIERS CAN FREEZE IN PLACE, PROTECTING THE ROCK BELOW

IMPORTANCE GLACIERS HAVE SHAPED MUCH OF THE LANDSCAPE OF EARTH'S TEMPERATE LATITUDES

Glaciers are well known for their ability to grind down mountains, and until recently this action was thought to prevent some mountain ranges from rising above a certain level. But new research suggests that, at least in some cases, they can have the opposite effect.

The enormous mass of ice contained within a glacier gives it tremendous erosive power, wearing down mountain sides and valleys into distinctive 'alpine' landscapes with u-shaped valleys, dish-like cirques, teardrop-shaped *roches moutonnées* and deep valley bottoms that fill with water to become finger lakes and fjords. High alpine mountains around the world (even close to the equator) bear the marks of glaciers at work.

Because of all this erosion, geologists had assumed that glaciers acted as a 'buzzsaw', preventing mountain peaks and ridges from rising to any significant height above the level of permanent glaciation. However, a new study of climate history and erosion patterns in the Patagonian Andes of South America indicates that, at least in high latitudes, extensive glaciers can actually protect rising mountain ranges from erosion, allowing them to rise above maximum heights predicted by the 'glacial buzzsaw' theory.

The coming of the ice

From 5 million to 3 million years ago, as Earth's climate cooled into the Quaternary Ice Age (see page 293) high-altitude glaciers began to appear across the world's major mountain ranges. Formed by the compression of snowcaps that persisted and grew thicker from year to year, the development of these glaciers has been held responsible for producing the characteristically jagged profiles of alpine mountains.

OPPOSITE The erosive power of glaciers relies not on the ice itself, but on the rock debris held within it. The lower surfaces of glaciers are like immense sandpapers, covered with rock fragments. The weight and motion of the glacier, driven by gravity, grinds away the surface of the rock beneath, cutting down even the toughest mountains.

Most mountain belts are linear portions of Earth's crust that have been squeezed by internal tectonic forces to such an extent that the crust has thickened beneath them, forming deep roots whose buoyancy lifts them upwards (see page 79). The height and width of any mountain belt are determined by the balance of the tectonic forces lifting it up and the surface processes, linked to climate, that are wearing it down. In theory, most mountain ranges should eventually reach a 'steady state' in which the two processes balance out over a million years or so. However, if the balance is disturbed – most likely by a climate change that either increases or decreases the rate of erosion, the mountain belt will either shrink or grow.

This theory has been difficult to prove because it is hard to determine whether climate change is the cause of a particular change in the dimensions of mountain belts, or its result. This is largely because increased uplift tends to produce heavier rainfall quite independently of climate change. What's more, since mountain belts are essentially sites of erosion, the evidence of their history is largely destroyed by the erosive processes themselves.

RIGHT The precipitous rock walls and sharply dog-toothed, glacial horns of the Torres del Paine in the Patagonian Andes owe their astonishing form to glaciation, but paradoxically seem to owe their elevation to protection from erosion afforded by those same glaciers.

Maintaining equilibrium

Some of the strongest evidence to support the idea that climate controls the height of mountain belts comes from what is known as the glacier equilibrium line altitude (ELA). This is the height at which the accumulation of snow is balanced by the rate of ice loss through melting, and coincides with the level at which glaciers erode bedrock most efficiently. Mapping of the ELA for the Andean Cordillera during the last glacial maximum some 20,000 years ago, when there was an extensive Patagonian ice sheet with numerous glaciers, produces a good match for the present height of much of the Cordillera.

However, this correlation does not apply to the heavily glaciated southern Patagonian Andes – here elevation increases away from the equator while the level of the ELA decreases. The development of this anomaly was originally blamed on the presence of a mid-ocean ridge that is being actively subducted below South America. The ridge's heat and buoyancy were thought to be responsible for the region's increased elevation.

The Patagonian anomaly

However, a recent investigation led by Stuart Thomson of the University of Arizona has produced a very different explanation. Thomson's team worked out the age of apatite grains in the exposed surfaces by measuring the number of 'fission tracks' (microscopic 'scars' within this phosphate mineral formed by the radioactive decay of uranium atoms). Apatite has an ability to heal itself when heated to temperatures of around 100°C (212°F), so only starts to accumulate fission tracks as it comes close to the surface and cools below this temperature. Because new tracks form at a steady rate, the number of tracks in a given sample indicates the time since it came to the surface.

The results of this technique, known as low-temperature thermochronology, show that surface rocks grow systematically older as one heads further south. Rather than being boosted in height by the subduction below, it seems that the Patagonian Andes have simply been eroding more slowly over the past 5 million years. This is the first field evidence to support a radical new suggestion that during extremely cold climates, mountain glaciers freeze onto the underlying bedrock instead of sliding over and eroding it. With no friction or erosion between glacier base and bedrock, the mountain peaks are protected rather than degraded by glaciers, allowing tectonic forces to raise them significantly higher than lower-latitude glaciated mountains under warmer climates. This kind of glacial protection may provide an explanation for otherwise puzzling high-altitude features such as the mountains surrounding the Tibetan Plateau, which have an average elevation of more than 5,000m (16,400ft), and extend across an area of 3,500 by 1,500km (2,200 by 930 miles).

'The height and width of any mountain belt are determined by the balance of the tectonic forces lifting it up and the surface processes, linked to climate, that are wearing it down.'

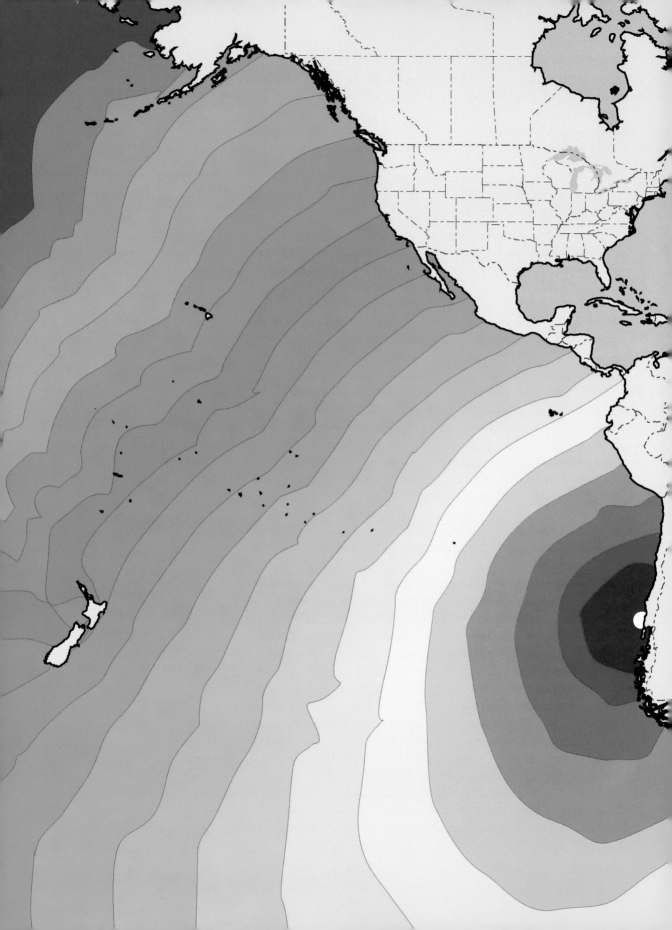

Earthquakes and plate tectonics

DEFINITION EARTHQUAKES ARE CAUSED BY ROCK FRACTURE AND ENERGY RELEASE AS EARTH'S CRUSTAL PLATES SLIP AGAINST ONE ANOTHER DEEP BELOW THE SURFACE

DISCOVERY SEISMIC STUDY OF EARTHQUAKE LOCATIONS WAS PIONEERED BY KIYOO WADATI IN THE 1920S

KEY BREAKTHROUGH HUGO BENIOFF IDENTIFIED THE LINK BETWEEN EARTHQUAKE ORIGINS AND TECTONIC 'SUBDUCTION ZONES'

IMPORTANCE A BETTER UNDERSTANDING OF EARTHQUAKE ORIGINS HELPS TO PREDICT WHICH REGIONS ARE PRONE TO THE WORST QUAKES, AND CAN ULTIMATELY SAVE LIVES

In May 1960, the most powerful earthquake ever recorded hit Valdivia in Chile with a magnitude of 9.5. The quake, and the devastating tsunami that followed, were the result of convergence between the Nazca Plate of the Pacific Ocean floor and the continental crust of the South American Plate.

The Pacific coast of South America is no stranger to 'geohazards' such as volcanoes and earthquakes, and today we know that both result from tectonic activity. Two tectonic plates converging on one another at a rate of 8cm (3.2in) per year might not seem very dramatic, but over geological time it amounts to 8km (5 miles) every 100,000 years. In the face of such inexorable forces, one plate or the other must ultimately yield, and in this type of collision it is always the oceanic plate that gives way, sinking beneath the oncoming continent in the tectonic process known as subduction.

The discovery of this process, and ultimately the formation of the entire theory of plate tectonics itself, is the result of a series of important discoveries between the 1920s and the 1960s. The seismic study of earthquakes – and especially the first detection of deep earthquakes by Japanese seismologist Kiyoo Wadati (1902–95) in the late 1920s – was a key factor, but it was not for another 30 years that their implications were more fully appreciated by American seismologist Hugo Benioff (1899–1968).

Benioff's breakthrough

In the 1930s, Benioff invented sophisticated new equipment for detecting and measuring seismic waves, and his innovations made possible the determination of precise earthquake wave-travel times and detailed analysis of their sources (see page 39). This equipment formed the basis for the first

OPPOSITE Originating deep beneath Valdivia, Chile, the intense magnitude 9.5 earthquake of 1960 was caused by the subduction of the Nazca Plate of oceanic crust below the South American Plate. The quake generated offshore tsunami waves that crossed the Pacific to Japan. The contours on this map show the travel time for the tsunami in hours.

World-Wide Standard Seismograph Network, set up after the Second World War to monitor cold war nuclear tests. The new data that this network provided about natural earthquakes revealed a remarkable concentration of earthquake epicentres (the point on Earth's surface directly above the quake's true origin or focus) around deep-ocean trenches. As a result, Benioff suggested in the 1950s that the deep-ocean trenches were fault zones, and proposed that the generally discredited theory of 'continental drift' was a possible explanation for their origin.

Benioff went on to show that the depth distribution of earthquake foci also followed clearly defined patterns, dipping from the bottom of the trenches down to some 700km (435 miles) beneath the continental margins. Now known as Wadati-Benioff zones, these regions define the seismically active part of a subduction zone, where the converging tectonic plates are slipping past one another.

The Great Quake

It was this kind of plate convergence that caused the Valdivia earthquake. Deep beneath the Pacific Ocean, the relatively dense ocean-crust rocks of the Nazca Plate are being subducted eastwards, down a 30-degree slope, below the thick piles of crustal rocks that make up the Andean mountains of South America. Huge amounts of pent-up energy are created by the subduction process, and friction between the rocks that are being pushed past one

BELOW When the tsunami triggered by the 1960 Chilean earthquake struck Easter Island in the southeastern Pacific, several of the famous Moai statues were knocked over and even transported inland.

another is released spasmodically in the form of earthquakes that propagate violent shaking forces in all directions. Although the rupture produced by the Valdivia quake was only some 20m (66ft) at most, it spread for 800km (500 miles) along the rupture zone at around 3.5km (2.2 miles) per second, releasing centuries of stored energy in a few catastrophic minutes. Measured at 9.5 on the moment magnitude scale (MMS), which measures the amount of energy released and replaces the older Richter scale, the quake devastated the towns of Valdivia and Puerto Montt and flooded coastal areas. Its various effects are estimated to have killed at least 1,500 people, and fatalities could have been much higher if people had not been warned by earlier minor quakes. Today scientists believe that earthquakes cannot get much bigger than this one, which amounted to something like 25 percent of Earth's total earthquake 'budget' for a century.

'Two tectonic plates converging on one another at a rate of 8cm (3.2in) per year might not seem very dramatic, but over geological time it amounts to 8km (5 miles) every 100,000 years.'

Wave of destruction

With its origin lying some 160km (100 miles) offshore at a relatively shallow depth of 33km (20 miles) below the Peru-Chile Trench, the Valdivia quake, like so many subduction zone quakes, generated tsunami waves. Rising to heights of 25m (82ft), they swept away coastal villages and carried ships a kilometre or so inland. Tsunami waves are notorious for their ability to cross oceans at hundreds of kilometres per hour with little effect on shipping, but then rise into huge destructive waves again as they approach coastal regions far away from their origin. Travelling at 350km/h (220 mph), this particular tsunami took 15 hours to reach Hawaii, where 10m (33ft) waves took some 61 lives, and continued on for another seven hours as far as Japan, some 17,000km (10,500 miles) from its origin, where another 140 people were killed.

Learning the lesson

From a geological point of view, the great magnitude of the 1960 quake prompted a new phase in the science of seismology. Most notably, studies in the aftermath provided unequivocal evidence that our entire planet 'rings' like a bell with the harmonic vibrations from such large shock waves.

The earthquake also prompted international collaboration that led to the establishment of a more advanced Pacific tsunami warning system, now based at Ewa Beach, Hawaii and Palmer, Alaska. Monitoring stations here detected the Indian Ocean tsunami of December 2004 and issued warnings, but they were not picked up in time by the regional authorities. The staggering death toll of 230,000 people in that event led to the establishment of a separate warning system for the Indian Ocean, which became active in June 2005. Even so, warnings of the July 2006 earthquake in Java and subsequent tsunami were not passed on to some of its remote coastal regions, and more than 600 people were killed in that event.

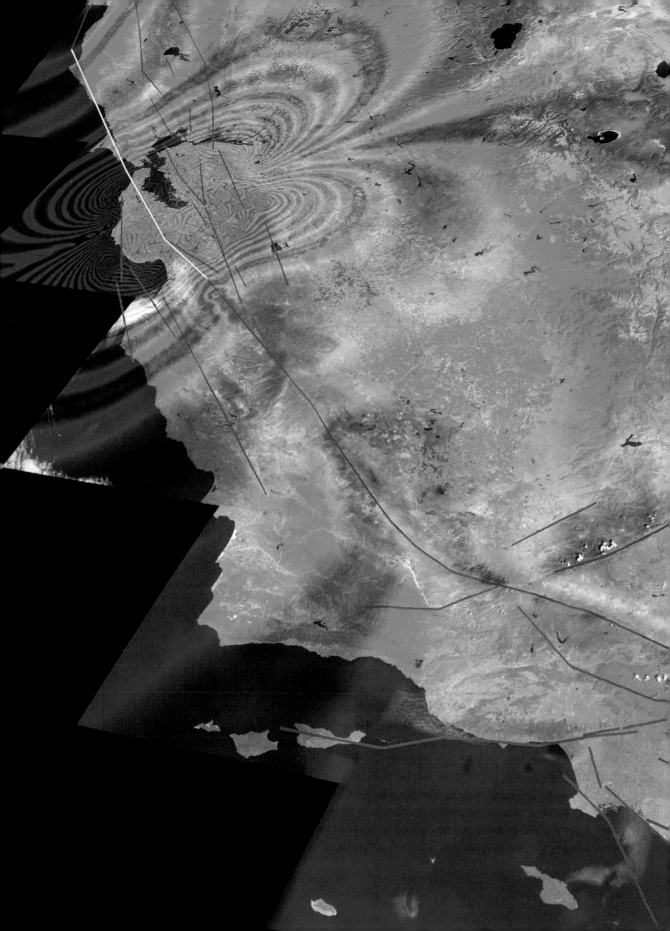

22 What makes faults slip?

DEFINITION A BUILD UP OF ENERGY WITHIN THE CRUST IS RELEASED BY FRACTURE, FAULTING AND SLIPPAGE OF BRITTLE ROCK

DISCOVERY FAULTS WERE WELL KNOWN TO MINERS, AND WERE MAPPED BY GEOLOGISTS DURING THE INDUSTRIAL REVOLUTION

KEY BREAKTHROUGH RECENT STUDIES SHOW THAT THE RATE AND FORM OF FAULT SLIPPAGE IS LINKED TO THE PRODUCTION OF A ROCK POWDER CALLED GOUGE ALONG THE FAULT SURFACE

IMPORTANCE BETTER UNDERSTANDING OF THE CONDITIONS IN WHICH FAULTS SLIP MAY ULTIMATELY HELP TO PREDICT EARTHQUAKES

Fracture planes through rocks, known as faults, are a widespread result of interactions between Earth's dynamic interior and its brittle outer crust. When rocks break and faults shift, they generate shock waves that we experience as earthquakes – but how and why does this happen?

The common geological phenomenon of fault-fractured rocks was first mapped and described several centuries ago, when miners encountered them below ground. Mineral veins, coal seams and other valuable deposits might run continuously for long distances, then abruptly come to a halt against a plane where the rock had clearly been dislocated. If the miners were lucky, the seam they were following might have suffered little displacement and could be picked up again on the other side of the fault, but when the movement across the fault plane was greater, the target layer might be lost altogether. Unsurprisingly, the first efforts to understand the direction and amount of displacement in such fractures were driven by economic interest.

Examinations of the fault planes soon revealed that the direction of displacement could usually be worked out from parallel scratches produced as the rocks ground against one another. Further studies revealed a host of other features that could help narrow the direction and type of movement, if not its extent. Geological mapping showed that rock fracture and dislocation happens on all scales, from a few millimetres to hundreds of kilometres.

The process of faulting was soon linked to other geological phenomena such as folding, mountain building and large-scale vertical movements of Earth's crust. It became clear that different types of fault were the product of different forces at work – tension pulling the rocks apart, compression

OPPOSITE This false-colour image models the patterns of seismic deformation and resulting ground displacement (shown in different colours) caused by a simulated earthquake on California's San Andreas Fault (yellow line). The results are overlaid on a satellite radar map of the state, with other fault lines represented in red.

pushing them together and horizontal forces tearing them apart in opposite directions. Mapping and detailed examination of faults and fault patterns became an integral part of attempts to unravel the geological history of any sequence or 'suite' of rocks, from those encountered in a single mine or quarry, to those found across an entire mountain belt.

Faults and earthquakes

It was also clear that faulting was linked to earthquake activity and that fault ruptures produce earthquakes through friction and the dissipation of energy. The initial point of rupture – the 'focus' of the earthquake – is also the point from which the fault propagates. In any one event the dislocation is only a matter of centimetres, or a few metres at most, but the 'fault plane' across which this motion occurs may range from less than 1 square metre (about 10 square ft) to thousands of square kilometres, depending upon the amount of energy that has to be dissipated. When the global distribution of earthquakes was mapped, it also became evident that the vast majority of earthquake epicentres lie along active fault zones – regions we now know to be the boundaries between tectonic plates.

Earthquakes are produced as a result of brittle rocks rupturing under pressure, and the ensuing friction generated by movement of the ruptured surface. However, in order to trigger an earthquake, the rock must reach some critical point at which it 'fails', losing its strength so that the two sides of the fault can slip past each other. Geologists are still divided on the nature of the mechanism that triggers this failure, but a better understanding of it could have major implications for our ability to predict imminent earthquakes.

Experimental faulting

Recent experiments carried out by Ze'ev Reches of the University of Oklahoma and David Lockner of the US Geological Survey have at last thrown some light on the weakening mechanism. When blocks of dry and brittle granite are ground against one another over short distances and at slow speeds (less than 1mm per second or 1in every 25 seconds), they are weakened by slow 'creep' over the contact surface, and the rolling of fragmented granules in a shear zone. At higher speeds, the friction is so great that rock surfaces will actually melt. Observations of both natural and artificially created fault planes show that they are typically covered with what is known as fault gouge – a fine-grained rock powder that coats the contact surfaces, often with a smooth, hard and glaze-like surface.

'In any one event the dislocation is only a matter of centimetres, or metres at most, but the "fault plane" across which this motion occurs may range from less than 1 square metre (about 10 square ft) to thousands of square kilometres.'

The new experiments show that gouge plays an essential role in the dynamics of faulting, and that its properties change with the speed, distance and heat of the fault motion. At relatively low slip rates of 10–60mm (0.4–2.4in) per second, the powder develops after a few millimetres of movement and produces a thin deforming layer that lubricates the fault, weakening it by a factor of two to three times. When slippage halts, the gouge rapidly hardens and the fault regains its strength in a matter of hours or days. For the fault to slip again, the gouge must be reactivated to repeat its lubricating function.

In contrast, with an accelerating slip rate from stationary to about 1m (40in) per second, the initial low-velocity movement produces only small amounts of fault gouge. As the slippage picks up speed, however, the gouge reaches a critical thickness, with low strength and rate of wear. Thereafter, the gouge becomes highly reactive, but may strengthen even during slip as a result of dehydration from frictional heat. The formation of fault gouge separates the fault blocks and lubricates the movement and acts as a common and highly effective mechanism, which controls earthquake stability in brittle crustal rocks.

A case study in volcanism

23

DEFINITION THIS 1980 ERUPTION WAS THE FIRST MAJOR VOLCANIC OUTBURST IN THE MAINLAND UNITED STATES FOR 65 YEARS

DISCOVERY SCIENTISTS WERE ALREADY MONITORING THE VOLCANO WHEN IT ERUPTED ON 18 MAY 1980

KEY BREAKTHROUGH THE CATASTROPHIC RESULTS ASSOCIATED WITH SIDEWAYS ERUPTION FOLLOWING THE FLANK COLLAPSE PROVED TO BE REMARKABLY DEVASTATING

IMPORTANCE STUDIES OF A MAJOR VOLCANIC ERUPTION IN PROGRESS HELP TO PREDICT THE BEHAVIOUR OF FUTURE VOLCANIC EVENTS

It is now more than 30 years since Mount St Helens volcano in Washington State blasted its way into the headlines in mid-May 1980. With plenty of warning, scientists were able to watch and measure the whole eruptive sequence in real time, revealing some real surprises.

Although there are many active volcanoes within the mainland United States, before 1980 the most recent activity had been the 1915 eruption of Lassen Peak in California. Mount St Helens provided an opportunity for experts to study the development of a major eruption in real time and recruited a whole new generation of volcanologists. Their efforts have produced a new level of insight into these violent expressions of Earth's internal power.

Blast off

The main eruption of Mount St Helens blasted away the top 400m (1,300ft) of the volcano, reducing its overall height from 2,950m (9,667ft) to 2,549m (8,363ft). The blast was triggered by an earthquake of magnitude 5.1 that originated roughly 1.6km (1 mile) below ground, but the outburst had been preceded by around 10,000 smaller earthquakes over the previous two months, steam venting from the flanks of the volcano and the injection of molten rock into its subterranean magma chamber. This caused the entire mountain to bulge upwards, particularly on its northern flank, which grew by some 150m (500ft) and became unstable. At 08.32 PST on Sunday, 18 May, an earthquake shook the volcano so violently that its entire north face slid away in a massive rock avalanche. One of the largest landslides on record, it crashed down the mountain side at 200km/h (more than 125mph), crested a ridge 10km (6 miles) away, and spread about 3 cubic km (0.7 cubic miles) of debris across 60 square km (24 square miles) of the landscape.

OPPOSITE A false-colour satellite image of Mount St Helens in October 2004 (with ash and rock coloured grey against the green of the surrounding forests), reveals the volcano's continuing restlessness, with a red glow showing where lava had broken through the newly formed lava dome within the caldera.

Pyroclastic killers

The collapse abruptly released pent-up pressure in the magma chamber, causing hot gases within to explode violently, pulverizing the surrounding rock and lava to form 'pyroclastic flows' of hot gas, ash and pumice. Accelerating to more than 1,000km/h (620mph) as they expanded, these flows overtook the rock avalanche and fanned out across some 600 square km (230 square miles) of forest. With temperatures of around 700°C (1,300°F), they incinerated all life in their path, including several thousand big game animals and some of the human victims of the eruption.

By the time the pyroclastic flows had travelled for 13km (8 miles), their energy had dissipated somewhat, and trees were just knocked down instead of being seared away altogether. Beyond a range of 30km (19 miles), trees were left standing but stripped of all foliage. Altogether there were some 17 separate pyroclastic flows on 18 May alone, and they continued intermittently for more than a year. When scientists saw the hummocky deposits and rounded knolls created in the landscape, they realized that such distinctive features might reveal where similar processes had happened in the past. Since the 1980s some 200 such deposits have been found – for example at Unzen in Japan, where a similar 'flank collapse' in 1792 led to the deaths of some 15,000 people.

'The blast was triggered by an earthquake of magnitude 5.1 that originated roughly 1.6 km (1 mile) below ground, but the outburst had been preceded by around 10,000 smaller earthquakes over the previous two months.'

Ash clouds and mud flows

As the pyroclastic flows were racing over the landscape below, a mushroom cloud of ash blasted its way 24km (15 miles) into the sky, lifting its load of ash above the lower layers of atmosphere into the stratosphere, from where it spread out across 11 states. High-altitude winds carried the ash east-northeast, reaching Yakima in Washington State, some 145km (90 miles) away, by 09.45, and dumping some 10cm (4in) of ash on the region. Altogether some 540 million tonnes of ash were spread over 60,000 square km (22,000 square miles), while the finest particles and gases were carried around the globe over the following two weeks by high-altitude winds. The weight of uplifted material was so great that parts of the ascending cloud collapsed back down onto the volcano flanks, generating further pyroclastic flows. As the flows crossed snow and water, they flashed explosively into steam, forming small craters and blasting ash up to 2km (1.2 miles) into the air.

Heat from the eruption also melted snowfields and glaciers around the volcano, generating huge muddy floods, charged with volcanic debris and meltwater, known as lahars. Initially flowing at 140km/h (90mph), they spewed down river valleys, swept away bridges and finally dumped vast amounts of debris into the lower reaches of the rivers. Such unpredictable flows are highly dangerous – five years after the Mount St Helens outburst, lahars generated by a small eruption of Nevado del Ruiz in Colombia killed more than 23,000 people as they engulfed mountainside villages.

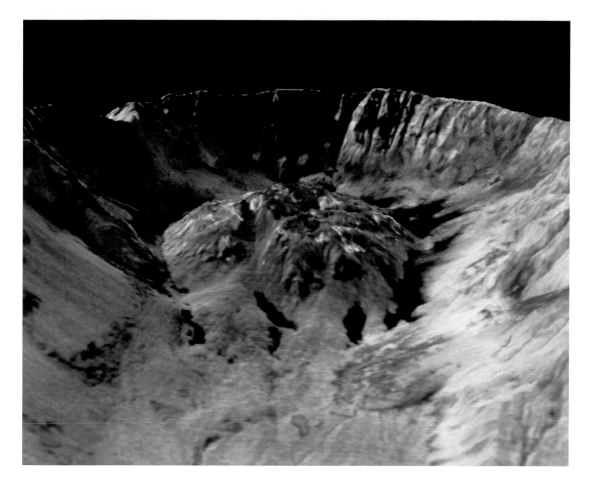

The volcano today

Today, Mount St Helens is still very much an active volcano, and since 1980 a series of lava domes have grown up within the new crater. Within its geological history there have been even bigger and longer-lasting eruptions than those of 1980, and these are likely to recur, though it will take some time for the magma chamber to recharge itself. Nevertheless, the volcano is closely monitored for any warning signs of eruptive activity.

Mount St Helens is just one of 13 potentially active volcanoes in the Cascade Range of the US Pacific Northwest, produced by the subduction of the oceanic Juan de Fuca Plate beneath the continental North American Plate. They tend to erupt explosively, with an average of some two eruptions per century, and since the eruption of Mount St Helens, all have been closely monitored. Particular attention is given to Mount Rainier, the tallest volcano in the United States outside of Alaska, since more than a million people within the Tacoma-Seattle area of Washington State live in its shadow. This 4,392m (14,411ft) volcano has produced at least ten eruptions in the last 4,000 years and is built on giant lahar deposits that are less than 1,200 years old.

ABOVE A false-colour image shows the new lava domes that have grown up within the Mount St Helens summit crater since its last major eruption in 1980. There have also been short-lived explosive events, with steam and ash rising some 12km (7.5 miles) into the air within minutes.

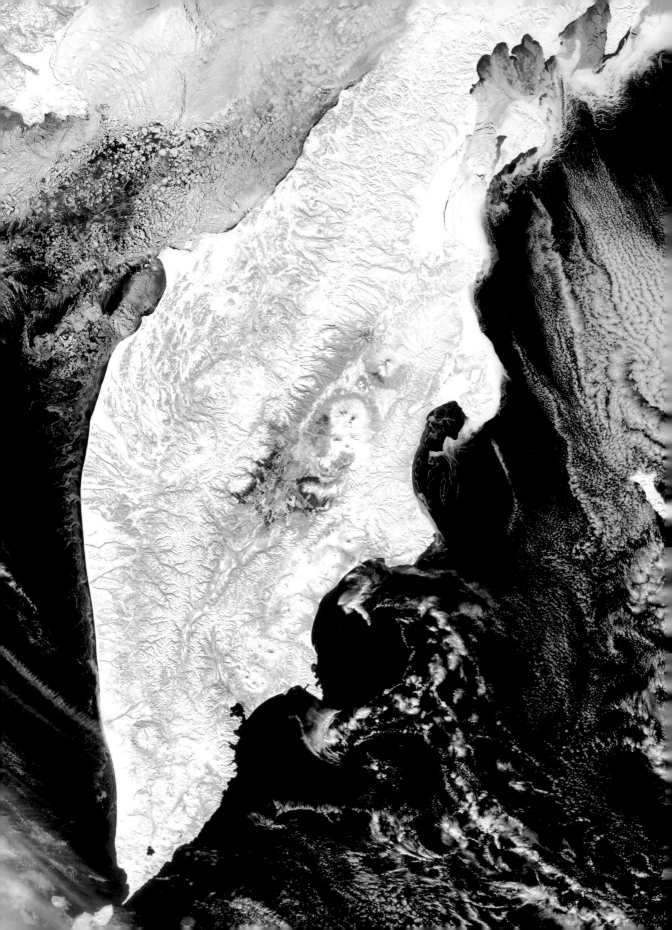

Island arc volcanoes

24

DEFINITION A CHAIN OF VOLCANIC MOUNTAINS PRODUCED WHERE
ONE PLATE OF OCEANIC CRUST IS SUBDUCTED BENEATH ANOTHER

DISCOVERY VOLCANIC ISLAND CHAINS ARE FOUND ALL AROUND THE
RIM OF THE PACIFIC OCEAN

KEY BREAKTHROUGH THE NATURE OF THE SUBDUCTION ZONES
BENEATH VOLCANIC ISLAND ARCS WAS IDENTIFIED BY WADATI AND
BENIOFF IN THE 1930s

IMPORTANCE ISLAND ARC VOLCANOES SUCH AS MOUNT FUJI PRESENT
A SIGNIFICANT THREAT TO LARGE CENTRES OF POPULATION

Well over half of all presently active volcanoes above sea level are strung out around the rim of the Pacific Ocean, forming what is known as the Ring of Fire. Many of them lie along island chains called volcanic island arcs and have a number of features in common.

As their name suggests, volcanic island arcs have a characteristic curvi-linear form – concave on the side that faces their nearest neighbouring continent, and with a curving deep-ocean trench running parallel to the islands themselves on their seaward side. These features indicate the presence of a subduction zone where one plate is disappearing below another, and geologists now know that they are created where two plates of oceanic crust converge and one, generally the older and slightly denser plate, is subducted. Such processes are particularly widespread in and around the Pacific, and island arcs account for some 65 percent of all Earth's active surface volcanoes.

The arcs tend to be significantly longer than they are wide – the Izu-Mariana, one of the best known, is more than 2,500km (1,550 miles) long, and stretches from Japan in the north to the island of Guam in the south. The 'angle of dip' of the subducted Pacific Plate creating this arc ranges from 50 degrees in the north of its range to a vertical 90 degrees in the south, and seems to be related to increasing age (ranging from 125 to 170 million years) and density of the oceanic crust. The Pacific Plate is a relatively old and cold segment of oceanic crust and therefore continues to descend steeply within the mantle – seismic surveys (see page 39) have traced it down to more than 1,000km (620 miles), and it probably reaches right down to the boundary with the core itself. As the plate has bent downwards, parallel cracks have developed across it and these are deeply penetrated by seawater.

OPPOSITE This true-colour image of Russia's Kamchatka Peninsula covered in snow was captured by the MODIS instrument aboard NASA's Terra satellite on 1 December, 2002. A line of highly active volcanoes running along the length of the peninsula are the result of the subduction of the Pacific Plate beneath the Eurasian Plate.

In the 1930s, American geophysicist Hugo Benioff and Japanese seismologist Kiyoo Wadati independently discovered that the foci of large earthquakes tend to occur along dipping planes that decline under a continental margin or island arc. Now known as Wadati-Benioff zones, the surface manifestation of these regions is a deep-ocean trench where the crust is bent downwards in the process of subduction. The dip of Wadati-Benioff zones typically varies from 25 degrees up to 70 degrees, but can exceptionally be as much as 90 degrees. The opposing motion of converging plates builds up intense pressures that are periodically released in the form of earthquakes and slippage. Most of these earthquakes originate within the upper 100km (62 miles) of the zones but some occur down at depths of more than 600km, (375 miles) especially directly below island arcs.

As the denser subducting crust descends below the more buoyant overlying plate, the heat released can trigger melting in the overlying crust, creating pockets of molten magma that in turn produce volcanoes on the surface. New continental crust is produced above the arc by a process called accretion, in which rising molten material adheres and builds up on the underside of the overlying plate, increasing its thickness and buoyancy so that the overlying plate starts to emerge above sea level, carrying the newly formed volcanoes with it. The process of subduction is responsible for a large majority (85 percent) of historic volcanic eruptions, including most of the catastrophically large explosive eruptions that have occurred throughout human history and prehistory.

'Fuji is a large composite volcano whose magmas are generated by the subduction process, and it poses a major threat to a very densely populated region.'

Forearcs and back-arcs

The region between the trench on the oceanward side and the volcanic islands themselves is known as the forearc. In this particular instance, its sediments, derived from erosion of the arc landmass, are subducted rather than accreted onto the arc as can happen in other subduction situations. As they are subducted, the sediments carry a considerable volume of seawater with them. At depths of around 100km (62 miles) this water is squeezed out of the subducted sediments and the descending plate itself, into the surrounding upper mantle where it lowers the melting point of some of the minerals. Partial melting generates magma, which eventually rises to the surface and erupts through the island volcanoes.

Over time the subducted 'incoming' plate retreats, while the upper 'opposing' plate continues to advance through a process known as slab roll-back. Stretching and thinning of the opposing plate allows the mantle to well up from beneath, creating what is known as a back-arc basin. As the pressure from above is relieved, the rising mantle material here also experiences partial melting to create magma, which erupts all along new spreading centres in the basin, parallel to and behind the island arc. As a result, the volcanic island arc can continue to regenerate itself for a considerable time.

Mount Fuji

The 3,776m (12,388ft) cone of Mount Fuji is the highest volcano in Japan and one of the most famous in the world. This sleeping giant, one of the largest peaks in the Izu-Mariana chain, has developed at the junction of the incoming and subducting Pacific Plate, the Philippine Sea Plate and the continental Eurasian Plate. Fuji is a large composite volcano whose magmas are generated by the subduction process, and it poses a major threat to a very densely populated region.

The last significant eruption, in 1707, followed a very strong magnitude 8.2 earthquake, whose epicentre lay some 50km (30 miles) south of the volcano and which seriously damaged the city of Osaka. The basaltic magma feeding the volcano itself was highly enriched with water, which may explain the violently explosive nature of this eruption. In contrast, a previous eruption in AD 864 also involved basaltic magma, but had a very different character, most likely because in that case the magma had lost much of its water content. Such contrasting behaviour in a single volcano makes it very difficult to build up an accurate assessment of the risks to those living nearby.

25

Hotspot volcanoes

DEFINITION A MAJOR CENTRE OF IGNEOUS ACTIVITY CREATED AS EARTH'S CRUST MOVES OVER A LONG-LIVED, HOT MANTLE PLUME

DISCOVERY THE 'HOTSPOT' MODEL WAS FIRST PROPOSED BY JOHN TUZO WILSON IN 1963

KEY BREAKTHROUGH IN THE 1970s W.J. MORGAN CONNECTED WILSON'S HOTSPOTS TO RISING PLUMES OF HOT MANTLE MATERIAL

IMPORTANCE HOTSPOTS BUILD VOLCANIC ISLAND CHAINS IN OCEAN CRUST, BUT CAN CREATE HIGHLY EXPLOSIVE SUPERVOLCANOES WHEN THEY BURN THROUGH CONTINENTAL CRUST

Almost any volcano can be potentially dangerous if you get close enough to the source, but at least most volcanoes seem to be confined to tectonic boundaries. However, the last few decades revealed evidence for another form of volcanism on a vast scale with potentially global implications.

In the mid-19th century, the German geologist Karl von Fritsch (1838–1906) was the first to identify a significant pattern – the progression in age among certain chains of volcanic islands. Based on the extent of erosion on the various Canary Islands, he concluded that, as one moved away from the continent of Africa and towards the mid-Atlantic, the islands became gradually younger. His conclusion was supported by the greater number of historic eruptions recorded on the most westerly islands. Later in the century, US geologist James Dwight Dana made similar observations about the Hawaiian islands, which grow steadily older towards the west-northwest.

However, it was not until 1963 that the underlying cause of this pattern became clear, when Canadian geologist John Tuzo Wilson (1908–93) used the theory of plate tectonics to formulate the 'hotspot' volcanism model. Wilson argued that there are specific locations on Earth's crust beneath which intense heat flow gives rise to melting in the lower crust and surface volcanic eruptions. While these sites seem to be anchored and persist for considerable periods of geological time, the overlying plates of the lithospheric crust are in more or less constant motion, so that as they pass over the hotspot, a series of volcanoes is created. In ocean settings, such as the Hawaiian and Canary Islands, the manifestation of such eruptions is a submarine volcano that grows and may eventually reach the surface, only to become extinct once it has passed over the hotspot.

OPPOSITE The volcanic Hawaiian Islands have developed in the Pacific Ocean above a mantle hotspot that originated some 85 million years ago. Movement of the Pacific Plate has carried the oldest volcanoes some 6,000km (3,700 miles) to the northwest. As each volcano is torn from its hotspot 'root' it becomes extinct, cools and gradually shrinks to form a submerged seamount.

RIGHT A false-colour satellite image shows the present-day lava-based terrain around Yellowstone Lake (coloured black in this image) and giant caldera in Wyoming, USA. The last major eruption of this enormous supervolcano occurred some 640,000 years ago.

Wilson's idea was developed in the 1970s by William Jason Morgan, who realized that the rising heat sources, which he called mantle plumes, were effectively columns some 100–150km (60–93 miles) in diameter, rising through the mantle. Not only do these plumes transport heat and generate melting, but they can eventually lead to plate fragmentation, and as a result are one of the driving mechanisms for tectonic motion.

The Hawaiian-Emperor chain

The Hawaiian-Emperor chain is now known as a classic example of hotspot development. Over some 85 million years, some 129 volcanoes have burned their way through the ocean crust as it has moved across the hotspot. They form a chain of islands and submerged seamounts some 6,100km (3,800 miles) long, stretching from Hawaii, west-northwest across the Pacific, with a distinct 'dogleg' dated to around 43 million years ago. Beyond this change in direction, the older submerged volcanoes of the Emperor seamount chain continue north towards the Aleutian island arc. Based on the distribution of the seamounts, geologists calculate that the Pacific plate has had an average speed of just over 7cm (2.8in) per year.

The volcanoes of Hawaii produce low-viscosity 'runny' basaltic magmas, which flow away from the vents and fissures, and accumulate as vast shield volcanoes with low slopes. Lacking unpredictable explosive activity, they are relatively safe to approach and attract huge numbers of tourists. But hotspot volcanoes in continental settings can be very different beasts.

Continental hotspots

Yellowstone volcanic caldera in the northwestern USA is the latest manifestation of a hotspot that began to form around 17 million years ago Throughout the hotspot's lifetime, the North American Plate has slowly drifted southwest at a relatively slow speed of around 2cm (0.8in) per year. Initially, rising heat set off a major pulse of extensive volcanism across Oregon, Idaho and Washington, producing the 16–17 million-year-old Columbia River lavas, along with the lavas of the Snake River Plain, which cover an area of 400 by 100km (250 by 62 miles). At first, the hotspot-generated magmas were predominantly basalts and therefore, like the Hawaiian eruptions, formed 'runny' lavas. However, more recent magma became contaminated by silica-rich minerals derived from the continental crust, increasing its viscosity and leading to far more explosive eruptions. As present-day Nevada moved across the hotspot, five major explosions produced large volumes of viscous 'rhyolitic' lava and ashfalls over a wide region. So much material was erupted that the floors of the volcanic craters collapsed to form vast sunken depressions or calderas ranging from 15 to 26km (9 to 16 miles) in diameter.

The history of the Yellowstone volcanic complex began around 2.1 million years ago, with the first of three more caldera-forming eruptions. Huge flows of granitic lava called ignimbrite, containing up to 2,500 cubic km (600 cubic miles) of magma, erupted at roughly 700,000 years intervals, and were among the largest volcanic explosions of the time. At 64 x 93km (40 x 58 miles), the Island Park Caldera is the oldest and largest, with the smaller, 1.2 million-year-old Henry's Fork Caldera lying within it. The Yellowstone Caldera itself, some 72 x 55km (45 x 34 miles) across, is the remnant of the most recent major eruption, some 640,000 years ago. Since then, there has been a succession of far less violent outbursts, including lava and steam-powered geysers fuelled by magma-related heating beneath the surface. Heated groundwater within these geysers becomes chemically enriched as it scavenges minerals from the surrounding rocks, which are hydrothermally altered and turned yellow in the process.

'Wilson argued that there are specific locations on Earth's crust beneath which intense heat flow gives rise to melting in the lower crust, and surface volcanic eruptions.'

Fortunately, studies of Yellowstone's eruptive material suggest that its explosive potential is probably now exhausted. However, the North American Plate is continuing to move, and in another million years or so, the hotspot could produce another supervolcano, probably somewhere close to the Canadian border. Any eruption on the scale of the previous major outbursts would be a catastrophic event for the life and environments of the North American continent and would have global consequences, so it's little wonder scientists still keep Yellowstone under close observation.

26 Earth's greatest eruptions

DEFINITION LARGE IGNEOUS PROVINCES ARE MASSIVE OUTPOURINGS
OF VOLCANIC LAVA COVERING HUGE AREAS

DISCOVERY THE SIGNIFICANCE OF THESE ERUPTIONS HAS ONLY BEEN
RECOGNIZED SINCE THE 1990S, ALTHOUGH THEIR SITES HAVE
BEEN KNOWN FOR MUCH LONGER

KEY BREAKTHROUGH RECENT DISCOVERIES HAVE SHOWN THAT SOME
'LIP' ERUPTIONS CAN INVOLVE EXPLOSIVE SILICA-RICH LAVAS

IMPORTANCE LIPS MAY HAVE PLAYED A MAJOR ROLE IN CAUSING
MASS EXTINCTIONS THROUGHOUT THE HISTORY OF LIFE ON EARTH

No historic eruptions have approached anything like the scale of the prehistoric super-eruptions known as Large Igneous Provinces. Their presence in the rock record shows that such exceptional outbursts have punctuated Earth history, often coinciding with major extinctions.

Fortunately for us, humans have very little experience of the truly immense scale and power of the largest volcanic eruptions. But around the world, geologists have found traces of some 18 known Large Igneous Provinces (LIPs) that have formed since Permian times, around 300 million years ago. Such eruptions occur on average every 17 million years or so, and since the last LIP erupted around 16 million years ago, another could begin at any time. What's more, many LIPs coincide with major extinctions, presumably through their effects on ocean and atmospheric chemistry and climate. A handful of historical eruptions, though on nothing like the same scale as true LIPs, can give us some idea of what to expect when the next one occurs.

Sizing up LIPs

A typical LIP event involves lavas with cumulative volumes of more than 100,000 cubic km (24,000 cubic miles) spewing out in tens or hundreds of eruptions across a few million years. The flows smother vast areas of Earth's surface, with dimensions ranging from thousands of square kilometres up to a million or more (the area of countries such as Iceland or Portugal). The largest known LIP is the Ontong Java Plateau, estimated to have originally covered some 2 million square km (770,000 square miles), or about the size of Alaska. Most of this area was (and still is) beneath the Pacific Ocean near the Solomon Islands, where perhaps as much as 100 million cubic km (24 million cubic miles) of magma were erupted onto the ocean floor between

OPPOSITE Time-lapse photography of volcanic eruptions at night can provide startling images of dramatic fire fountains. However, thanks to their rarity, no human has ever witnessed the true immensity of some of Earth's largest eruptions.

125 and 119 million years ago, building up a thickness of some 30km (19 miles) in places.

LIP lavas

Recent research has shown that LIPs involve eruptions of two main types of lava – basaltic and silica-rich – producing equally vast volumes of lava with very different styles of eruption. The basaltic flood or plateau lavas, found in LIPs such as the North Atlantic Tertiary Igneous Province, the Columbia River Basalts and India's Deccan Traps are the most easily recognizable. The Deccan lavas have an estimated volume of 8 million cubic km (1.9 million cubic miles), and erupted between 67 and 60 million years ago, coinciding with the extinction event at the end of the Mesozoic Era (see page 261).

'The flows smother vast areas of Earth's surface, with dimensions ranging from thousands of square kilometres up to a million or more (the area of countries such as Iceland or Portugal).'

These basaltic eruptions are similar in style to those currently occurring on a smaller scale in the Hawaiian Islands of the Pacific, where the low-viscosity magma is hot and free-flowing. Such eruptions are relatively quiet, without large-scale violent explosions. Lava tends to run freely out of fissures and down the nearest slopes for considerable distances before cooling enough to solidify. Eruptions may last for years or decades, releasing magma at rates of between 10,000 and 1 million tonnes per second.

As a result, these Hawaiian-style eruptions build up extensive flat layers of basalt lavas, with individual flows up to 100m (330ft) thick and between 100 and 200 cubic km (24 and 48 cubic miles) in volume. The magma is derived directly from upwellings in the mantle, mostly by a process known as decompression melting in which reduction of pressure on mantle minerals changes their state from a hot solid to a mobile liquid that migrates upwards.

Explosive eruptions

Recently geologists have realized that silica-rich eruptions can also occur on a similar scale to basaltic ones. Their discharge rates are virtually unknown, but some geologists have estimated they could be up to ten times greater than for basaltic lavas. These silica-rich eruptions are found exclusively in continental settings, where mantle-derived magmas can be chemically altered by the addition of melted material from the continental crust. Thanks to these additions, the lavas run less freely and erupt more explosively, probably forming what are known as pyroclastic fountains.

With discharge rates of 1 million to 100 million tonnes per second (similar to that of the 1991 eruption of Mount Pinatubo in the Philippines at its peak), such eruptions produce flows of incandescent gaseous pumice and ash that re still not enough to fuse together into a rocky layer known as a welded gnimbrite once the flow has stopped.

The 132 million-year-old Paraná-Etendeka Traps (found in parts of Brazil and southwestern Africa) are a typical example of a silica-rich LIP, formed from an estimated 5,000 cubic km (1,200 cubic miles) of magma. Despite their tremendous rate of discharge, even eruptions on the scale of Pinatubo would take many months to create an LIP on such a scale.

The simple geometry and structure of Paraná-Etendeka's 100m (330ft) flows suggest that they were erupted over short periods, perhaps a matter of days. However, eruptions were probably less frequent – every 100,000 years within silica-rich LIPs compared with 1,000 to 10,000 years of the basaltic LIPs. There is evidence for at least nine major eruptions during the formation of the Paraná-Etendeka LIP, all within about a million years. These would have required the storage of thousands of cubic kilometres of silica-rich magma in the lower crust.

Many questions about LIPs remain to be answered – particularly how such large volumes of magma can be generated, stored and then erupted from the magma reservoir. What's more, while basaltic and silica-rich LIPs were previously thought to be generated by separate processes, recent evidence has emerged for some LIPs involving both types of eruption – geologists are still investigating exactly how this could happen.

BELOW The strange columnar rocks of the Giant's Causeway in Northern Ireland are a result of the cooling and shrinking of thick lava flows that erupted over the region some 55 million years ago as the North Atlantic rifted apart. Originally the basalt lavas covered some 1.8 million square km (700,000 square miles), forming the Thulean Igneous Province. Remnants of this mighty eruption are now found from Northern Ireland to Greenland, the Faroe Islands and western Norway.

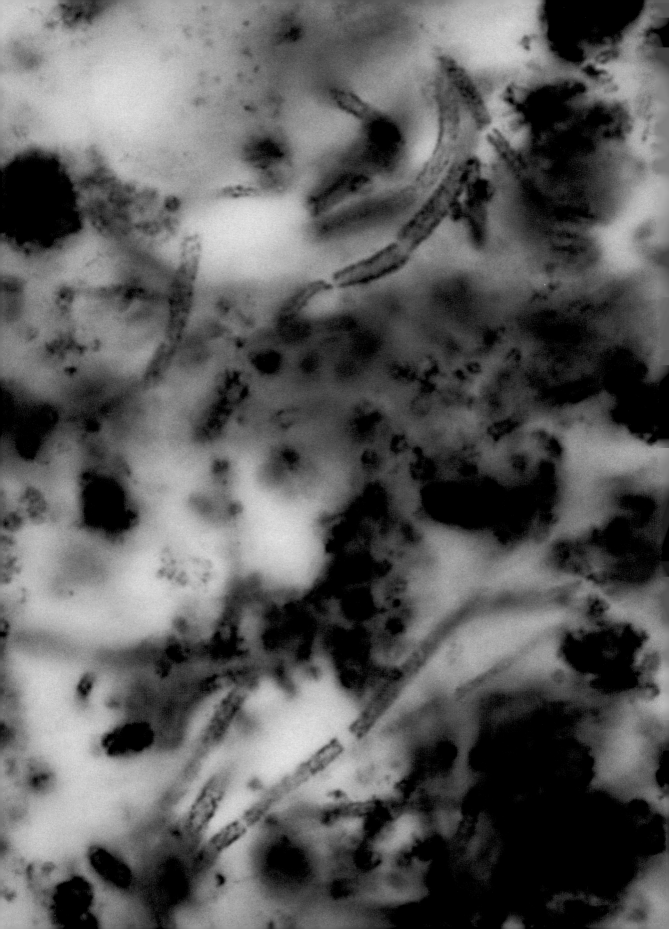

27

First fossils?

DEFINITION ORGANIC FOSSIL REMAINS OF THE EARLIEST SINGLE-CELLED ORGANISMS TO EXIST ON EARTH

DISCOVERY THE FIRST FOSSILS OF PRECAMBRIAN ORGANISMS WERE DISCOVERED IN CANADA'S GUNFLINT CHERT IN THE 1950S

KEY BREAKTHROUGH DISCOVERIES FROM SOUTH AFRICA'S AGNES MINE PUSH THE EARLIEST KNOWN FOSSILS BACK TO 3.2 BILLION YEARS AGO

IMPORTANCE THESE FOSSILS POSE MANY QUESTIONS ABOUT THE ORIGINS OF LIFE AND THE FIRST MICROBIAL ORGANISMS

Organic microfossils estimated to be around 3.2 billion years old discovered in a South African gold mine have been hailed as the earliest fossil evidence for life on Earth – but what exactly do these fossils tell us about the first forms of life?

Throughout the late 19th century, following the publication of Charles Darwin's *On the Origin of Species*, there was an expectation that the fossil remains of primitive organisms would soon be found in the ancient rocks that predated the appearance of more complex life forms. By 1859, when the *Origin* was published, the fossil record of life had already been traced back to early Cambrian strata, and so there was a widespread hope that more primitive fossils would be found in the rocks of the Precambrian Era. But it was to be almost a century before this expectation was fulfilled, and even today, fossil traces of Earth's earliest life forms remain hard to track down.

The main reason that early life is hard to find in fossil form is largely due to the mechanics of fossilization itself. Preserving the bodies of living or recently dead organisms in rock is not easy, and even when an organism has physically and chemically tough tissues such as bone, shell or wood, the processes of decay, burial and chemical transformation into rock generally filter out all but the most resistant remains. Since Precambrian life was entirely soft-bodied (see page 181), it's little wonder that these generally smaller organisms stand far less chance of being preserved. On top of this, of course, the older rocks are, the less chance of their surviving since formation without suffering extensive reworking or distortion.

OPPOSITE Microbes from around 2 billion years ago are preserved in the silica minerals of the Gunflint Chert of Ontario, Canada. Viewed with a microscope at a magnification of 400 times, this thin section of a chert sample reveals a variety of submillimetre-sized organic remains, including spheroidal colonies and filaments similar to those of living blue-green algae.

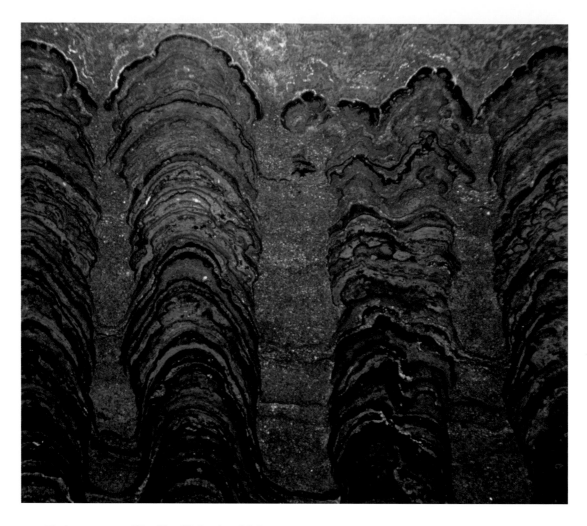

ABOVE The first
genuine indication
of life in Precambrian
strata came from the
discovery of laminated
dome structures
called stromatolites.
These formed
where thin sheets of
cyanobacteria (blue-
green algae) grew over
the seabed in quiet
and shallow tropical
waters. Deposition
of fine sediment
prompted new growth
of the light-seeking
algae and repetition of
the process produced
the laminated
structures.

The Gunflint microbiota

In the 1950s, the first genuine Precambrian fossils finally came to light in
the Gunflint Chert – a group of banded iron formation rocks (see page 125)
that lie exposed along the north shore of Lake Superior in western Ontario,
Canada. Even then, it took another ten years before the microscopic details
of these single-celled organisms were published. The fossils proved to be
mostly tiny spheres, rods and filaments, less than 0.1mm (1/250in) long,
and today are generally considered to represent photosynthesizing microbes
called cyanobacteria (see page 25). They are remarkably well preserved in
silicate-rich sedimentary chert rocks that must have formed very quickly,
encasing and preserving the delicate cellular membranes of the organisms
before they had time to decay on the ancient shallow seabed. The quality
and antiquity of the 'Gunflint microbiota', now dated to around 1.88 billion
years old, astonished the scientific world – they preserve so much detail that
some biologists at first thought they must be more recent contaminants. And
while the Gunflint fossils are no longer the oldest known, they are still some
of the most diverse.

South African acritarchs

More recent discoveries have pushed the early fossil record much further back in time, to an astonishing 3.2 billion years ago. A remarkable treasure trove of microfossils has been recovered in mud rocks drilled from the walls of South Africa's Agnes gold mine, some 600m (2,000ft) below ground. Extracting the organic (carbon-based) remains is a delicate process – inorganic elements in the rock samples have to be dissolved away with hydrofluoric and hydrochloric acids to release the organic material from the surrounding 'matrix'. Single fossils are then extracted by hand and examined with an array of imaging techniques.

The microfossils consist mostly of simple flattened spheres, up to 0.3mm (1/80in) across. With no surface ornamentation or discernible internal structure, they represent the largest, as well as the oldest, micro-organisms discovered so far. Generally classed as 'acritarchs' (members of an extinct group of organic-walled microbes), these Agnes Mine fossils are somewhat larger than any other Precambrian acritarchs so far recorded, but are comparable to some found in much younger rocks.

Problematic affinities

The meticulous protocols observed in collecting and preparing the Agnes Mine fossils allows palaeontologists to be very confident of their authenticity and age, and there is no known non-biological process capable of producing such structures. But even if they are undeniably the remains of living organisms, working out their biological relationships is still a huge challenge, even at the basic level of trying to decide which of the three great domains of life – Archaea, Bacteria or Eukaryota – they might fit into. The fossils are bigger than any known archaean, and although some bacteria do grow to this size, their individual cells do not produce tough organic walls that would survive in this way. The photosynthesizing, colonial cyanobacteria, on the other hand, can develop thick-walled cysts, but again there are no known fossil cyanobacteria of this size. One possibility it that the Agnes Mine acritarchs were ancestors of the cyanobacteria, which developed toughened organic walls to their colonies.

'The quality and antiquity of the "Gunflint microbiota", now dated to around 1.88 billion years old, astonished the scientific world – they preserve so much detail that some biologists at first thought they must be more recent contaminants.'

One final option is that the fossils are the remains of protists – single-celled life forms with a far more complex 'eukaryotic' cell structure. But apart from their size, they display none of the features that would support such a link, and the first eukaryotic cells are believed to have developed much more recently. The oldest confirmed protist fossils are around 1.8 billion years old and come from China's Changcheng strata. Although they are of similar size to the Agnes Mine fossils, the Changcheng fossils have a complex multiwalled structure that reveals their eukaryote nature beyond doubt.

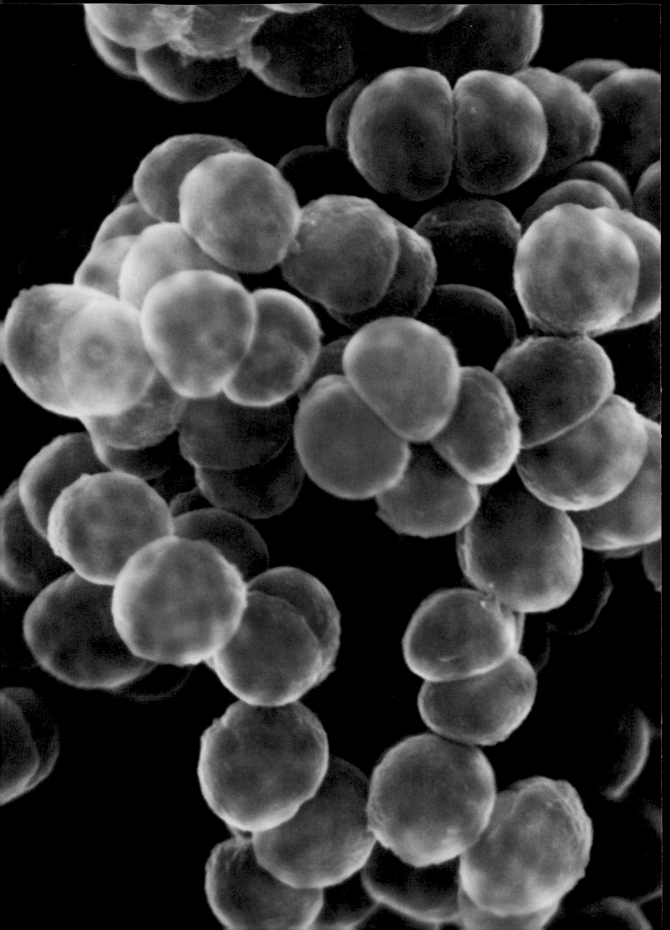

LUCA

28

DEFINITION THE MOST RECENT COMMON ANCESTOR SHARED BY ALL LIFE ON EARTH

DISCOVERY IN THE 1980S, MOLECULAR GENETICS ALLOWED BIOLOGISTS TO REVISE THEIR VIEW OF RELATIONSHIPS BETWEEN LIVING ORGANISMS

KEY BREAKTHROUGH A 2010 ANALYSIS CONFIRMED THAT ALL LIVING ORGANISMS PROBABLY SHARE A COMMON ANCESTRY

IMPORTANCE A COMMON GENETIC HERITAGE RAISES MANY QUESTIONS ABOUT HOW ORGANISMS EVOLVED AND DIVERGED THROUGH HISTORY

Concluding his revolutionary book *On the Origin of Species*, Charles Darwin (1809–82) suggested that 'probably all the organic beings [...] have descended from some one primordial form'. Today, this hypothetical organism is known as the last universal common ancestor, or 'LUCA'.

Darwin's cautious tone was perfectly reasonable given the cultural and scientific climate of the 1860s. Quite apart from the general hostility to the very idea of evolution, there was little scientific evidence to support the idea of a common ancestor – and indeed, the known fossil record seemed to lack any traces of early primitive life on Earth. As Darwin admitted, this apparent absence of fossils from the most ancient rock strata was a considerable problem for his theory, although today that issue has largely been resolved, with the discovery of fossil stromatolites, acritarchs and more complex organisms (see pages 25, 115 and 149). Nevertheless, the classic theory of 'evolution by descent through modification' still requires that everything alive today, from humans and whales, through insects and worms, to giant redwood trees, mushrooms and bacteria, has evolved from that single common ancestor.

The categories of life

The 19th century had seen a boom in the collection of plants and animals, presenting a considerable organizational challenge to biologists. A scientific method of classification had been put in place a century earlier by Swedish botanist Carl Linnaeus (1707–78) – a hierarchical scheme using the species as the basic unit. But this was essentially a means of ordering and grouping like with like, with no immediate evolutionary implications. For instance, humans, chimps and monkeys were grouped together in the Order Primates on account of certain common features or 'characters'. The

OPPOSITE Miniscule prokaryote bacteria, whose ancestry lies close to the origin of life, are still very much with us, including *Staphylococcus*, shown here magnified 4,000 times with an electron microscope. The numerous *Staphylococcus* species inhabit skin and mucous membranes and can cause a number of diseases and infections.

primates in turn were grouped along with other orders into the Class Mammalia. All the classes of animals formed the Kingdom Animalia, and all the plants the Kingdom Plantae, but by the mid-19th century it was already clear that some organisms – especially many microscopic, single-celled ones – could not easily be assigned to one group or the other, and so the Kingdom Protista was created.

The common characters and hierarchy of categories in the Linnaean scheme proved very useful to evolutionists, but the definition of the categories depended very much on which characters were deemed more or less significant – a matter of opinion that could vary from expert to expert. Nevertheless, the system was refined and elaborated, introducing many more subdivisions as closer relationships between groups of organisms were recognized. But at the same time, cracks began to appear in the system, revealing problems with the major divisions and their evolutionary relationships.

'Theobald's tests do not necessarily mean that life emerged just once on Earth, some 3.5 billion years ago – in fact, it's perfectly plausible for it to have arisen on several separate occasions.'

Revising the Tree of Life

In 1937, Swiss biologist Edouard Chatton (1883–1947) recognized that all life could be divided into two apparently fundamental groups, the eukaryotes and prokaryotes, depending on whether or not an organism's cells have a membrane-bounded nucleus that contains the bulk of its DNA. In 1959, US biologist Robert Whittaker (1920–80) took this further, proposing a five-kingdom 'Tree of Life', with four eukaryote divisions (Animalia, Plantae, Fungi and Protista) and a single prokaryote group, Monera.

Whittaker's system seemed to settle the issue for a while, but since the 1980s, increasing use of DNA data has seen a revolution in the way that life is classified – based not on external characters, but on 'molecular systematics' – the recognition of universal protein sequences in living organisms. This has led to some major revisions in the 'Tree of Life', and renewed interest in the LUCA problem. Today, three main lineages (called domains) are recognized – the eukaryotes as before, and two distinct types of prokaryote – the bacteria and the newly recognized 'archaebacteria'. In the 1990s, the archaebacteria were renamed Archaea to emphasize their uniqueness, since they include some highly unusual organisms known as extremophiles that can grow in conditions too hostile for other forms of life (see page 337).

Testing our ancestry

But even this new classification does little to 'root' the Tree of Life or address the question of its origins and Darwin's 'primordial form'. Is it possible that life has always been divided into three entirely incompatible strands, or do all living organisms have a common ancestor? In 2010, biochemist Douglas Theobald of Brandeis University, Massachusetts, carried out a statistical test of the theory, analysing the most likely ways in which 23 proteins in

a wide array of different life forms had evolved. The tests showed it was overwhelmingly likely that all life is indeed unified in its descent from a common ancestor.

However, Theobald's tests do not necessarily mean that life emerged just once on Earth, some 3.5 billion years ago – in fact, it's perfectly plausible for it to have arisen on several separate occasions. It is impossible to apply DNA analysis to extinct fossil species, so all the molecular evidence can really tell us is that all of today's extant life is derived from a single ancestor – one which need not necessarily have been the very first organism on Earth.

What's more, there may well have been extensive 'lateral gene transfer' among early life forms. This process, in which one organism incorporates DNA from another that is not its parent, is relatively common among micro-organisms, especially Bacteria and Archaea, and may have played a major role in the development of the complex eukaryotic cell (see page 129), but it also 'blurs the edges' between different groups of organism, and confuses the question of ancestry still further.

ABOVE *Pyrococcus* is an extremely primitive life form known as a microscopic prokaryote archaean, which lives in extreme conditions within hot springs at temperatures of 70–100°C (158–212°F). It has an unusually simple respiration system that could have been a precursor to the more complex mode of respiration found in higher organisms.

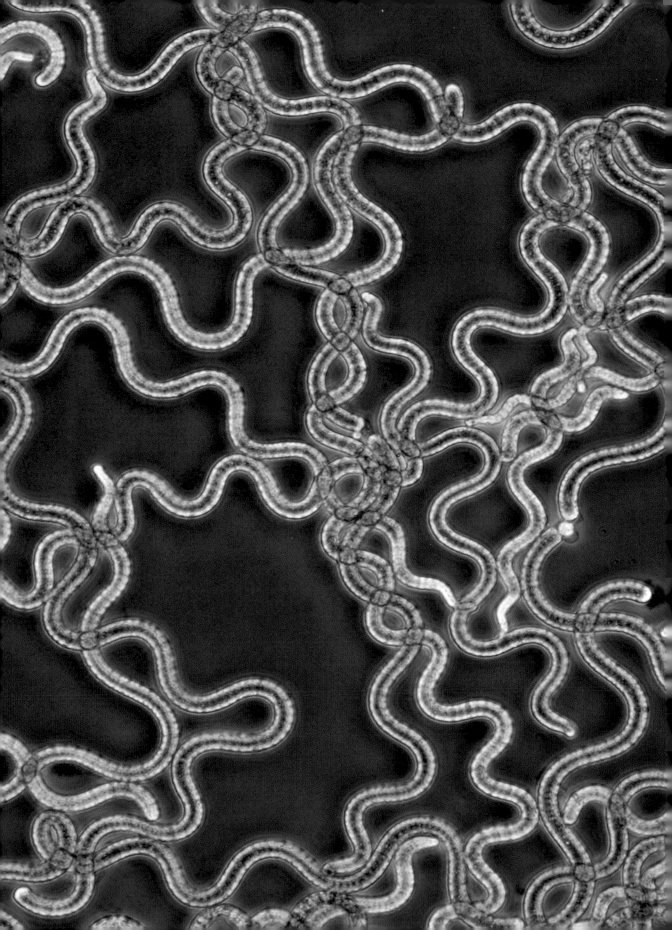

29

Controversial origins

DEFINITION PRESERVATION OF ANY REMAINS OF PRIMITIVE LIFE IN ANCIENT ROCKS IS INHERENTLY UNLIKELY

DISCOVERY THREAD-LIKE STRUCTURES RESEMBLING CYANOBACTERIA COLONIES WERE DISCOVERED IN 1993

KEY BREAKTHROUGH SPANISH SCIENTISTS SUCCEEDED IN REPRODUCING THE STRUCTURES USING INORGANIC TECHNIQUES

IMPORTANCE THESE CONTROVERSIAL FOSSIL TRACES SHOW HOW DIFFICULT IT CAN BE TO IDENTIFY THE EARLIEST LIFE

In recent years, geologists working in Western Australia claim to have discovered bacteria-like micro-organisms in 3.5 billion-year-old strata and organic chemicals from 2.7 billion-year-old rocks. But are they really what they seem?

Biologists generally agree that the first life on Earth would have been microscopic, water-based and single-celled. Aside from rare exceptions such as the 'acritarchs' recently found in South Africa (see page 113), such micro-organisms do not have any hard parts of the kind that are normally preservable in the fossil record. As a result, the hopes of finding bodily fossils from the very earliest life forms are remote – geologists must rely either on exceptional conditions of preservation, or on detecting some indirect signature of their existence left in the rock.

Fossil cyanobacteria?

In 1993, the microscopic remains of thread-like structures resembling photosynthetic cyanobacteria colonies were found in 3,465 million-year-old chert rock within the Warrawoona strata of Western Australia. These microfossils were proclaimed as by far the oldest apparent remnants of Earth's primitive life, and seemed to originate from a time well before the atmosphere and oceans became enriched with oxygen. Their appearance was taken to mark the very beginnings of biological oxygen production by photosynthesis.

However, a decade later, further analysis of the original fossils and other similar specimens threw doubt on their biological origin. While the carbon composition of the fossils could be biological in nature, it could also be

OPPOSITE Living cyanobacteria such as *Spirulina*, shown here, often form distinctive chains and build up into microbial mats. While the stromatolite structures formed by cyanobacteria colonies are preserved in the fossil record back to very ancient times, fossil evidence for the organisms themselves is far more elusive and controversial.

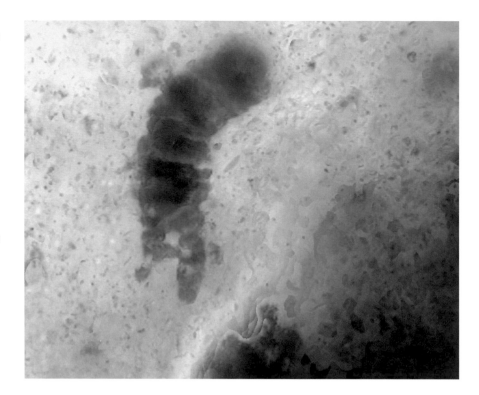

caused by the action of hydrothermally heated water on pre-existing minerals to form amorphous graphite (mineralized carbon). In fact, Spanish geologist Gael-Garcia Ruez and colleagues at CSIC-University of Granada succeeded in manufacturing just these kind of structures in the laboratory.

Furthermore, some biologists argued that the so-called fossils are not typical of filamentous bacteria or cyanobacteria, since they show features (such as branching) that are unknown in these organisms today. As a result of this reassessment, the Agnes Mine acritarchs (see page 113) must remain the oldest undisputed bodily fossils, at least for the moment.

Chemical fossils

However, even simple single-celled organisms are composed of complex organic molecules that do not just disappear. Upon death, these molecules degrade and disintegrate into simpler molecules, losing volatile components such as water but leaving behind a carbon-based residue that can be remarkably persistent, with the potential to form 'chemical fossils' within ancient sedimentary rocks.

In 1999, geologists recovered 2.7 billion-year-old chemical fossils of this kind from the Pilbara strata of Western Australia. On close analysis, they proved to contain a diverse range of naturally occurring fatty molecules (so-called 'lipid biomarkers'), whose origins are normally biological. These biomarkers included specific carbon compounds called hopanes and steranes, linked to the chemical composition of cell membranes formed by cyanobacteria and complex eukaryotic cells. Their discovery appears to push the origin of both groups back into late Archaean times, close to the first appearance and rise of Earth's oxygen supply. However, it also raises some serious problems – in particular the huge gap between the first chemical signatures of the eukaryotes and their earliest known appearance as bodily fossils (see below).

The hydrocarbon problem

One particular problem with reliance on chemical fossils is the danger that source rocks have been contaminated by more recent hydrocarbons – organic compounds that are notoriously mobile within Earth's crustal rocks. Since the Pilbara rocks are known to have suffered some metamorphism around 2.2 billion years ago, it is certainly possible that the biomarkers were introduced at this time. Nevertheless, supporters of Pilbara's importance have argued that the abundance of the biomarkers and their distribution indicates that they are indigenous to the rock. However, there is also another chemical clue that all may not be quite right – the ratios of carbon isotopes (atoms of the same element with different weights) in the biomarkers do not match those from other (non-biological) carbon residues called kerogens found in the surrounding sedimentary rocks. This suggests that the biomarkers could indeed be younger contaminants.

'These microfossils were proclaimed as by far the oldest apparent remnants of Earth's primitive life, and seemed to originate from a time well before the atmosphere and oceans became enriched with oxygen.'

In 2008, Birger Rasmussen of Australia's Curtin University of Technology reanalysed the Pilbara chemical fossils in an attempt to resolve these issues. He confirmed the disparity between the 'background' kerogen hydrocarbons and the extracted biomarkers, showing that the biomarkers were most likely introduced to the rocks at some later stage, probably after 2.2 billion years ago. However, other scientists, including Harvard biochemist Andrew Knoll, kept the debate open with claims to have found similar biomarkers in South African rocks contemporary to Pilbara.

If the Pilbara findings can be dismissed, then the oldest fossil evidence for cyanobacteria reverts to Canada's Belcher Islands, where body fossils dating back to 2.15 billion years ago have been found. The oldest confirmed eukaryotes, meanwhile, are fossils from around 1.8 billion years ago, found in China's Changcheng strata (though see page 115 for more on the Agnes Mine fossils).

If there is one lesson to be learned from these various controversies surrounding the earliest traces of life on Earth, it is that scientific investigations pushing the boundaries of knowledge – especially those in areas of research where far-reaching conclusions are drawn from limited samples of scientific data – always needs to be regarded sceptically, and ideally reassessed when there are contradictions in the results.

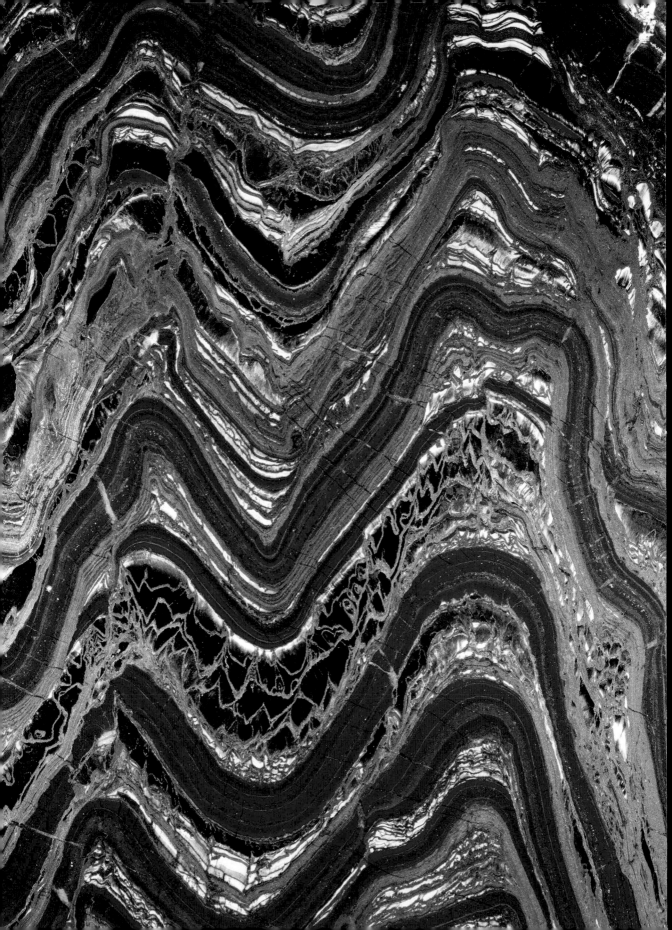

30

Rising oxygen levels

DEFINITION FLUCTUATIONS IN ATMOSPHERIC AND OCEANIC OXYGEN
IDENTIFIED THROUGH ANALYSIS OF BANDED IRON FORMATIONS

DISCOVERY THERE WAS A STEADY RISE IN OXYGEN LEVELS EVEN
PRIOR TO THE 'GREAT OXIDATION EVENT' 2.4 BILLION YEARS AGO

KEY BREAKTHROUGH ISOTOPE LEVELS IN THE BANDED FORMATIONS
CAN REVEAL OXYGEN LEVELS IN THE ANCIENT ATMOSPHERE

IMPORTANCE OXYGEN IN THE ATMOSPHERE IS AN IMPORTANT
INDICATOR FOR THE PRESENCE OF EARLY AEROBIC LIFE

The appearance and disappearance of the iron-rich sediments known as banded iron formations, which come and go from the rock record between 3.7 billion and 2.4 billion years ago, reveals a complex story of changing oxygen levels in Earth's ocean waters.

The Pilbara region of Western Australia is famous for its colourful red landscape, with hills composed almost entirely of iron-rich minerals interlayered with shale and silica-rich chert. These remarkable and distinctive deposits, known as banded iron formations (BIFs), were originally laid down on the seabed during early Precambrian times, more than 2.4 billion years ago, and have since been lifted above sea level by the powerful forces of tectonics. These cratonic rocks (see page 65) are economically important and are commercially mined for iron ore, both in Australia and in the United States, where similar rocks outcrop around Lake Superior. However, more importantly for geologists, the BIFs can also tell us about the chemistry of Earth's early oceans and atmosphere.

Scientists have long debated the process through which Earth's early atmosphere and oceans were gradually transformed by the introduction of large amounts of oxygen. There is little doubt that, more than 2 billion years ago, oxygen levels in Earth's early atmosphere were vanishingly small (less than 1 percent), and that the oceanic oxygen levels were equally low. Oxygen was apparently being generated in ocean surface waters by photosynthesizing microbes (see page 121), but was rapidly captured by chemical reactions with dissolved iron that was plentiful in the seas of the time, and so could not build up in sufficient quantities to oxygenate the water.

OPPOSITE Alternating bands of red haematite with siliceous chert and argillaceous shale are characteristic of the banded iron formations that occur in early Precambrian marine strata. Today, they are a major source of iron ore, but their origins reveal a time when ocean waters were deficient in oxygen.

The Great Oxidation Event

Recent geochemical analysis of early Precambrian seabed deposits, focused especially on the distinctive, widespread and abundant BIFs, supports this idea, but the rapid disappearance of the BIFs around 2.4 billion years ago led to the identification of a major change in Earth's environment known as the Great Oxidation Event (GOE). For the first time, this saw a significant rise in atmospheric oxygen, coinciding with the end of iron-rich, oxygen-poor conditions in ocean waters. The rapid rise in oxygen wiped out much of the microbial life of the time, which had evolved to survive and thrive in an oxygen-free environment. It is widely believed that they were supplanted by an explosion of photosynthesizing 'aerobic' cyanobacteria, which inhabited shallow waters and were ultimately responsible for the rise in oxygen.

However, the exact nature and timing of these rising oxygen levels and the link between atmospheric and oceanic oxygen are still contentious. This is particularly because the BIFs continue to appear in the geological record for some time after the GOE supposedly made such things impossible. One such event occurred around 1.9 billion years ago and a final one happened just 750 million years ago. Inspired by this puzzle, a new geochemical approach to the problem has revealed that the process of oxygenation was more subtle and less straightforward than previously thought.

BELOW
Characteristically red-coloured strata of the banded iron formation are exposed over vast areas of the Pilbara Craton in Western Australia. Lying at the heart of this ancient continental landmass, the flat-lying strata have been relatively undisturbed by folding or metamorphism for over 2 billion years, but they are deeply weathered and eroded.

Chromium isotopes

This new technique relies on the measurement of chromium isotopes (atoms of the same element with different masses) whose form can act as a tracer for the atmosphere's oxygen content. The elements chromium and manganese both formed fairly common minerals in the rocks of Earth's early crust and were locked away within them for as long as the atmosphere remained oxygen-free. But rising oxygen levels caused the manganese minerals to oxidize, reacting with chromium so that it also oxidized and became more chemically 'mobile'. Rainwater would then tend to leach the heavier chromium isotope (^{53}Cr) from the soil in preference to the lighter and more common form (^{52}Cr), carrying it to ocean waters where it reacted with iron and was immobilized again in deposits such as the BIFs. As a result of this complex chemical pathway, the quantities of the heavier chromium isotope measured in BIF deposits can be put to work as a useful indicator of atmospheric oxygen at the time they were laid down.

'The rapid rise in oxygen wiped out much of the microbial life of the time, which had evolved to survive and thrive in an oxygen-free environment.'

A 2009 study using this technique, carried out by an international team of scientists led by geochemist Robert Frei of the University of Copenhagen, produced some intriguing surprises. Firstly, there was an indication that oxygen levels began a long, slow rise well before the GOE 2.4 billion years ago. Atmospheric oxygen, it seems, was gradually building up through the photosynthetic work of marine microorganisms from around 2.7 billion years ago, suggesting that the cyanobacteria already had an evolutionary foothold.

However, there was a sting in the tail, since the later chromium record shows that roughly 500 million years after the GOE atmospheric oxygen rapidly disappeared, falling back to almost pre-GOE levels. Perhaps unsurprisingly, this event coincides neatly with a drop in oxygen dissolved in seawater and the reappearance of BIFs around 1.9 billion years ago.

Thereafter, other minerals can help to tell the story – atmospheric oxygen levels rose again, prompting a significant increase in the chemical weathering of iron sulphide minerals that were abundant in Earth's surface rocks. The accelerated weathering process delivered huge volumes of sulphate chemicals into deep, oxygen-poor ocean waters, where sulphur-eating bacteria processed them into hydrogen sulphide. As a result of this feeding frenzy, the sulphur bacteria flourished and may have gone through a period of evolutionary diversification. Finally, 750 million years ago, the BIFs make their last appearance during the widespread glaciations of the 'Snowball Earth' period (see page 137), when the oceans would have been cut off from supplies of atmospheric oxygen. The retreat of the glaciers was soon followed by another significant rise in oxygen levels throughout the atmosphere and ocean, and the end of BIF creation.

From simple to complex life

DEFINITION A MAJOR TRANSITION IN THE HISTORY OF LIFE, MARKED BY THE FORMATION OF THE FIRST COMPLEX EUKARYOTIC CELL

DISCOVERY IN THE 1960S, LYNN MARGULIS SUGGESTED THAT THE EUKARYOTIC CELL DEVELOPED BY INCORPORATING SIMPLER PROKARYOTE CELLS FOR DIFFERENT FUNCTIONS

KEY BREAKTHROUGH A RECENT STUDY SHOWS THAT THE APPEARANCE OF MITOCHONDRIA WOULD ALLOW A HUGE LEAP IN THE CELL'S AVAILABLE ENERGY AND POTENTIAL GENETIC COMPLEXITY

IMPORTANCE THE ORIGINS OF EUKARYOTIC LIFE HELP US TO BETTER UNDERSTAND THE NATURE OF OUR OWN CELLS

There is a fundamental division in the organization of life forms between the prokaryotes, with small and relatively simple cells, and the larger eukaryotes with cell nuclei. All complex life forms are eukaryotes, and they probably evolved in a single event from prokaryotes around 4 billion years ago.

For much of the 20th century, biologists believed that life was fundamentally divided into prokaryotes and eukaryotes, although this division was largely based on negative evidence (prokaryotes being defined by their lack of the cellular nucleus seen in eukaryotes). In 1977, however, Carl Woese proposed a fundamental subdivision *within* the prokaryotes based on differences in their structure and genetics.

Today, Woese's threefold division of life into eukaryotes, prokaryote bacteria and prokaryote archaeans has been upheld by further molecular studies (see page 117). Typically, prokaryotes are microscopic in size, lack eukaryote features, such as a membrane-bounded nucleus or subunits called organelles, and reproduce solely by asexual budding or fission (splitting apart to form two genetically identical offspring). Despite this simplicity, the prokaryotes are so successful that, according to some estimates, a typical animal body probably contains more prokaryotes living within it than it does cells.

Prokaryotes occupy most Earth environments where there is some modicum of water – bacteria are arguably the most successful organisms on the planet, and archaeans can survive extreme conditions where no eukaryote can exist (see page 337). In the circumstances, it's interesting to ask why more complex prokaryotic life forms never developed, and a new hypothesis argues that the difference is all a matter of energy.

OPPOSITE Mitochondria are the 'power plants' of eukaryotic cells, oxidizing sugars and fats to produce energy as part of the process known as respiration. Their development is thought to have marked a key stage in the origins of the complex eukaryotic cells. Here, mitochondria are coloured blue in a transmission electron micrograph (TEM) of a fat cell magnified 20,000 times.

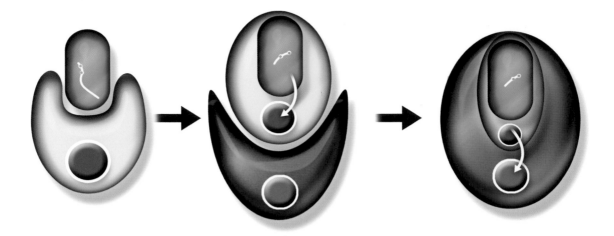

Eukaryotic life

Typically, eukaryotes are larger and more highly structured than the prokaryotes, with a membrane-bound nucleus and specialized organelles. Unsurprisingly, eukaryotes also have a much bigger genome than prokaryotes, containing more detailed genetic information to produce all these structures. Today, complex eukaryotic life is subdivided into half a dozen groups with names familiar only to biologists. For instance, animals, fungi and some other multicellular organisms belong in a major eukaryote division called the Opisthokonta. The plants and green algae, meanwhile, belong in another major division, known as the Archaeplastida.

Organelles called mitochondria are a particularly distinctive and important feature of the eukaryote cell – not only are they involved in generating the cell's chemical energy, but they also possess many prokaryote features. In the 1960s, American biologist Lynn Margulis suggested that the mitochondria originally developed from prokaryotic bacteria that 'invaded' an ancestral eukaryote cell and became incorporated into it.

Another important eukaryote innovation is a versatile mode of reproduction, involving two distinct mechanisms. In 'mitosis', the cell divides into two genetically identical daughter cells through a process that is superficially similar to the asexual reproduction of prokaryotes. In contrast, 'meiosis', used in sexual reproduction, involves the exchange and recombination of genetic material from two different parent cells, resulting in an offspring that has its own unique recombination of the parents' genetic material.

Conveniently for biologists, however, certain cell structures only ever reproduce by mitosis. In particular, the mitochondria in every living eukaryotic organism are derived from the mother (since they are contributed by the female egg cell, not the male sperm). While DNA from both parents is mixed within the nucleus, mitochondrial DNA (mtDNA) is essentially unchanged from one generation to the next, except for random mutation or 'genetic drift' that accumulates over time. The measurement of this drift in mtDNA forms the basis for the modern genetic classification of life, since the degree of difference in the mtDNA of two organisms reveals the time since they diverged from their common maternal ancestor.

Prokaryote constraints

In recent decades, almost every significant eukaryote feature, from genetic recombination and nucleus-like structures to internal membranes, large size, predation and parasitism, has been found in prokaryotes as well. It seems that over the vastness of evolutionary time, the prokaryotes have tried out almost every pathway towards increased complexity, but have never succeeded in making it stick.

A recent proposal from Nick Lane of University College London and William Martin of Dusseldorf University suggests that, while the move from simple prokaryotes to complex eukaryotes was a unique event, it was not the result of natural selection acting in the usual way (on gradually accumulated mutations of individual prokaryotes) but was instead the result of a sudden 'bioenergetic leap' in genome capacity. They argue that the presence of mitochondria within almost all eukaryotes shows that both cell and mitochondrion share a common origin in the same event.

Indeed, the very presence of the mitochondrion may have prompted this important evolutionary leap forward. Commonly described as the cell's 'power plants', mitochondria generate most of a cell's chemical energy and are also involved in other vital cell functions, such as growth, differentiation and death. They enable a highly efficient series of chemical reactions that oxidize nutrients to produce a molecule called adenosine triphosphate (ATP), which in turn supplies energy to the cell's metabolism.

'Despite their simplicity, the prokaryotes are so successful that, according to some estimates, a typical animal body probably contains more prokaryotes living within it than it does cells.

According to Lane and Martin, the opening of this 'metabolic pathway', and the sudden boost in energy available to the cell, promoted an astonishing expansion in the size of the genome, by a factor of 200,000 relative to prokaryotes. With additional energy suddenly available to manufacture specialized structures, cells with mitochondria 'on board' rapidly developed new gene families and increased the range and complexity of their proteins. Traditionally, biologists had assumed that genetic complexity came first, as a prerequisite to the acquisition or development of mitochondria, but the new approach emphasizes the mitochondrion's role in increasing the 'bioenergetics' of the cell and therefore the possibilities for the genome. Thus it was only after the acquisition of the mitochondria that 'energy-expensive' multi-gene processes and functions, such as the cell cycle, sex and the development of multicellular organisms, could evolve.

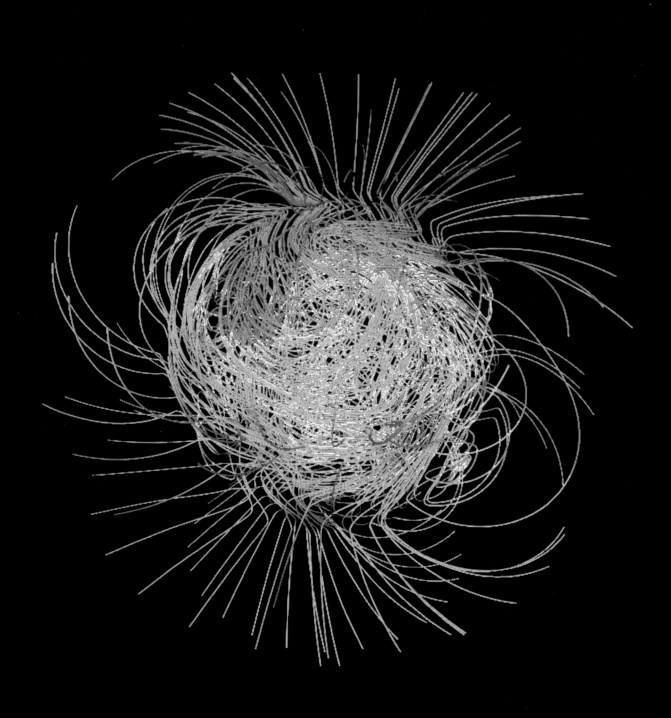

32 Rodinia

DEFINITION THE EARLIEST KNOWN SUPERCONTINENT, FORMED IN
THE LATE PRECAMBRIAN

DISCOVERY FIRST SUSPECTED FROM PALAEOMAGNETIC
MEASUREMENTS AND STUDIES OF ANCIENT MOUNTAIN BELTS

KEY BREAKTHROUGH NEW DETAILED MAPS OF THE SUPERCONTINENT
OUTLINE THE PRECISE ARRANGEMENT OF THE MAJOR CRATONS

IMPORTANCE THE BREAK-UP OF RODINIA COINCIDED WITH MAJOR
DEVELOPMENTS IN THE HISTORY OF LIFE

Rodinia was a late Precambrian supercontinent that saw most of the world's continents united in a single landmass. However, recent attempts to 'rebuild' the billion-year-old Earth push the current techniques of tectonic reconstruction to their limits.

The discovery that Earth's magnetic field can be 'fossilized' in certain ancient rocks has revolutionized our reconstruction of Earth's geological history. Iron-rich mineral grains align to the prevailing magnetic field before solidifying into rock and therefore preserve the original orientation relative to magnetic north, and even their latitude, thanks to the angle of magnetic 'dip' that varies with distance from the poles. What is more, the remnant magnetism of sea-floor rocks retains detailed evidence for the way in which the oceanic crust has developed and moved over the last 180 million years or so (see page 61). Alongside plate tectonic theory, this has allowed geologists to model the past movements of the many jostling crustal plates with great precision.

One slight drawback of this 'palaeomagnetic' technique, however, is that ancient continental rocks alone can only provide magnetic data about their latitude – establishing their longitude, especially before 180 million years ago when the help provided by the oceanic crust begins to peter out, requires other data. Studies of the geological history and palaeontology of the continental landmasses and surrounding shelves have helped to fill this gap and now allow fairly accurate plate reconstructions back to the beginning of Cambrian times, more than 540 million years ago. However, the further back in time we go, the more problematic and debatable these reconstructions become.

OPPOSITE The orientation of Earth's geomagnetic field, shown here as a fountain of field lines emerging from close to the planet's poles, can be measured in ancient rocks and used to show their original latitudinal position. This is essential for the reconstruction of ancient plate configurations, such as the Rodinian supercontinent.

Reconstructing the late Precambrian

The last few decades have seen increasing efforts to extend the techniques of tectonic reconstruction further back in time, to around a billion years ago. This time frame encompasses the late Precambrian world in which complex multicellular life first appeared, as well as the puzzling 'Snowball Earth' glaciations (see page 137). So far back in time, there is little reliable palaeomagnetic data, and still fewer fossils to help with the task – instead, reconstructions are based on the geological 'matching' of rock formations from different plates. At first, this largely relied on studying 'structural fold belts' (the remains of ancient mountain ranges), but new techniques based on chemical similarities between rocks, have proved much more precise.

However, Precambrian rocks are only exposed on Earth's surface in limited patches within ancient cratons in the heart of major landmasses, and the further back we want to look, the rarer appropriate rocks become. Useful geological data for reconstructing the world of the late Precambrian is largely restricted to shallow marine sediments deposited on the continental shelves of the ancient cratons and some igneous rocks, especially fragments of ancient ocean floor brought to the surface during orogenic (mountain-building) tectonic events and preserved by chance. As a result, there are many different ways of analysing the scattered fragments of data, and many different configurations of the ancient cratons have been suggested.

However, most recent reconstructions have agreed that around a billion years ago many, if not most, of the ancient cratons were clustered together in a supercontinent similar to the later Pangea (see page 146). This huge landmass, named Rodinia from the Russian word for 'motherland', lay mostly in the southern hemisphere. It seems to have coalesced around 1,200 million years ago, and finally began to break apart around 850 million years ago.

'Precambrian rocks are only exposed on Earth's surface in limited patches within ancient cratons in the heart of major landmasses, and the further back we want to look, the rarer appropriate rocks become.'

Putting Rodinia together

One of the most recent reconstructions of Rodinia, produced by John H. Stewart of the United States Geological Survey, uses the available palaeomagnetic data along with detailed geological information. This suggests that the ancient Laurasian continents of Baltica, Greenland and Laurentia (most of modern North America) came together with South America (itself composed of the even older Amazonia, Rio de la Plata and São Francisco cratons), creating a mountain belt and triggering volcanic activity down the west coast of South America. The east coast, meanwhile, butted against the four cratons making up western Africa (West Africa, Trans-Sahara, Congo and Kalahari).

The association of South America and Africa with the other continental cratons of Arabia, India, Antarctica and Australia to the east, was an

early manifestation of the great southern landmass generally known as Gondwana (see page 145). Within Gondwana, 'magmatic arcs' (where subduction and disappearance of ocean floor triggered volcanic activity in the overlying craton) mark the junctions of the Trans-Sahara and Congo cratons with Arabia and East Africa, while other collision zones mark the attachment of southern Africa, eastern India, Antarctica and western Australia.

The break-up of Rodinia began around 870 million years ago and continued until around 740 million years ago, by which time the main constituents had separated. Thereafter, however, the motion of the plates reversed and the continents began moving towards one another, resulting in further subduction and the production of new volcanic arcs (see page 101), especially between 600 and 500 million years ago.

Given the lengthy geological history of these plates over some 400 million years, it is not surprising that reconstructing their distribution and relationships to one another is the subject of so much speculation. Some attempts have been made to reconstruct the world in even earlier times, but the results are even more speculative.

ABOVE The coast of Cape Breton Island in Canada's Maritime Provinces exposes sedimentary and volcanic rocks that are over a billion years old. They formed during the collision of the continental crustal plates of Laurentia and South America to form the supercontinent of Rodinia.

33 Snowball Earth

DEFINITION A DRAMATIC PHASE OF LATE PRECAMBRIAN HISTORY, DURING WHICH EARTH WAS COVERED BY EXTENSIVE ICE

DISCOVERY SUGGESTED BY SIR DOUGLAS MAWSON AS EARLY AS 1949 AND INDEPENDENTLY REVIVED MORE RECENTLY

KEY BREAKTHROUGH DISCOVERY OF GLACIAL DROPSTONES AND OTHER GLACIAL FEATURES IN LOCATIONS THAT WOULD HAVE BEEN AT LOW LATITUDES DURING THE LATE PRECAMBRIAN

IMPORTANCE THE SNOWBALL EARTH PHASE MAY HAVE BEEN CRUCIAL TO TRIGGERING THE FLOWERING OF COMPLEX LIFE

Over recent decades, geologists have found increasing evidence that Earth was engulfed with ice down to equatorial latitudes several times during its remote past. Known as Snowball Earth, this hypothesis is still the subject of fierce debate.

In 1949, Australian geologist and Antarctic explorer Sir Douglas Mawson (1882–1958) published a scientific paper that put forward a controversial idea. Based on ancient glacial deposits found in South Australia, he suggested that late Precambrian times saw an ice age on an unprecedented scale, in which glaciation extended to tropical latitudes. Since Mawson's time, of course, the discovery of plate tectonics has shown that Australia has shifted its location considerably through tectonic movements since Precambrian times, but despite this, the evidence from the glacial deposits remains strong – we now know that 640 million years ago, at the end of Earth's 'Cryogenian' Period when the deposits were laid down, Australia actually lay just a few degrees north of the equator.

British geologist Brian Harland (1917–2003) independently proposed a global ice age in the 1960s, based on evidence that ancient glacial deposits called tillites on the Arctic island of Svalbard had been formed in low latitudes. Harland's evidence for Svalbard's original latitude came from palaeomagnetism – measuring the orientation of magnetized grains or crystals preserved from the time when they were laid down around 650 million years ago.

However, it was another 30 years before the idea now called Snowball Earth really hit the headlines, when Harvard geologist Paul Hoffmann described

OPPOSITE This artist's impression of Earth's Snowball phase depicts a world covered in ice down to equatorial latitudes, but scientists are still debating whether these glaciations were indeed entirely global, or whether large regions of open ocean persisted without sea ice.

the late Precambrian rock succession from Namibia, southwest Africa. Here, spectacular 'dropstones' from melting icebergs and covered in ice striations form glacial deposits that are overlain or 'capped' by carbonate rocks laid down in warm tropical waters relatively soon afterwards – strong support for the idea that glaciation extended to near-equatorial latitudes.

The Snowball Earth glaciations

Over the past decade, geologists have subjected these ancient glacial deposits to intensive study and there is now little doubt that there were some major glaciations in the Precambrian. However, debate still rages about the true extent of these ice ages and how close they really came to freezing our entire planet. The youngest of these events (known as the Gaskiers glaciation) occurred around 580 million years ago in the late Ediacaran, but its global extent is doubtful. The Marinoan glaciation of Cryogenian times (660–635 million years ago) is much more secure, with good evidence from South Australia and Oman. Glacial deposits from Oman in particular (then at a latitude of around 13 degrees North) are overlain by muds and sands containing chemical evidence for abundant marine life, including the earliest known animals (see page 153).

RIGHT Some of the best evidence for Snowball Earth comes from Namibia, where boulders up to 1m (40in) across were dropped into the seabed sediments from floating sea ice during the Sturtian Ice Age around 726–660 million years ago. Above these 'dropstones' lie carbonate rocks that formed in warm waters, showing that seawater temperatures recovered quickly after the glaciation.

Another Cryogenian glaciation, the Sturtian, is even older and dates to somewhere between 726 and 660 million years ago. This was first identified in Namibia, and has been backed up by discoveries in Canada's Yukon Territory, which then lay at a latitude of around 10 degrees South. Radiometric dating suggests this episode lasted around 5 million years, and the strata again provide chemical evidence for abundant marine life throughout this period. There is also evidence for far older glaciations, such as the 2.25 billion-year-old Huronian glacial deposits.

But how could life survive under a global ice sheet? It would seem that, however extensive the ice became, there must have been regions with sufficient light and warmth to support life. Supporters of Snowball Earth argue that sea ice is never totally static, flowing and thinning in some places to create patches of open water that may have allowed life to cling on.

Snowball problems

However, a more fundamental problem with the Snowball hypothesis is the nature of the mechanism that might trigger such an event. Some climate models suggest that, if glaciers and ice sheets extend to within 30 degrees of the equator, Earth's increased reflectivity could produce a 'feedback loop', with rapid cooling and growth of ice to the equator.

But what process could initiate such a widespread glaciation in the first place? Large-scale volcanic eruptions, belching out gas and ash that cooled the atmosphere, are a strong contender, and the Sturtian glaciation certainly coincides with the formation of a Large Igneous Province (see page 109) that stretches across northern Canada. However, changes in Earth's orbit and axial tilt, or the Sun's energy output, could also have caused drastic declines in global temperatures.

'Some climate models suggest that, if glaciers and ice sheets extend to within 30 degrees of the equator, Earth's increased reflectivity could produce a "feedback loop", with rapid cooling and growth of ice to the equator.'

Another major question is the security of the palaeomagnetic evidence that points to low-latitude glaciation. Ancient rocks such as the ones that provide most of the evidence are prone to suffering later changes that may 'reset' the rock magnetism, and it is also possible that Earth's magnetic field has not always had the same dipolar, north–south, structure that we know today. Dating the glacial strata is also contentious, since the glacial strata tend to have few rocks that are directly datable by radiometric means. Furthermore, the mechanism that formed the overlying limestones (which often 'cap' the glacial deposits) is also disputed – unusual features suggest they may result from some profound change in ocean chemistry that has not recurred since Precambrian times. Finally, the mechanism by which Snowball events come to an end is also disputed – whatever it was, it seems to have occurred quite rapidly (within a million years or so), but modelling suggests that melting the ice in such a geologically brief period would require unreasonably high levels of carbon dioxide.

The phosphorus effect

DEFINITION A MASSIVE INCREASE IN THE AMOUNT OF PHOSPHATE NUTRIENTS DEPOSITED IN EARTH'S OCEANS AT THE END OF EACH SNOWBALL EARTH GLACIATION

DISCOVERY THE IMPORTANCE OF THE PHOSPHORUS CYCLE WAS FIRST IDENTIFIED BY HEINRICH HOLLAND AND OTHERS IN THE 1980S

KEY BREAKTHROUGH PHOSPHATE-RICH ROCKS FROM POST-SNOWBALL DEPOSITS SHOW THAT GLACIATIONS HELPED TO STOCKPILE AND THEN RELEASE PHOSPHORUS

IMPORTANCE PHOSPHORUS IS A VITAL NUTRIENT FOR LIFE ON EARTH TODAY, AND IT SEEMS IT WAS ALSO IMPORTANT TO EARLY LIFE

Why did complex life suddenly 'take off' in a burst of diversification around 700 million years ago? New research suggests that a hitherto overlooked mechanism – the phosphorus cycle – may have played an important role.

In Darwin's day, knowledge of the fossil record seemed to show that life did not make its first significant increase in abundance and diversity until the beginning of Cambrian times, some 542 million years ago. Since then, however, a whole new group of mysterious soft-bodied organisms, known as the Ediacaran biota (see page 149), has been discovered in ancient strata dating back around 610 million years. We now also know that life originated as far back as 3.5 billion years ago in the form of sea-dwelling marine prokaryote microbes, taking billions of years to progress to more complex but still essentially microscopic eukaryote organisms, and finally to multicellular (metazoan) organisms.

Various triggers have been suggested for the diversification of metazoan life. Since the discovery of evidence for the Snowball Earth glaciations (see page 137) some scientists have wondered whether their conclusion, around 700 million years ago, released some pressure that had hitherto restrained life from evolving beyond relatively simple forms.

Certainly, there was a significant increase in atmospheric oxygen at the time, which could have literally fuelled the growth of metazoan marine organisms dependent upon oxygen for respiration. The rise in atmospheric oxygen would also have built up the layer of ozone (molecules of oxygen consisting of three individual atoms, O_3, as opposed to the more common O_2) in the

OPPOSITE During periods of intense glaciation, the growth of glaciers and ice sheets across landscapes effectively shuts down the cycling of water from the oceans to the land and back to the oceans. Moisture from the oceans is precipitated as snow on the land, where it remains locked up until the end of the glaciation. As a result, the normal cycling of soluble chemicals, such as phosphorus compounds that are essential to life, is also cut off during glaciation.

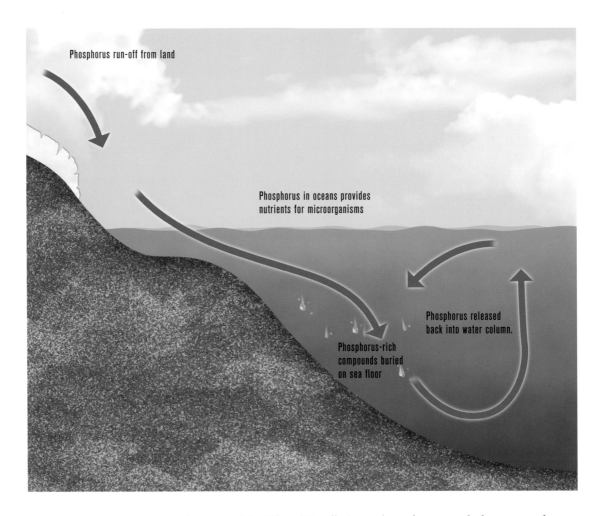

Phosphorus run-off from land

Phosphorus in oceans provides
nutrients for microorganisms

Phosphorus released
back into water column.

Phosphorus-rich
compounds buried
on sea floor

ABOVE When
glaciers retreat,
rocky landscapes
charged with fresh
rock debris are
re-exposed to
processes of
precipitation and
run-off. Huge volumes
of meltwater carry
an abundance of
chemicals, such as
phosphorus salts, into
the sea, where their
nutrient value can
stimulate the growth,
diversification and
evolution of marine
organisms.

stratosphere roughly 30km (19 miles) up. As we've recently become only too aware (see page 378), ozone plays a major role in protecting Earth's surface against penetrating ultraviolet radiation from the Sun, which can inflict severe damage on delicate individual cells. But was the oxygen increase directly linked to the end of the Snowball Earth glaciations, and if so, how?

The role of phosphorus

Evidence for just such a link has been provided through a new analysis of phosphate levels in the oceans of the time. Phosphorus is an essential nutrient for plants and animals, and forms the basis for global biological productivity, contributing to energy storage within cells and playing a crucial role in the construction of the complex nucleic acids (DNA and RNA) that store and transfer genetic information. As with several other crucial chemicals in the environment (such as water and carbon), there is a well-defined 'phosphorus cycle' in which the element is transferred from the rocky lithosphere, through the watery hydrosphere into the living biosphere, and from there back into the lithosphere through the decay and burial of dead organisms on land and in the sea.

Although phosphates have been used as fertilizers since the 19th century, it was only in the 1980s that scientists including Heinrich Holland of Harvard University developed an understanding of how phosphorus is recycled over tens of thousands of years through rock weathering, transport, biological absorption and burial (especially in marine deposits). As a result, measurements of ancient marine phosphorus can now be used as a proxy measure for biological activity. These show that phosphorus levels were relatively constant through Earth's first few billion years but then, 700 million years ago, shot up to unprecedented levels. This late Precambrian peak coincides with the end of the Snowball Earth glaciations and was sustained for tens of millions of years. So what was going on?

Origins of the phosphorus cycle

Throughout the first few billion years of Earth's history, the continental landscapes were radically different from those of today – there was no plant cover and hence few if any soils. Instead, the overwhelming processes were geological – tectonic uplift, mountain building, erosion and destruction, alongside high levels of volcanic activity. This led to a relatively rapid cycling of material from solid rock through fragmentary 'regolith' into sediments transported by gravity, water and wind down to lower altitudes and ultimately into the sea. As a result any phosphorus in minerals near the surface was rapidly stripped from the rocks and delivered to the oceans.

However, during the Snowball Earth glaciations, the flow of phosphorus to the oceans would have been held back, or even stopped, as the delivery system of rainwater run-off and upland rivers froze solid. When the glaciations ended, rocky landscapes emerged from their icy cover and rebounded (see page 321). The processes of weathering and erosion accelerated, and the release of meltwaters from the ice flushed debris from land to sea with renewed vigour, carrying with it a new supply of phosphorus. This 'revved up' phase of erosion and phosphorus cycling is likely to have continued throughout the early Palaeozoic, only slowing down with the development of soils and rooted land plants during the Silurian Period around 470 million years ago (see page 181). According to some estimates, the post-glaciation rate of phosphorus erosion and recycling could have reached as much as ten times its present-day level.

The effect of so much nutrient entering the waterways and oceans would have been similar to that of modern farmland fertilizer run-offs, which can cause runaway algal blooms (see page 381). As the thriving single-celled organisms died, their carbon-rich remains would have been deposited into sediments at an increased rate, leading in turn to increased release of oxygen into the atmosphere and oceans, where it boosted the evolution and radiation of the oxygen-hungry metazoan animals (see page 185).

'The "revved up" phase of erosion and phosphorus cycling is likely to have continued throughout the early Palaeozoic, only slowing down with the development of soils and rooted land plants during the Ordovician Period around 470 million years ago.'

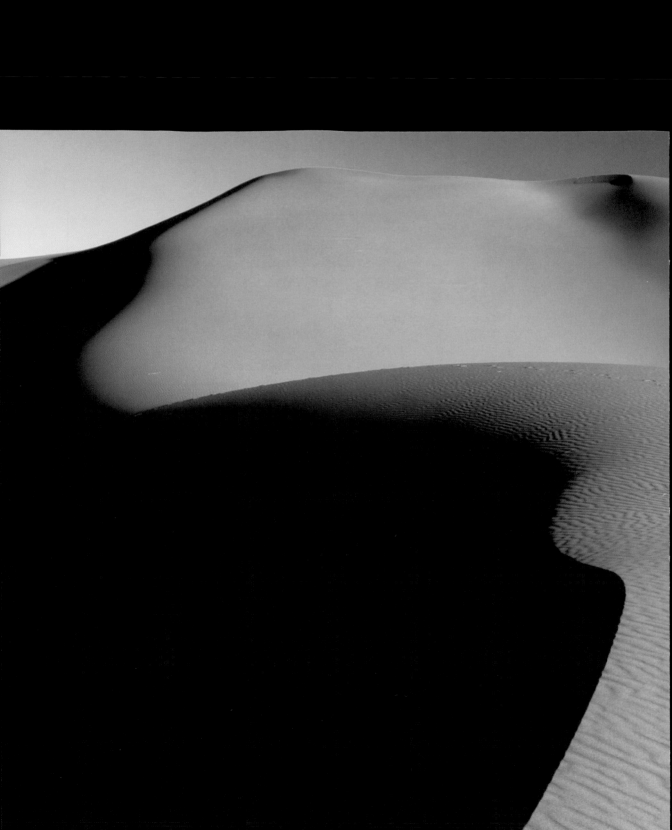

35 Gondwana and Pangea

DEFINITION GONDWANA WAS A LONG-LIVED SOUTHERN HEMISPHERE SUPERCONTINENT THAT UNITED WITH OTHER CONTINENTS TO FORM PANGEA IN LATE PALAEOZOIC TIMES

DISCOVERY THE EXISTENCE OF GONDWANA WAS FIRST HYPOTHESIZED BY EDUARD SUESS IN 1861

KEY BREAKTHROUGH WEGENER'S THEORY OF CONTINENTAL DRIFT CLAIMED THAT CONTINENTS HAD BEEN DISPLACED OVER TIME

IMPORTANCE GONDWANA AND PANGEA PROVIDED HABITATS FOR THE DEVELOPMENT OF COMPLEX MULTICELLULAR LIFE

550 million years ago, the slow drift of Earth's tectonic plates produced a supercontinent named Gondwana that dominated the southern hemisphere. This huge landmass survived for some 400 million years and had profound effects on the evolution and distribution of life.

Earth's history has been punctuated with large-scale rearrangements of the continental landmasses as oceans have formed, grown and ultimately been destroyed. Occasionally, these processes inevitably bring the continents together, forming massive supercontinents such as Rodinia (see page 133). Gondwana was the first of these huge landmasses to be identified, by Austrian geologist Eduard Suess (1831–1914) in 1861, although Suess explained the presence of similar plants on widely separated continents through a change in sea levels rather than a wholesale rearrangement of continents. Fifty years later, Alfred Wegener explained Gondwana using his 'continental drift' theory, and co-opted it as part of an even larger landmass, Pangea, which we now know dominated the late Palaeozoic and Mesozoic world.

Gondwana began to come together in late Precambrian times through the amalgamation of two previously assembled clusters of ancient cratons. Eastern Gondwana was formed from the African, Indian, Antarctic and Australian cratons, while western Gondwana consisted of the amalgamated South American cratons. The process of formation was a lengthy business, and its story is told in the ancient mountain belts formed as the cratons converged (rather like the modern Himalayas – see page 241). The final phase of assembly around 550 million years ago saw India and Australia converge on eastern Africa, while the North American craton (Laurentia) began to drift away from Gondwana as a southern polar sea called the Iapetus Ocean

OPPOSITE When continental masses amalgamate to form supercontinents such as Pangea, their vast interior landscapes are far removed from the sea and develop typical continental interior climates, especially in the subtropics. As they receive little or no rain, they become harsh, arid desert environments where it is difficult for life to survive.

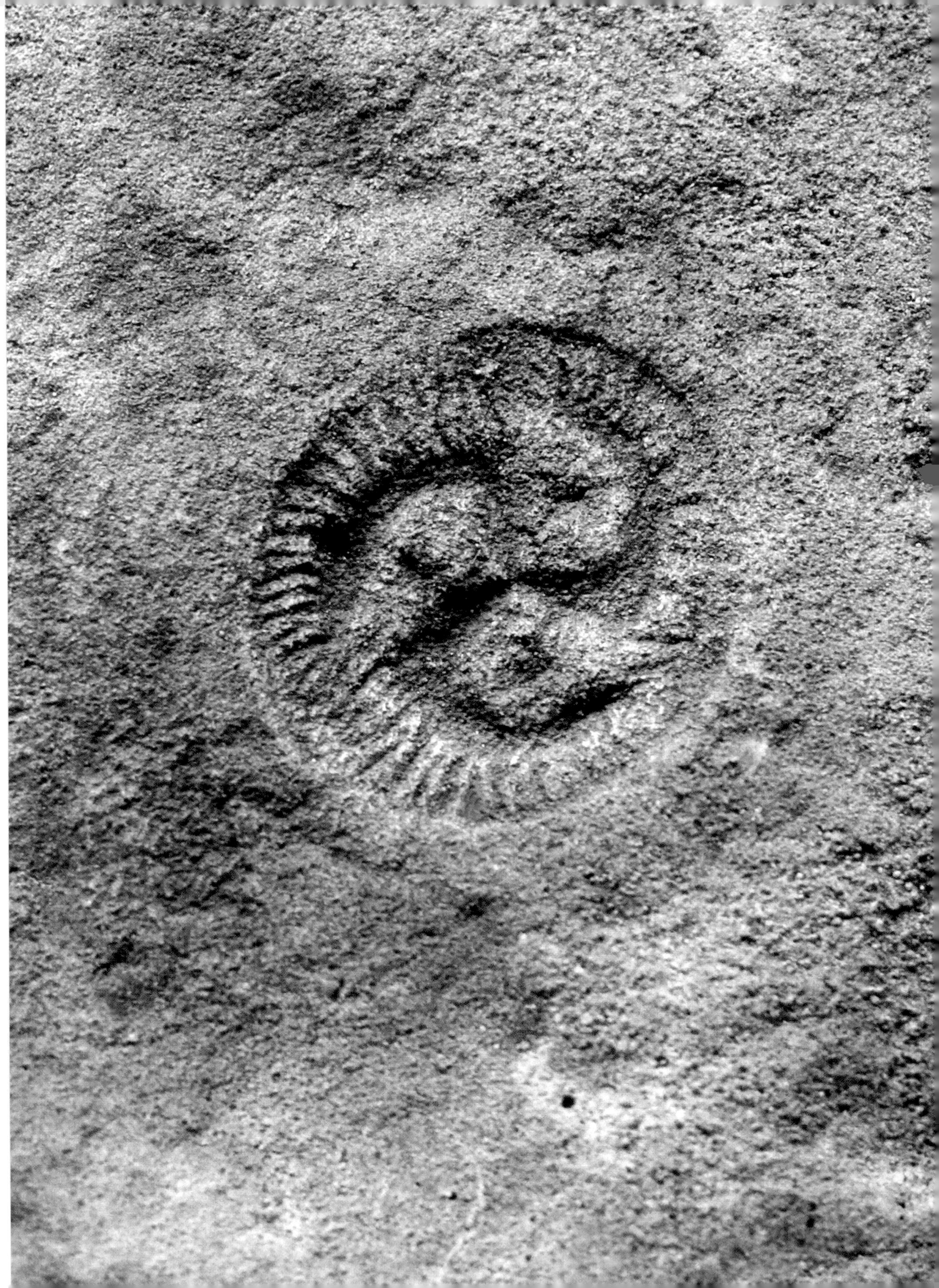

36

Mysterious Ediacarans

DEFINITION THE EARLIEST KNOWN ECOSYSTEM OF LARGE MARINE
ORGANISMS, APPEARING IN THE LATE PRECAMBRIAN

DISCOVERY FIRST DISCOVERED IN AUSTRALIA'S FLINDERS RANGE BY
GEOLOGIST REG SPRIGG IN 1946

KEY BREAKTHROUGH IDENTIFICATION OF SPECIES FROM CHARNWOOD
FOREST, UK, ESTABLISHED THEIR PRECAMBRIAN ORIGIN

IMPORTANCE AS THE FIRST LARGE, PROBABLY MULTICELLULAR,
ORGANISMS, THE EDIACARANS ARE A VITALLY IMPORTANT ELEMENT
IN THE STORY OF LIFE, BUT ARE STILL POORLY UNDERSTOOD

During late Precambrian times, life blossomed into a variety of large,
soft-bodied marine organisms known as the Ediacarans. But what kind of
organisms were they – giant protists, relatives of the jellyfish, ancestors
of today's worms, molluscs and arthropods, or unique extinct life forms?

Between 635 and 542 million years ago, during the Ediacaran Period of
the late Precambrian, Earth's seas were home to many kinds of strange
organisms, few of which have living representatives. All were soft-bodied,
and are only fossilized thanks to the impressions their bodies made in
surrounding sediments. They range in size from a few millimetres up to
a couple of metres in length, and typically have flattened and sometimes
curiously 'quilted', mattress-like body forms. Some lived on the sea floor,
others were mostly buried within sediment and some frond-like forms were
rooted in the seabed. First discovered in the late 1940s, they are collectively
known as the Ediacaran biota, and their biological and evolutionary
relationships are still hotly disputed.

Discovering the Ediacarans

A few scattered fossil remains of creatures that are now classed as Ediacarans
have been known for nearly 150 years, but since fossils were not generally
believed to have appeared in the rock record until the beginning of Cambrian
times, these were generally ignored or considered to be deceptive patterns
formed by other processes. Modern investigations began around 1946, when
Australian geologist Reg Sprigg (1919–94) found numerous well-preserved
remains impressed in the shallow seabed sandstones of the Ediacara Hills in
South Australia's Flinders Range. However, even Sprigg believed that his
strange fossils were Cambrian in age, and it was another decade before their

OPPOSITE The
sandstone mould of
Tribrachidium was one
of the first fossils to
be found in the late
Precambrian strata of
the Flinders Range of
South Australia in the
1940s. Because of its
shape it was thought
to belong to a jellyfish,
but it is now seen
as one of numerous
problematic marine
organisms in the
Ediacaran biota.

37

The first animals

DEFINITION SPONGES ARE THE SIMPLEST MULTICELLULAR ANIMALS
SURVIVING TODAY AND AMONG THE FIRST MULTICELLED ANIMALS
IN THE FOSSIL RECORD

DISCOVERY RECENT FINDS OF SPONGE-RELATED 'CHEMICAL FOSSILS'
IN OMAN DATE TO AROUND 635 MILLION YEARS AGO

KEY BREAKTHROUGH NEWLY DISCOVERED BODY FOSSILS FROM
AUSTRALIA MAY SHOW AN EARLY FORM OF PRIMITIVE SPONGE

IMPORTANCE SPONGES MARK A KEY STAGE IN THE EVOLUTION OF
ANIMAL LIFE, SINCE THEY ARE FORMED FROM GROUPS OF
COOPERATING EUKARYOTIC CELLS

The remains of fossil sponges have recently been discovered in
635 million-year old strata from Oman and Australia. Simple though
these organisms may be, they nevertheless represent the first definite
animals, and therefore a major development in the story of life.

Charles Darwin, whose theory of evolution by natural selection transformed
our view of the natural world in the mid-19th century, was only too well
aware that the fossil record, as it was known at the time, did not provide
much evidence for the basic evolutionary idea that multicelled 'metazoan'
organisms developed from smaller single-celled ancestors. Instead, the
available fossils seemed to show that abundant and diverse animal forms,
ranging in complexity from sponges to arthropods, appeared quite suddenly
in the earliest Cambrian times, around 542 million years ago, without any
record of the expected precursors in the late Precambrian.

It was another hundred years before palaeontologists began to discover
the first remains of late Precambrian life, and even then, the soft-bodied
organisms of the Ediacaran biota (see page 149) did not seem to fit in with
Darwin's neat conception of animal evolution. But while the Ediacarans
may have presented taxonomists with new problems, they did at least
show that large, probably multicellular, creatures existed at the time.
And now, the very recent discovery of both chemical and body fossils
seemingly related to the most primitive animals is helping to bridge the
evolutionary gap. These 635 million-year-old fossilized traces of early
sponges provide two lines of independent evidence that the first metazoans
not only evolved, but became widespread and successful, amid the climate
extremes of Snowball Earth (see page 137).

OPPOSITE Sponges
are one of the most
successful of the
ancient and primitive
animal groups found
in ocean waters. A
biological curiosity,
their remarkably
complex structures,
supported by various
skeletal materials,
are built through
the cooperation of
numerous individual
cells, but are lacking
specialized tissues such
as nerves.

Traces from Oman

The common image of fossils is one of shells and bones embedded in rock strata, and most fossils are indeed formed by the hard parts of organisms that persist after all the soft parts have rotted away. However those soft tissues, while composed mostly of water, also contain a wide variety of organic (carbon-based) chemicals, some of which can be preserved in rocks for vast spans of time. Although recovering them and proving their origin is difficult, such chemical fossils provide invaluable evidence of life in Precambrian strata that were laid down before the evolution of hard body parts. What's more, these chemical signatures can be linked to the very group of simple animals whose presence Darwin might have predicted in these ancient rocks – the sponges.

Although fundamentally composed of soft-bodied cells, sponges often support themselves on elaborate skeletons made up of spiny needles and star-like shapes called spicules. The composition of these spicules varies between the three major sponge groups – they are made from silica in the hexactinellids or glass sponges, from calcium carbonate in the calcareans or stony sponges, and from a variety of materials including the organic protein spongin in the demosponges. Skeletal remains of glass sponges are well known from early Cambrian strata, and there may be demosponge-like spicules in rocks dated at around 750 million years old (although their identity is not completely secure). There are also suggestions that the Doushantuo embryos (see page 161) record the early stages of glass sponge growth.

However, it was only in 2009 that scientists from Massachusetts Institute of Technology discovered numerous chemical fossils typical of living demosponges, providing definite evidence for the appearance of primitive metazoan animals more than 635 million years ago. Geobiologist Roger Summons and his team analysed samples from wells in Oman that tap into some of the oldest known oil reserves on Earth. Here they found some very resilient and distinctive hydrocarbon 'biomarkers' – most notably a 30-carbon steroid molecule known as 24-isopropylcholestane. This complex molecule is only known to occur naturally in living demosponges, and its presence in numerous oil samples from Oman suggests that these sponges were thriving in the ancient seabed environments, even though none of their body fossils have so far been recovered. The oil reservoirs can be precisely dated thanks to intervening layers of volcanic ash, and crucially, the samples and surrounding rocks have been rigorously tested to make sure that the molecules are not the product of contamination from younger strata (in contrast, for example, to the controversial and much earlier biomarkers found in Australia's Pilbara region – see page 121).

'Even 100 million years before the first shelled creatures began appearing in the Cambrian Period, it seems that the shallow seas contained enough dissolved oxygen to sustain this early animal life.'

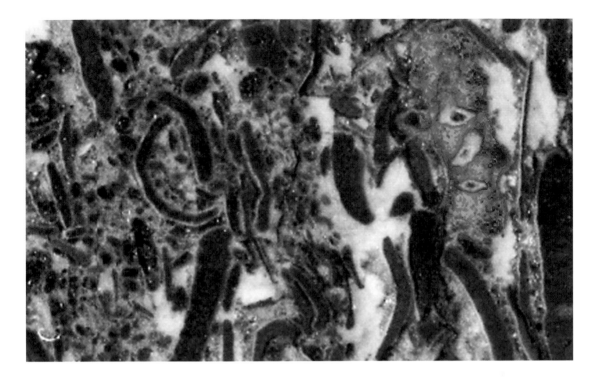

Bodies from Australia

In 2010, however, a team led by Adam Maloof of Princeton University went one better than the Oman discovery, excavating sponge-like body fossils from the Cryogenian strata of South Australia's Flinders Range. These fossils can be dated to around 5 million years earlier than the Omani oil deposits. Several millimetres across, they were found within stromatolite limestones (see page 25) and have been reconstructed as irregularly shaped organisms with a series of interior chambers and canals. Problematically, the Australian fossils lack the characteristic large 'exhalant opening' through which sponges expel water, and also appear to lack spicules (although there are some traces of silica that might represent spicule bases), so they do not fit the normally accepted sponge body plan. However, they could be the remains of primitive sponges in which this pattern had not yet fully evolved. If this is the case, then it's also possible that the Omani biomarkers come from these same sponge precursors.

All this evidence suggests that sponges were well on the road to establishing themselves throughout the latter part of the Snowball Earth period, and were able to survive the global Marinoan glaciation 660–635 million years ago, and the later Gaskiers event (see page 138). Even 100 million years before the first shelled creatures began appearing in the Cambrian Period, it seems that the shallow seas contained enough dissolved oxygen to sustain this early animal life. The search for even older chemical and body fossils will undoubtedly continue, and hopefully new kinds of biomarker molecule will help to further refine the dating of our first animal relatives.

ABOVE Centimetre-sized fossils (coloured red in this thin section) found in South Australia's Flinders Range are the oldest known calcified fossils, and have been interpreted as possibly the oldest known sponges. They occur in limestones that predate the Marinoan glaciation of 635 million years ago, but lack some features that are associated with modern sponges.

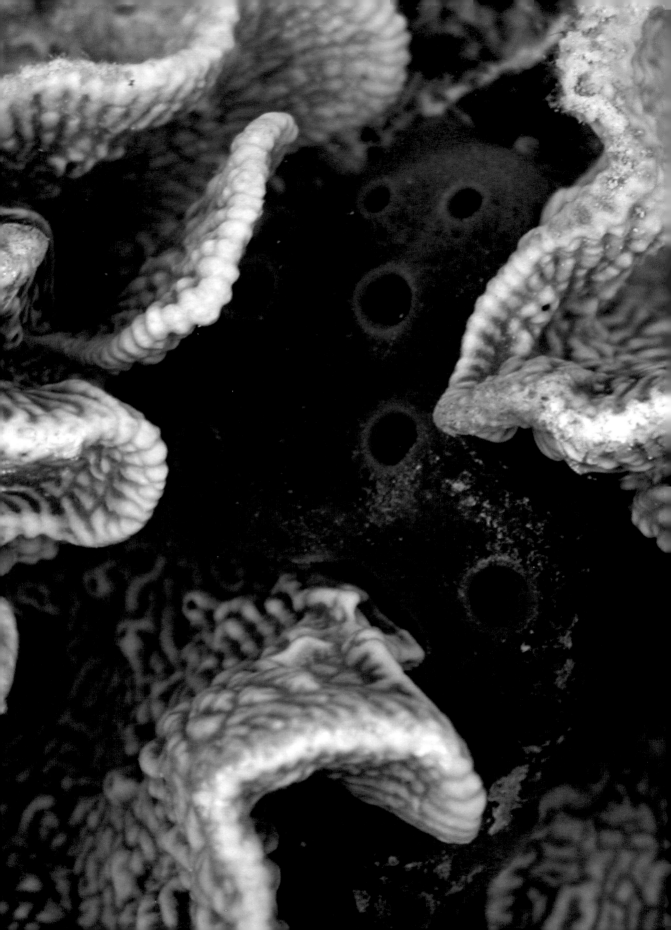

Secrets of the sponge genome

DEFINITION THE FIRST SEQUENCING OF THE GENETIC CODE FOR A SIMPLE SPONGE, FOR COMPARISON WITH MORE COMPLEX ANIMALS

DISCOVERY THE DRAFT GENOME OF THE SPONGE *AMPHIMEDON QUEENSLANDICA* WAS COMPLETED IN 2010

KEY BREAKTHROUGH THIS SIMPLE ANIMAL CONTAINS A MORE COMPLEX GENETIC 'TOOLKIT' THAN EXPECTED, SUGGESTING THE SAME FEATURES WERE PRESENT IN THE VERY EARLIEST ANIMALS

IMPORTANCE THE SPONGE GENOME CAN TELL BIOLOGISTS A GREAT DEAL ABOUT GENETIC CODE THAT IS SHARED BY ALL ANIMALS

Amphimedon queenslandica is a small and seemingly insignificant sponge that lives on Australia's Great Barrier Reef. However, it has leapt to biological fame thanks to the sequencing of its genome, completed in 2010, which has revealed some real evolutionary surprises.

Sponges seem to be remarkably simple and ancient organisms, with a fossil record that extends back at least 635 million years (see page 153). They represent the earliest branch from the common ancestor of all animals (including ourselves). Sponges thus occupy a pivotal position in the evolution of complex animals from single-celled organisms.

The name 'sponge' is still familiar enough today, although few of us ever encounter the real thing – instead (and fortunately for the sponges themselves) we have to make do with artificial plastic substitutes. Biologically, sponges are especially interesting, as they bridge the gap between single-celled eukaryote organisms (unicells or protists) such as amoebas and foraminiferans, and more complex multicellular animals (metazoans) with specialized tissues and body organs, such as insects, dinosaurs and humans.

Real sponges are made from aggregates of cells that combine to produce characteristic body forms, which are typically vase- or globe-shaped. They filter water through their porous walls to obtain tiny food particles before passing it out through a single larger opening. Today, there are some 10,000 different sponge living across all water environments from ocean depths to fresh waters. They can grow to substantial sizes of 1m (40in) or so, and can be reinforced with various skeletal materials such as tough organic proteins, calcium carbonate and silica. All of this is achieved with little in the way of cell

OPPOSITE The sponge *Amphimedon queenslandica* was first discovered in 1998 and is just one of many species that are native to Australia's Great Barrier Reef. It has achieved international status as the first sponge to have its genome sequenced. Although sponges are very primitive animals the genome of this sponge is showing that it possesses remarkably complex genes.

specialization or organization such as muscles or nerve cells. Sponges with resilient skeletal frameworks have a well-established fossil record, including evidence of reef formation, that extends back into late Precambrian times (see page 155).

According to the recent finds of early fossil sponges, it was well over 600 million years ago that tiny single-celled ocean organisms gave rise to the first multicellular animals with bodies capable of growing to a significant size. The huge organizational and functional gap between unicells and metazoans – even simple ones such as sponges – seems to have been bridged remarkably quickly, and biologists have been eager to understand how this was achieved. In 2010, an international team of scientists led by Mansi Srivastava of the University of California completed the sequencing of a sample sponge genome with more than 18,000 individual genes, revealing the surprising complexity of these apparently simple animals. Further analysis of this genome helps to clarify exactly what it takes to make an animal and suggests that the sponges may have a more complex evolutionary history than previously thought.

'Biologically, sponges are especially interesting as they bridge the gap between single-celled eukaryote organisms such as amoebas and foraminiferans, and more complex multicellular animals such as insects, dinosaurs and humans.'

Sampling the sponge genome

Amphimedon's genome is made up of more than 18,000 individual genes (compared to 23,000 in the human genome). While the Human Genome Project (begun in 1990) found that we had far fewer genes than expected, the sponge genome team were surprised by the complexity of *Amphimedon*'s genetic 'toolkit'. In both human and sponge, however, the genome provides a basis for fundamental processes that allow cells to cooperate in an organized fashion to form a larger organism rather than competing with one another, and yet also enable them to recognize and repel 'foreign' and potentially threatening cells. The sponge genome also includes genes that seem to be implicated in the formation of nerve and muscle cells – specialized tissues that are absent in sponges themselves, but which together form the all-important neuromuscular systems of more advanced animals.

Ancestral problems

Various lines of evidence, including body fossils, chemical traces and molecular genetics, now seem to indicate that there was a relatively small window of opportunity – just 100 million years or so – in which this elaborate genetic toolkit was assembled and the first metazoan animals appeared. Now the sponge genome is telling us that even the simplest of modern animals have many of these essential genetic precursors in place. The implication is that either the more complex genes evolved independently within the sponges themselves but were never put to use, or more likely that sponges evolved from a simpler ancestor that already had these basics too – though what form this ancestor could have taken remains unknown.

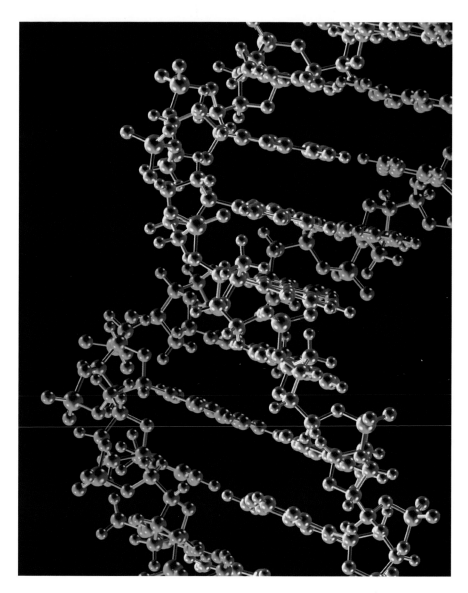

A further unexpected and intriguing discovery from the sponge genome is the apparently deep-rooted connection between the evolution of metazoan life and the occurrence of cancers. Some parts of the sponge genome are associated with the crucial mechanism by which animal bodies control or suppress rogue cancerous cells (cells that have a tendency for runaway multiplication, which can impair the functioning of the organism as a whole). Clearly the cancer mechanism is as ancient as the evolution of multicellular animals themselves – in fact, it may be an inevitable side-effect of the more advanced cell division and replication processes required to make specialized tissues. It certainly appears to have presented a significant challenge to even the simplest metazoans – enough to spur the evolution of controlling mechanisms of 'apoptosis', or cell suicide, inherited by even the simplest of modern multicellular animals.

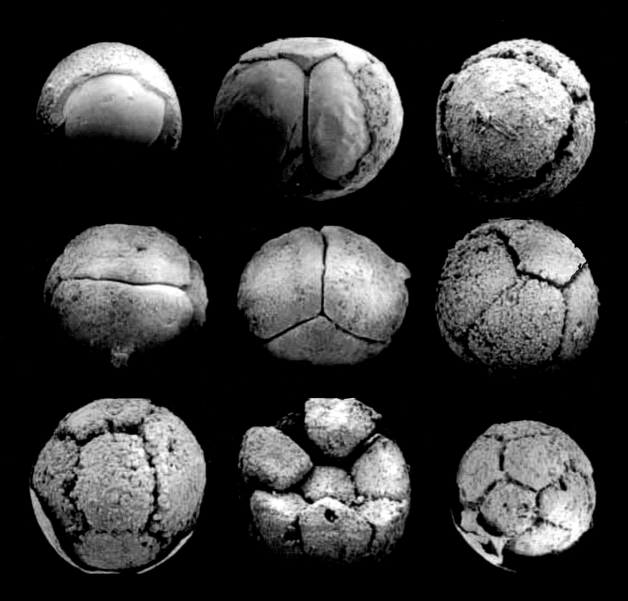

39

The Doushantuo embryos

DEFINITION A COLLECTION OF MULTICELLULAR 'EMBRYOS' THAT MAY INCLUDE THE EARLIEST EVIDENCE OF ADVANCED ANIMALS

DISCOVERY FIRST FOUND IN CHINA'S GUIZHOU PROVINCE IN 1998

KEY BREAKTHROUGH SOME SCIENTISTS CLAIM THE EMBRYOS SHOW EVIDENCE THAT THEY BELONG TO EARLY 'BILATERIAN' ANIMALS

IMPORTANCE THE DOUSHANTUO EMBRYOS OCCUPY A PIVOTAL PLACE AT THE BEGINNING OF ANIMAL EVOLUTION, BUT THEIR TRUE STATUS IS HOTLY DISPUTED

Tiny spheres of mineral phosphate less than a millimetre wide, found in southern China and dated from 580 to 551 million years old, have proved to be fossilized animal embryos. They record the early evolution of primitive metazoan animals as our planet emerged from its Snowball Earth phase.

The so-called 'Doushantuo embryos' first hit the scientific headlines in 1998, after they were discovered during commercial mining of phosphate-rich seabed deposits in China's Guizhou Province. These phosphates were originally laid down in shallow coastal lagoons with low oxygen levels, high salinity and little in the way of sediments running off the land. There is growing evidence that they were associated with a peak in the phosphorus cycle at the end of a major glaciation (see page 141).

Under a high-powered microscope, these beautifully preserved, tiny spheres reveal a clearly divided structure that closely resembles the pattern of cell 'cleavage' seen in living animal embryos during their early stages of development. As a result, scientists from Harvard and the University of Beijing claimed that the spheres represented the earliest known fossil embryos, preserving all stages of development from the initial two-cell stage, up to clusters of more than 3,000 cells.

Other researchers have since claimed that embryos from a variety of distinct organisms – including sponges, corals and algae – can be recognized among the spheres. There have even been claims that some of the embryos show bilateral symmetry – the presence of distinct left and right sides indicative of more advanced 'bilaterian' animals such as worms, arthropods and vertebrates.

OPPOSITE At more than 550 million years old and less than a millimetre in size, these spherical fossils from the Doushantuo phosphate deposits resemble textbook illustrations of the early stages of cell cleavage in a developing embryo. However, questions were soon asked about the details of their structure and preservation.

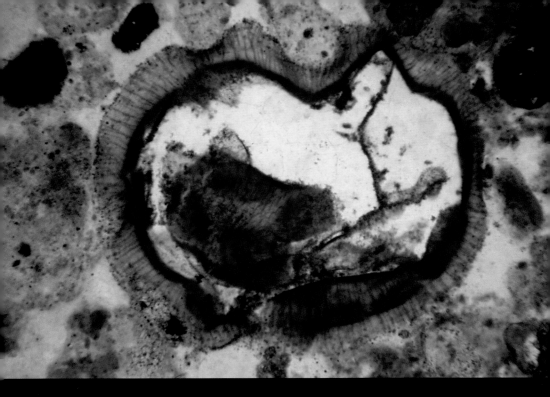

ABOVE A thin section through one of the Doushantuo embryos reveals a structured outer casing quite unlike anything found in animal embryos, resembling instead the organic wall of the cyst-like fossils known as acritarchs.

Indeed, in 2004 a Chinese/American team led by Jun-Yuan Chen of the Nanjing Institute of Geology and Palaeontology described a new species *Vernanimalcula*, based on its embryonic evidence alone, and controversially proposed that it was the oldest bilaterian animal.

But are these claims overambitious? Some critics pointed out that even those fossils containing large numbers of cells do not seem to have any well-defined structures, and therefore are unlikely to represent organisms more advanced than sponges. Others argued that such small embryos could not be fossilized by any known process, and that fossil embryos are not preserved in this way from anywhere else in the fossil record. What's more, the fossils are surrounded by wall-like structures not seen in living embryos. One alternative explanation was that, rather than being embryos, the Doushantuo fossils are in fact the remains of giant sulphur bacteria, such as the present-day *Thiomargarita*, which are known to divide repeatedly and grow into multicelled spheres up to a millimetre or so in size.

Reviving the embryos

However, further discoveries and analyses over the last few years have resolved many of the doubts surrounding the Doushantuo embryos. Laboratory experiments have shown that complex and rapid processes of 'phosphate fossilization' in the right conditions can preserve soft tissues of this kind and fossil embryos have also now been found in younger Cambrian strata

Furthermore, close examination has confirmed the embryonic nature of the fossils, and some rare specimens have revealed complex three-dimensional cell structures with features that are normally found only in the embryos of living bilaterian animals. However, even if these embryos do indeed represent fossil bilaterians, they cannot be assigned to any living group and probably represent early forms that subsequently became extinct.

Some of the embryo-like fossils are preserved inside larger ornamented spherical structures known as acritarchs – an extinct group of microscopic organic-walled fossils with unknown biological affinities (see page 115). In particular, some Doushantuo embryos have a close resemblance to the acritarch genus *Tianzhushania*, whose internal chambers are up to 0.7mm (0.03in) in diameter. Fossil embryos that have advanced to the 16-cell stage have now been found within these *Tianzhushania*-like acritarchs.

Although some modern and fossil bacteria have relatively simple structured coatings, none display the elaborate ornamentation seen in acritarchs. As a result, it's generally assumed that the acritarchs are a form of 'microbial cyst' – a dormant stage in the life cycle of various eukaryote organisms including some animals. Embryonic cysts are a form of 'suspended animation', in which an organism becomes metabolically inactive but more robust. This increases the cyst's chances of surviving adverse conditions in the environment such as falling oxygen levels or the temporary drying up of a lagoon or lake. Among living organisms, several different groups of water-dwelling invertebrates, including some relatively advanced arthropods, can pass through this stage.

'Other researchers have since claimed that embryos from a variety of distinct organisms – including sponges, corals and algae – can be recognized among the spheres.'

Doushantuo rocks, glaciations and evolution

What can we learn about the environment in which the Doushantuo embryos evolved? New investigations of the strata where they were discovered have revealed an interesting correlation with the late Precambrian glaciations. The phosphorite rocks in which the embryos are embedded have been dated at between 580 and 551 million years old. They first appear after the end of the Gaskiers glaciation of Ediacaran times, and may have formed as a result of sea-level changes around that time.

However, if the suspected association between the embryos and the fossil acritarch *Tianzhushania* can be fully confirmed, then the range of these organisms will be extended much further back in time – since *Tianzhushania* is found in strata from as far back as 632 million years ago. Interestingly, this date is within 3 million years of the global Marinoan glaciation. So, as elsewhere in the fossil record, it seems that a significant evolutionary development in early multicellular life was triggered by the end of a major ice age, and can perhaps be linked to the associated rise in oxygen levels within ocean waters (see page 185).

40

Life in the Cambrian seas

DEFINITION CAMBRIAN TIMES SAW A SUDDEN EXPLOSION IN THE
AMOUNT, SIZE AND VARIETY OF MARINE LIFE

DISCOVERY THE 520 MILLION-YEAR-OLD FOSSILS OF THE BURGESS
SHALE WERE DISCOVERED BY CHARLES WALCOTT IN 1909

KEY BREAKTHROUGH IN THE 1970S, A TEAM OF CAMBRIDGE
SCIENTISTS REASSESSED THE FOSSILS AND DISCOVERED THAT THEY
SHOWED REMARKABLE DIVERSITY IN CAMBRIAN SEAS

IMPORTANCE THE CAMBRIAN EXPLOSION GAVE RISE TO THE
ANCESTORS OF ALL TODAY'S MAJOR ANIMAL GROUPS, INCLUDING
VERTEBRATES SUCH AS OURSELVES

In recent decades, our understanding of the complex animal life that first appeared in Cambrian times has developed enormously, thanks to the discovery and excavation of fossil sites that partially preserve soft tissues from a diverse range of organisms that lived in the ancient seas.

The Cambrian Period of Earth's history was first identified in 1835 by the Reverend Professor Adam Sedgwick (1785–1873). This Cambridge academic, who had tutored Charles Darwin in the techniques of geological field work and mapping, named the period after a Roman term for North Wales, where rocks of the age are exposed. Sedgwick had difficulty in finding many fossils in these particular strata, and delayed describing those he had found, but it soon became clear from other rocks of the same age that even the earliest Cambrian sediments contained a variety of fossils.

The nature of the Cambrian fossils immediately presented a problem for Darwin and his theory of evolution, which predicted that the earliest fossils should represent primitive ancestral forms. In apparent defiance of the theory, however, the Cambrian fossils actually showed the evolutionary divergence of a number of major living groups such as sponges, molluscs and arthropods. The more primitive and less specialized ancestors of these animals should be found in older and deeper strata, but their fossils remained hidden until well after Darwin's time.

The wonders of Mount Burgess

The fossil-bearing strata of the 520 million-year-old Burgess Shale, high in the Canadian Rockies, were first discovered in 1909 by American palaeontologist Charles Doolittle Walcott (1850–1927), but it was another two years before

OPPOSITE This flattened but otherwise remarkably well-preserved fossil reveals a 520 million-year-old arthropod called *Marrella.* Just 2cm (0.8in) long, this is the most common fossil in the Burgess Shale. Apart from the tough exoskeleton, some soft tissues, such as antennae and legs, have also been preserved, showing that fossilization occurred rapidly after the animal's death.

his taxing job as Secretary of the Smithsonian Institution allowed him to begin publishing descriptions and illustrations of what he had found. He was so impressed by the richness and significance of the site that he returned several times, and by 1924 had collected 65,000 fossils. However, only a few of these were ever formally described, and when Walcott died in 1927, the fossils of the Burgess Shale became a near-forgotten curiosity. Despite the strangeness of the creatures he had discovered, Walcott did his best to shoehorn them into known biological groups such as the arthropods and the annelid worms, inadvertently understating the importance of his finds.

As a result, it was not until the 1960s that British palaeontologist Harry Whittington (1916–2010) and his colleagues began a new and systematic attempt to describe and classify the Burgess fossils. It soon became clear that a significant number defied any attempt to fit them into known taxonomy, and classification was made even more difficult by the dissociation of parts of the larger organisms during the complex process of burial and fossilization.

Since 1975, the Royal Ontario Museum has reopened excavations in the Burgess Shale and collected even more specimens than Walcott, some of which have helped to resolve the taxonomic problems. It has now been possible to reassemble some of the fossil fragments into whole creatures such as *Anomalocaris*, a 1m (40in) swimming predator with a pair of large grasping claws and a circular mouth structure that had been described as a jellyfish by Walcott in 1911. Another pair of giant spiky appendages are now known to belong to a related 50cm (20in) predator called *Hurdia*. From sponges and worms to a huge range of arthropods, the Burgess Shale fossils reinforce the view that marine life was already hugely diverse by mid-Cambrian times.

'Despite the strangeness of the creatures he had discovered, Walcott did his best to shoehorn them into known biological groups such as the arthropods and the annelid worms, inadvertently understating the importance of his finds.'

New finds from Chengjiang

Since the 1980s, fossils from the Maotianshan Shales of China's Yunnan Province, discovered by Chinese geologist Hou Xian-guang, have rivalled the Burgess Shale and provided many new insights into Cambrian life. Known as the Chengjiang biota, these 525 million-year-old fossils predate the Burgess Shale by 5 million years, and the conditions of preservation here reveal details not seen in the Burgess fossils. As in Canada, there is a huge diversity of forms (more than 185 known species), with representatives of major animal groups from sponges and jellyfish, through soft-bodied worm relatives and arthropods, to the first known echinoderms and vertebrates. It seems clear that anomalocarids were the top predators of Cambrian times, but the most important fossils from a human point of view are those which illustrate our remotest vertebrate ancestry.

Tiny eel-like creatures just 2–4cm (0.8–1.6in) long, such as *Myllokunmingia*, already show some important vertebrate features such as paired eyes, gills and gonads, and parallel sets of muscles that helped the animal to swim by throwing its body into a series of sideways, wave-like movements. Both *Myllokunmingia* and the superficially similar *Pikaia* from the Burgess Shale were primitive filter feeders somewhat similar to the living lancelet, with no bones in their bodies and a barely developed head, but they represent a clear beginning of evolution towards the more familiar fish-like vertebrates that appeared around 70 million years later.

Most recently, Chengjiang has also provided the first evidence for collective behaviour by animals. Palaeontologists from Yunnan University have discovered 14cm (5.5in) chains consisting of up to 22 individuals of a crustacean-like arthropod species, *Waptia*, apparently firmly connected head to tail. This behaviour is unknown in any living arthropod or other extinct form, but may be connected to migration.

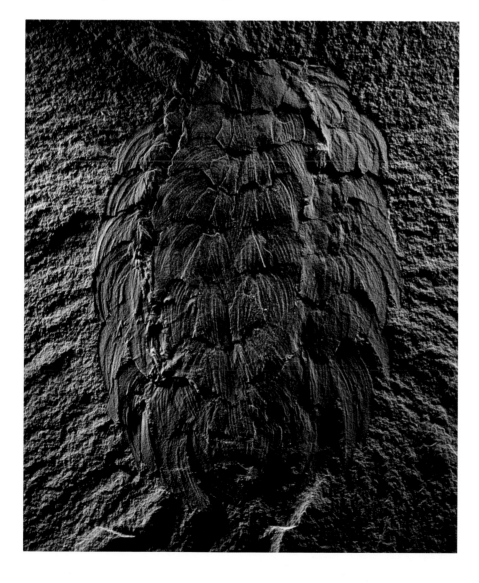

LEFT Growing to some 5cm (2in) long, *Wiwaxia* has a characteristic covering of scale-like structures and is one of the main biological puzzles found in both the Burgess Shale and the older Chengjiang strata. Isolated scales and spines from similar species have now been found in rocks from Australia and Siberia, showing that the seabed-living wiwaxids had a worldwide distribution.

Soft bodies from the Ordovician

41

DEFINITION A SERIES OF FOSSILS OF SOFT-BODIED ANIMALS
DISCOVERED IN MOROCCO'S DRAA VALLEY

DISCOVERY THE FIRST FOSSILS WERE FOUND IN THE LATE 1990s BY
MOROCCAN COLLECTOR MOHAMMED OU SAID BEN MOULA

KEY BREAKTHROUGH UNUSUAL CREATURES THOUGHT TO HAVE
BECOME EXTINCT AT THE END OF THE CAMBRIAN HAVE BEEN
FOUND SURVIVING WELL INTO THE ORDOVICIAN PERIOD

IMPORTANCE THE DRAA VALLEY SHOWS HOW MUCH THE FOSSIL
RECORD CAN BE SKEWED IN TERMS OF THE MATERIAL PRESERVED

The fossil record is highly biased towards those organisms with readily preservable hard parts, such as shell, bone and woody tissues, but is our view of evolution equally skewed by this preservational bias? New discoveries from Morocco suggest that it might be.

Much of our detailed information about the life of the past comes from a relative handful of fossil sites where special environmental conditions have allowed preservation of the soft body tissues normally lost through decay early in the process of fossilization. Although several such sites, most famously the Burgess Shale of British Columbia (see page 165), reveal the rich diversity of marine life in Cambrian times between 542 and 488 million years ago the subsequent Ordovician Period (488–433 million years ago) has not been so fortunate. But recently intriguing sites have been discovered in Morocco, revealing a range of soft-bodied and shelled marine organisms from the early Ordovician.

The greatest surprise is that the Moroccan rocks preserve a number of survivors from Cambrian times – organisms that were thought to have died out in an end-Cambrian extinction similar to those which mark the end of other major phases in the history of life. We can now see that their descendants did in fact persist, and apparently flourish, alongside new kinds of organisms that evolved in what is known as the Great Ordovician Biodiversification Event.

A Moroccan bonanza

Among fossil collectors, Morocco is best known for the abundance of its trilobites and the polished lumps of cephalopod limestone that are quarried

OPPOSITE At just 1cm (0.4in) long, this cheloniellid fossil is thought to be a distant relative of the trilobites and was recently found in 470 million-year-old early Ordovician strata from Morocco. It is by far the oldest representative of its group, which were previously known from Devonian strata between 416 and 360 million years old.

commercially and appear for sale in practically every market with a rock and fossil stall. Less spectacular but much more interesting are the marine fossils that have now been found around the Draa Valley, north of Zagora in the southeast of the country. So far, more than 1,500 fossils have been recovered from 40 localities spread over some 500 square km (190 square miles). The fossils were originally found by Mohammed Ou Said Ben Moula, one of Morocco's many local commercial collectors, and it appears that the full richness of the Draa Valley's fossil record still awaits discovery.

Life in Ordovician seas

The Draa Valley discoveries reveal a considerable range of marine life, with representatives of some 75 different kinds of organisms (technically known as taxa). Most of them seem to have lived on the seabed, and about two-thirds of them were entirely soft bodied, including a wide range of annelid worms. The remainder include representatives of fairly familiar groups such as molluscs, echinoderms and trilobites, all of which had shells. This kind of variety and relative abundance of soft-bodied and shelled organisms is much closer to the reality of marine life since mid-Cambrian times than what is normally represented by the fossil record. However, few of the individual creatures would be recognizable to anyone without specialist knowledge of marine life more than 470 million years ago.

'The Draa Valley fossils have opened a new window onto life in early Ordovician seas and shown that the apparent differences between Cambrian and Ordovician marine life... may well be just an artefact of preferential preservation in the fossil record.'

Many of the organisms belong to what is known as the Lower Palaeozoic biota, much of which became extinct at the end of Palaeozoic times. Trilobites – those distinctive, superficially horseshoe crab-like marine arthropods – are perhaps the best-known members of this ancient extinct fauna, and species from at least seven different trilobite groups have been found in the valley. Other extinct groups are less familiar – such as the strange graptolites, with their curiously geometric linear skeletons of organic protein. One group that was evolving rapidly at the time were the echinoderms, familiar today as sea urchins, sea lilies (crinoids) and starfish. While starfish and sea lilies are represented in these Ordovician rocks, they were accompanied by a number of less familiar echinoderms that are now extinct.

Cambrian survivors

With normal processes of fossilization, in which only hard parts are preserved, two-thirds of this fauna would be missing entirely. What's more, most of the 50 or so soft-bodied creatures from the Draa Valley have been discovered here for the first time in the Ordovician fossil record. Intriguingly, they include a number of formerly iconic Cambrian organisms, such as primitive sponges, armoured lobopods and delicate marellomorphs. Additionally, there are some of the first known examples of other organisms, such as horseshoe crabs (xiphosurids).

LEFT Fossils in the early Ordovician Fezouata biota of Morocco preserve soft tissues in much the same way as those from Chengjiang and the Burgess Shale. As a result, the Ordovician biota includes marine animals closely related to the earlier Cambrian ones, such as this 1.5cm (0.6in) marrellomorph.

In fact, horseshoe crabs are the most abundant fossils found in the entire valley. Several hundred specimens belonging to two quite different taxa have been recovered – a diversity that suggests these remarkably persistent life forms were already well established in the early Ordovician, and probably originated in earlier Cambrian times. Surprisingly, however, the Moroccan deposits have so far revealed no sign of early vertebrates, already known from the older Burgess and Chengjiang deposits (see page 165).

Considered as a whole, the Draa Valley fossils have opened a new window onto life in early Ordovician seas and shown that the apparent differences between Cambrian and Ordovician marine life, and the impression that there was a dramatic turnover between the two, may be nothing more than an artefact of preferential preservation in the fossil record. Furthermore, the considerable diversity found among some 'new' members of the early Ordovician fauna suggests that they probably had common ancestors further back than previously thought. Yet again, the fossil record shows its potential to throw up new surprises that rewrite evolutionary history.

42

Avalonia

DEFINITION AVALONIA WAS A TECTONIC REGION FORMED OFF THE
COAST OF THE SOUTHERN CONTINENT OF GONDWANA, WHICH
EVENTUALLY UNITED WITH LAURENTIA

DISCOVERY MOUNTAINS ASSOCIATED WITH THE COLLISION WERE
IDENTIFIED IN NORTH AMERICA BY WILLIAM LOGAN IN THE 1860s

KEY BREAKTHROUGH PALAEOMAGNETIC MEASUREMENTS HAVE SHOWN
THAT THE NORTHWESTERN BRITISH ISLES WERE ONCE UNITED
WITH GONDWANA

IMPORTANCE WETLANDS FORMED AT THE EQUATORIAL HEART OF
THE COLLISION SAW THE EMERGENCE OF LIFE ONTO LAND

The British Isles did not exist as a separate geological entity until the opening of the North Atlantic around 55 million years ago. However, their geological history and development in much earlier times is quite extraordinary, and has only been revealed over recent decades.

During the scientific revolution that swept across Europe and beyond from the late 18th century onwards, British geologists were in the vanguard of geological mapping. As early as 1815, the first detailed geological chart of England, Wales and parts of Scotland was published by William Smith. He also produced a vertical section across the country, drawn from London to North Wales, showing a sequence of strata that we now know were deposited over 500 million years. A geological map of Scotland was published posthumously by John MacCulloch (1773–1835) in 1836, and was soon followed by Sir Richard Griffith's (1784–1874) map of Ireland in 1838.

However, none of these pioneers could have had any inkling of the true geological history of this part of northwest Europe, which was not fully revealed until the revolution in understanding brought about by the discoveries of palaeomagnetism and plate tectonics. Although some scientists had tentatively pointed out similarities between the geological history and structure of the British Isles and those of North America, there was no possible mechanism to explain such connections. Now that we can 'wind back' the tectonic clock, closing up the Atlantic and removing Iceland from consideration, the case for a geological link is inescapable. What's more, it now seems that England, Wales and southern Ireland began their geological life as part of North Africa, while Scotland and Donegal were part of North America – and all lay in the southern hemisphere.

OPPOSITE A satellite image of the northern British Isles reveals the striking geological features associated with their formation in Ordovician and Silurian times. A series of faults runs across the county separating northern and southern landmasses, the most striking of which, the long, straight Great Glen Fault, which originated with the formation of the ancient Caledonian Mountains in the late Silurian, but has moved several times since.

A curious history

The story of the British Isles begins around 1 billion years ago when, during the formation of the supercontinent Rodinia (see page 133), the South American craton of Amazonia converged with that of Laurentia, encompassing much of modern North America. The resulting collision pushed up an ancient range of now-eroded mountains known as the Grenville mountain belt, whose rock remnants were first identified at Grenville in Quebec by the great Canadian geologist William Edmond Logan (1798–1875) in the 1860s, and have since been found in Scotland and Ireland. After the break-up of Rodinia some 800 million years ago, the southern landmass of Gondwana briefly coalesced with Laurentia and Baltica (encompassing much of modern Scandinavia and eastern Europe) to form another short-lived supercontinent, called Pannotia, from 620 million years ago. Along the West African and Amazonian edge of Gondwana, a region of active subduction created volcanic island arcs (see page 101), and ultimately gave rise to an Andean-style mountain belt called Avalonia. This was where southern Britain and Ireland began to form as geological entities, some 60 degrees south of the equator.

The break-up of Pannotia began between 600 and 580 million years ago, along Laurentia's eastern margin. With Scotland and northwestern Ireland still firmly attached, the North American craton began to drift from around 45 degrees South into higher latitudes, opening up the Iapetus Ocean between the two landmasses, while Gondwana plus Avalonia stayed close to the South Pole. During early to mid-Ordovician times (480–460 million

BELOW The mountains of northern England's Lake District were originally laid down as volcanic and sedimentary strata on the margins of Avalonia in Ordovician times. The Avalonian microplate was moving towards Laurentia, subducting the Iapetus ocean crust in front of it and generating intense volcanic activity.

years ago) however, Avalonia began to rift away from the edge of Gondwana margin, creating a new ocean known as the Rheic.

Closing Iapetus

Carried along by the growth of the Rheic Ocean at the rift to its south, the microcontinent of Avalonia moved north towards Laurentia. As it did so, the Iapetus sea floor was subducted beneath its leading edge, generating intense volcanic activity that is preserved in England's Lake District. By early Silurian times, Avalonia began to converge on Laurentia, and the two halves of Britain and Ireland were brought together for the first time. Their collision was marked by formation of the long, straight Caledonian mountain belt, whose roots are still seen in upland Scotland and Ireland. Aligned from northeast to southwest, this mountain chain can also be traced in Newfoundland and New England, where it is known as the Acadian orogeny. By mid-Silurian times, around 425 million years ago, the craton of Baltica also converged on Greenland, and the resulting collision extended the Caledonian mountains northeast through Scandinavia.

Today, the Ordovician and Silurian rocks of Wales and the Welsh Borders, the Lake District, Scotland's Southern Uplands and their equivalent tracts in Ireland mark the remains of marine sediments that were deposited along Avalonia's continental shelf as it was pushed towards Laurentia. As the gap between the two closed in Silurian times, the seas became shallower, and land began to rise in the English Midlands. These hot and humid coastal environments were the first to be colonized by small upright-growing vascular land plants and their accompanying animal communities (see page 181). The converging plates squeezed together the Ordovician and Silurian sediments, thickening the crust so that the folded and faulted rocks were pushed upwards and exposed to increased erosion (see page 353).

'Rivers and lakes were invaded by the first freshwater communities, including primitive jawless fish. Nearby wetlands, meanwhile, were colonized by rapidly evolving land plants.'

Laurussia

Continued uplift throughout Devonian times (416–359 million years ago) eventually created Laurussia, an extended continent that stretched from the Baltic, across northwest Europe and Greenland into northeastern Laurentia. Historically, this region has been known as the Old Red Sandstone because of the prevailing colour of its sedimentary strata. Its barren hills and mountainous landscapes were crossed by powerful rivers that pooled in huge lakes on their way to the distant sea, far to the south. These rivers and lakes were invaded by the first freshwater communities, including primitive jawless fish. Nearby wetlands, meanwhile, were increasingly colonized by rapidly evolving land plants, which grew into the first tree-sized plants by late Devonian times. The fossil remains of these first land communities are preserved in the Silurian and Devonian rocks of Wales, Scotland (see page 193) and New York State.

43 Ordovician boom and bust

DEFINITION THE ORDOVICIAN SAW A RAPID DIVERSIFICATION OF LIFE, FOLLOWED BY A MAJOR MASS EXTINCTION

DISCOVERY METEORITES IN ORDOVICIAN STRATA SUGGEST EARTH WAS BOMBARDED WITH FRAGMENTS OF A SHATTERED ASTEROID

KEY BREAKTHROUGH A NEW ASSESSMENT SUGGESTS THE IMPACTS COULD HAVE PLAYED AN IMPORTANT ROLE IN CREATING DIVERSITY

IMPORTANCE THE ORDOVICIAN BOMBARDMENT SHOWS HOW IMPACTS FROM SPACE CAN HAVE BENEFICIAL AS WELL AS NEGATIVE EFFECTS

Life in Ordovician times seems to have gone through a huge boom in diversity, only to suffer an equally devastating 'bust' at the end of the period. The causes of this rise and fall are not entirely clear, but extraterrestrial impacts and a major ice age may both have had a role to play.

The period of Earth's history from 488 to 443 million years ago is today known as the Ordovician – a name taken from an ancient Welsh hill tribe known as the Ordovices, and invented by British palaeontologist Charles Lapworth (1840–1920) in 1879. Originally, this new geological period was largely intended as a compromise to fill the gap between rocks of the better-defined Cambrian and Silurian ages, whose boundary had become the subject of a bitter dispute between geologists Adam Sedgwick and Roderick Murchison (1792–1871).

Since then, however, the Ordovician has been recognized worldwide, and geological discoveries have given it a character entirely its own. Most notably, the Ordovician is now known to contain the second most important explosion of early metazoan life – the Great Ordovician Biodiversification Event. Around 470 million years ago, this evolutionary boom (especially clear among marine invertebrates) produced some irreversible changes in the biological make-up of life on the world's seabeds.

Relatively few major new groups of life evolved, but the existing groups that had appeared in the earlier 'Cambrian explosion' tripled their diversity by the end of the period, with the recognized number of families (the standard measurement of evolutionary diversification) rising from around 500 to 1,500. This burst in evolution is especially well seen in familiar groups such

OPPOSITE The surprising discovery of fossil meteorites from Swedish mid-Ordovician strata is thought to be related to the fragmentation of a single asteroid around 470 million years ago, whose impact upon Earth has been linked to four known craters. Intriguingly, this impact seems to be linked to a diversification of seabed life, rather than a large-scale extinction.

as the cephalopods (squids and their relatives) and echinoderms (starfish, sea urchins and others), as well as in groups that are less familiar today such as the twin-shelled brachiopods and the extinct conodonts.

A helpful bombardment?

The reason for this sudden radiation of new life forms is not known, but recently a detailed study has linked it, remarkably, to a disruption in the asteroid belt, caused by the largest known asteroid break-up in the solar system's recent history. From Earth's point of view, this cataclysm resulted in frequent impacts from kilometre-sized asteroids. Numerous smaller 'fossil' meteorites have now been found embedded in mid-Ordovician strata, and impact craters from this age provide good evidence that Earth was indeed bombarded by an unusual amount of extra-terrestrial debris.

As early as the 1960s, meteorite scientists discovered that a particular class of meteorites, called the L-type chondrites, frequently had a 'shock age' (measured from changes linked to their arrival on Earth) of 450–500 million years. This suggested that they originated from the break-up of a single asteroid, whose debris rained down on Earth and seems to account for some 20 percent of all known L-type chondrites. More recent refinements have dated the asteroid's disruption to 470 million years ago – almost exactly the same age as Ordovician strata from southern Sweden and south-central China within which the fossil meteorites have been discovered. Sweden also has four known impact craters of this age.

'Relatively few major new groups of life evolved, but the existing groups that had appeared in the earlier "Cambrian explosion" tripled their diversity by the end of the period.'

In 2007, scientists from Scandinavian and US universities published the details of a comprehensive survey of fossil remains from seabed organisms (especially brachiopods) in the mid-Ordovician strata of southern Sweden. Their study clearly shows that diversity increased markedly immediately after an influx of meteorite-derived chondrite grains. The researchers, led by Birger Schmitz of the University of Lund in Sweden, suggested that the frequent impacts destabilized seabed communities, allowing invasive species to take over and displace the previously established organisms. The resulting mix of increased competition and random 'culling' by the bombardment would have promoted rapid creation of new species shaped by changing evolutionary pressures. While impact events are normally associated with the destruction of life (see page 261), it seems that here, the many relatively small impacts and abundant showers of chondrite particles entering the marine environment actually prompted diversification.

An ice age and an extinction event

Of course, this sudden rise in the diversity of life was followed by an equally dramatic fall. The end-Ordovician extinction event, around 443.7 million years ago, was the third largest extinction in the history of life – about half of all marine life was wiped out, including around 100 invertebrate families,

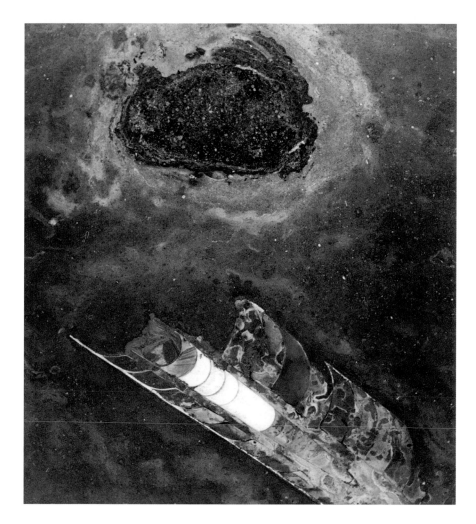

with brachiopods and bryozoans particularly affected. Detailed studies have revealed that there were actually two closely spaced extinction events, roughly a million years apart, apparently coinciding with the beginning and end of a severe ice age.

Firm geological evidence for glaciation at this time is found in strata from Morocco in North Africa, where perhaps five separate 'pulses' of glacial expansion and retreat are recorded. Here, geologists have found remarkably well-preserved ice-scratched boulders and rock surfaces, alongside ice-rafted 'dropstones' within seabed deposits. At the time, North Africa was positioned close to the South Pole alongside the northeastern part of South America, with the two continents both forming part of the great southern supercontinent of Gondwana (see page 145). With so much ocean water locked up in ice sheets on top of this huge landmass, sea levels fell and the wide continental shelves around the world's landmasses were exposed, becoming dry land. This exposure is thought to have been the main cause of extinction, since it decimated so many seabed habitats and their life forms.

Living on the land

DEFINITION FOSSILS INDICATE THAT THE FIRST LIFE EMERGED FROM
THE SEAS AND FRESH WATERS ONTO DRY LAND IN SILURIAN TIMES

DISCOVERY EARLY LAND PLANTS CALLED *COOKSONIA* WERE
DESCRIBED BY W.H. LANG IN 1937

KEY BREAKTHROUGH FOSSILS OF THE FIRST LAND ARTHROPODS HAVE
BEEN FOUND ALONGSIDE *COOKSONIA* AND AT SITES IN SCOTLAND

IMPORTANCE THE MOVEMENT OF PLANTS AND INVERTEBRATES
ONTO LAND PAVED THE WAY FOR THE EMERGENCE OF THE FIRST
TETRAPODS IN THE DEVONIAN PERIOD

The Silurian period saw a major development in the history of life, with the evolution of the first terrestrial plants and the 'greening' of Earth's landscapes. Alongside these early plants, the first invertebrate animals also made their way onto land.

Throughout most of geological time, Earth's landscapes have been barren wastelands of rock and sediment, lacking anything much in the way of visible life from the time they were first formed more than 4 billion years ago, right through to mid-Silurian times, around 428 million years ago.

Tiny, 1cm (0.4in) long fossils of simply forked, leafless stems, just a millimetre wide and preserved in black carbon, are nearly all that remains of the first true land plants. Technically known as vascular land plants, these fossils, called *Cooksonia*, have been found in Silurian strata along the borders of England and Wales since the late 19th century, and were formally described in 1937 by botanist William H. Lang (1874–1960). Only an expert would recognize them as the remains of plants, and their delicate fossils require careful preparation and powerful microscopes to reveal their features.

The trouble with life on land

Moving onto land and abandoning the natural buoyancy provided by water required major adaptations for plant life. Basic necessities included a stronger stem and a water-conducting or 'vascular' cell structure that allowed them to defy gravity. Land plants need to grow upright in order to expose their surfaces to the sunlight that powers their photosynthesis, and also need to carry water and dissolved nutrients up their stems from the soil in which they grow. They require waterproof surfaces in order to avoid drying out

OPPOSITE The modern millipede is a familiar land-living arthropod, but its ancient relatives were among the first creatures to move from marine waters into fresh waters in Ordovician times, and then onto land in the Silurian. Their tough, waterproof covering protected them from desiccation on dry land.

45 Oxygen booms and growth spurts

DEFINITION EARTH'S OXYGEN LEVELS HAVE GONE THROUGH SEVERAL 'SPIKES' DURING THE DEVELOPMENT OF COMPLEX LIFE

DISCOVERY A NEW TECHNIQUE USES MOLYBDENUM AS A MEASURE OF ATMOSPHERIC OXYGEN IN THE DEEP PAST

KEY BREAKTHROUGH THE MEASUREMENTS REVEAL A STEADY RISE IN OXYGEN FROM THE LATE PRECAMBRIAN INTO CAMBRIAN TIMES, AND A SHARP PEAK IN THE DEVONIAN

IMPORTANCE OXYGEN IS VITAL TO ANIMAL LIFE, AND THE OXYGEN BOOMS MAY BE LINKED TO THE DEVELOPMENT OF HARD PARTS AND LARGER ANIMALS

Oxygen is the essential fuel of aerobic life, including the vast majority of animals. What's more, the bigger and more active the animal, the more oxygen it requires. So could the evolution of large body size be related to changing oxygen levels in the oceans and atmosphere?

The story of how Earth's atmosphere and oceans developed from their early, oxygen-free state to a situation in which oxygen accounts for 21 percent of the present-day atmosphere, has tended to focus on the very slow and prolonged early rise in atmospheric oxygen and its implication for the evolution of relatively simple life forms (see page 129). It is now widely accepted that life's slow evolution during early Precambrian times was due in large part to a lack of oxygen in ocean waters. But following the Great Oxidation Event (see page 125), around 2.4 billion years ago, the next major step-change in oxygen levels did not occur until around 550 million years ago, during late Ediacaran times. This rise in oxygen coincides with a significant increase in the size and diversity of seabed-dwelling organisms, and the appearance of the first hard shells and skeletons in the fossil record – but despite this, most marine life was still limited in size to a few centimetres. Now, new research has linked an even more recent jump in oxygen levels during early Devonian times to another major increase in body size, and the appearance of the first truly gigantic animals.

Essential oxygen

Oxygen is essential fuel for the processes of cellular respiration and consequently the survival of all aerobic organisms, including nearly all animals, most fungi and some bacteria. In vertebrates, oxygen passes through membranes such as gills and lungs, binds to haemoglobin molecules

OPPOSITE We tend to take the oxygen in Earth's atmosphere and waters for granted but it has not always been present. There is geological evidence that oxygen levels did not begin to rise to any significant levels until around 2.4 billion years ago. Its increase was the death knell for many anaerobic organisms, but eventually promoted the rise of the aerobic life that dominates our planet today.

within the red blood cells and is then transported around the circulatory system for delivery to body tissues. As a rule, the larger the animal, the greater its oxygen needs – an adult human at rest consumes between 1.8 and 2.4g (0.06 and 0.08oz) of oxygen a minute and altogether the human population requires some 6 billion tonnes of oxygen a year. Most of this free atmospheric oxygen (some 70 percent) is produced by photosynthesis (the light-powered splitting of water molecules) in green algae and cyanobacteria, while the remaining 30 percent is produced by terrestrial plants.

Measuring past oxygen levels

But while it's relatively easy to measure the levels and sources of oxygen in Earth's atmosphere and water today, tracking the past history of oxygen levels is considerably harder. Oxygen is very reactive and has a tendency to undergo chemical reactions with many other materials, and it can be difficult to find an appropriate signal and measure of past oxygen levels in the rock record. However, a recent extensive survey of the geological record has revealed a real surprise in the history of oxygenation. The new survey, carried out by an international team led by Tais Dahl of Harvard University, relies on a 'proxy measure' of atmospheric oxygen, based on the chemistry of the element molybdenum.

'This rise in oxygen coincided with a significant increase in the size and diversity of seabed-dwelling organisms, and the appearance of the first hard shells and skeletons in the fossil record.'

Molybdenum has a number of different natural isotopes (atoms of the same element that have different masses), and its heaviest stable form (^{98}Mo) is also the most common, accounting for roughly 24 percent of all atoms found in molybdenum ore. However, this element's geochemistry depends upon the availability of dissolved oxygen and hydrogen sulphide in ocean water – as a result, it can be preserved in sediments that are rich in the sulphide mineral iron pyrite (familiar as so-called 'fool's gold') and in organic matter. Under oxygen-free 'anaerobic' conditions, such as those that existed before the Great Oxidation Event, molybdenum is immobile, but in the presence of oxygen, water-soluble molybdate ions can leach out of soils into rivers and eventually into ocean waters, where they ultimately accumulate in sediments. Due to its slightly greater mass, heavy ^{98}Mo tends to leach from the soil more readily than lighter forms such as ^{95}Mo, and so with increasing availability of oxygen, the molybdenum composition of ocean waters and sediments becomes artificially enriched with heavy molybdenum, above the normally expected 24 percent level.

The new study reveals that heavy molybdenum stayed at its natural level until around 2.7 billion years ago, when it began to increase until the time of the Great Oxidation Event around 2.4 billion years ago. A dip between 2.4 and 2.2 billion years ago may be due to an early Snowball Earth-type glaciation cutting off the enrichment process, but this was followed by a steady increase from 1.4 billion to 550 million years ago, alongside the rise and diversification of the seabed-dwelling metazoan animals (see page 165).

Similar ratios persisted until around 400 million years ago, in early Devonian times, when there was another, even bigger hike in ocean oxygen.

Devonian sea monsters

This Devonian oxygen boom coincides with a sudden increase in the amount of charcoal in the sedimentary record, and probably reflects a significant increase in the occurrence of wildfires among the first terrestrial woodlands. The first upright-growing vascular land plants had evolved in earlier Silurian times but grew to no more than 1m (40in) or so in height. It wasn't until Devonian times that the first tree-sized land plants evolved, producing large amounts of dry, wood-like tissue that was highly flammable in the presence of abundant oxygen. At the same time, high levels of oxygen in the oceans are also purported to be responsible for the sudden evolution of the first large predatory fish – creatures such as the spiny acanthodian 'sharks', which grew to 3m (10ft) or so in length, and placoderm armoured fish such as the gigantic *Dunkleosteus*, which grew to 10m (33ft) long.

After Devonian times, the molybdenum isotope record becomes quite sparse, but what few measures there are seem to point to a gradual increase in atmospheric oxygen up to modern levels. Although a general long-term and global link can be made between oxygen in the air and in the oceans, the sedimentary and isotope record reveals significant 'anoxic' events in ocean waters during the last 150 million years – periods when oxygen levels were seriously depleted. These show that the association between atmosphere and oceans at any one specific time is complicated, and can be affected by a number of other factors, including ocean circulation patterns and global climate.

46 Four feet on the ground

DEFINITION THE EARLIEST TETRAPODS EVOLVED AND MADE THEIR WAY ONTO LAND DURING EARTH'S DEVONIAN PERIOD

DISCOVERY A NUMBER OF FOSSILS SHOWING THE KEY STAGES OF TETRAPOD EVOLUTION HAVE BEEN RECOVERED FROM GREENLAND AND CANADA

KEY BREAKTHROUGH A RECENT DISCOVERY OF FOOTPRINTS IN POLAND SUGGESTS THAT ALL THE FOSSILS SO FAR FOUND ACTUALLY POSTDATE THE FIRST TETRAPOD ANIMALS

IMPORTANCE EARLY TETRAPODS WERE THE ANCESTORS OF ALL LIVING AMPHIBIANS, REPTILES, BIRD AND MAMMALS

The appearance of four-limbed vertebrate animals, known as tetrapods, was an important development in the history of life. The possession of two pairs of jointed, muscular limbs gave vertebrates the basic structural equipment to move out of the water and begin their conquest of the land.

Ever since fish such as lungfish and coelacanths, with paired muscular fins, were first described in the 19th century, scientists have realized that this 'lobefin' anatomy provided a template for the development of tetrapod limbs seen in a huge variety of land animals from reptiles to mammals to birds.

The discovery of extinct fossil lobefins (technically known as sarcopterygians) in Carboniferous and late Devonian strata suggested that the first tetrapods probably evolved around this time, and in the late 1890s fossil bones of possible early tetrapods were found in late Devonian rocks from Greenland. However, the remoteness of the sites meant that no more specimens were recovered until the 1930s, when Swedish geologist Gunnar Säve-Söderbergh (1910–48) described *Ichthyostega*, a 1.5m (5ft) tetrapod with a vaguely crocodile-like, flattened bony skull.

It was a further 20 years before Erik Jarvik (1907–98), another Swedish expert on fossil lobefins, made the first complete reconstruction of the 365 million-year-old *Ichthyostega*. Jarvik worked on the creature's anatomy for another 30 years, portraying it as a salamander-like amphibious animal that was in the process of emerging from the water onto land. Although it could walk on dry land, *Ichthyostega*'s long, sideways-flattened tail, well adapted for swimming, showed that it was still very much at home in an aquatic environment.

OPPOSITE In order to survive on land even small vertebrate animals like this salamander require a number of special adaptations. For instance, getting about requires muscular movements of the body and limbs that reduce friction with the ground to the bare minimum; seeing requires eyes whose surfaces must be kept moist; breathing requires a permeable and moist membrane for gas exchange; and the skin either has to be kept moist, or must become semi-waterproof to reduce water loss from the body.

47 The fossils of East Kirkton

DEFINITION DISCOVERIES FROM A QUARRY IN SCOTLAND PRESERVE
THE DETAILS OF AN EARLY LAND COMMUNITY

DISCOVERY FOSSIL HUNTER STAN WOOD DISCOVERED THE FIRST
SIGNIFICANT FOSSIL, 'LIZZIE', IN THE LATE 1980s

KEY BREAKTHROUGH AT FIRST BELIEVED TO BE AN EARLY REPTILE,
LIZZIE IN FACT REPRESENTS A VERY REPTILE-LIKE TETRAPOD

IMPORTANCE EAST KIRKTON SHOWS THAT LIFE ON LAND UNDERWENT
A RAPID EXPLOSION IN DIVERSITY

The fossil remains of one of the first complex land communities come from a limestone quarry at East Kirkton, near Edinburgh, Scotland. Here, detailed investigation has opened a window on life around a tropical lake in the shadow of active volcanoes some 328 million years ago.

While plants and invertebrates began to colonize the land in the Silurian Period and the first tetrapods followed them in Devonian times, it was not until the early Carboniferous, more than 320 million years ago, that plant and animal life began to extend its range beyond the watery edges of rivers, lakes and coastal lowlands, as some plants adapted to living in better-drained upland soils. Despite this, the fossil record mostly preserves the remains of life around large water bodies, where sediments were deposited and organic remains were more likely to be preserved.

The limestones around Edinburgh formed in just such environments, and a single quarry at East Kirkton, near Bathgate, has revealed a huge number and diversity of fossils, ranging from plants and arthropods to fish and some of the earliest known reptile-like animals.

'Lizzie the lizard'

East Kirkton first hit the headlines in the late 1980s, when professional fossil collector Stan Wood discovered the well-preserved remains of a reptile-like animal some 20cm (8in) long in its 328 million-year-old rocks. 'Lizzie the lizard', as the creature was nicknamed, was hailed as the oldest known reptile in the world by some 14 million years (see page 197) and as a result became a highly saleable fossil. Eventually, a consortium of Scottish buyers including Westlothian Council bought the specimen so that it could stay in Scotland.

OPPOSITE Today, primitive sphenopsid 'horsetail' plants, such as Equisetum, grow to only around 1m (40in) high, but their early relatives grew into 10m (33ft) trees that were common members of the understorey in wet areas of the Carboniferous forests.

However, when experts conducted a detailed analysis of the animal's anatomy, it became evident that *Westlothiana lizziae* (as the species was named) combined reptilian features with more primitive tetrapod characteristics. As a result, *Weslothiana* is today generally regarded as a reptiliomorph, part of a sister group to the reptiles themselves. It was probably not a true amniote (capable of laying shelled eggs on dry land) and would therefore have had to return to the water to complete its breeding cycle.

Tetrapods and invertebrates

However, Lizzie was not the only important tetrapod to be found at East Kirkton. Following Stan Wood's initial discovery, the quarry was systematically taken apart by an international team of geologists and palaeontologists in an ambitious effort to get a wide-ranging picture of the life and environments of the early Carboniferous.

It soon became clear that by the time of the East Kirkton deposits, terrestrial ecosystems had already evolved and diversified to a considerable extent, producing predators such as the 50cm (20in) salamander-like *Balanerpeton*, which is the most common tetrapod found here. More than 30 specimens of *Balanerpeton* have been discovered, including a larval stage that provides the first evidence for the evolution of this kind of development (seen in modern amphibians such as frogs and salamanders). *Balanerpeton* belongs to the extinct temnospondyls, a primitive group of tetrapod amphibians. Like living salamanders, it could pull its eyeballs down into its mouth to aid swallowing. It also had adaptations for living on land, such as a 'tympanic' inner ear structure that would have allowed it to hear high-pitched sounds in air, and eyelids that could have provided protection and kept the cornea moist. As with other primitive tetrapods, however, it would have been land living only until it came time to reproduce – breeding and larval development would have taken place in the water.

The far rarer top predator, *Eucritta*, was a significantly larger crocodile-like animal, 1.5m (5ft) long with a 30cm (12in) skull full of sharp piercing teeth. It belonged to another group of amphibian-like tetrapods known as the baphetids, and the form of its ribs suggests that it breathed by buccal pumping (rhythmic movements of the mouth) as seen in living amphibians.

In addition, there were two other small (35cm or 14in long) anthracosaurs (reptile-like amphibians) called *Silvanerpeton* and *Eldeceeon*. The aïstopod *Ophiderpeton*, meanwhile, was a snake-like limbless amphibian that lived like the present-day caecilians in the leaf litter of the forest floor and grew up to 1m (40in) long. The aquatic ecosystem of the lake and river waters was also well developed, with a variety of small extinct fish that ranged from plankton-feeding acanthodians, to ray-finned actinopterygians, including predators such as *Elonichthys* and the shell-crushing *Eurynotus*. There were even some small (40cm/16in long) hybodont sharks.

Interesting and important finds have also come from the numerous invertebrates, which included extinct eurypterids (sea scorpions), many-legged myriapods, an opiolinid (harvestman) and terrestrial scorpions. *Pulmonoscorpius*, one of the oldest known arachnids, was a fully terrestrial and predatory scorpion, up to 80cm (32in) long, with a 'booklung' that allowed it to breathe air. The tiny opiolinid *Brigantibunum*, just 1cm (0.4in) long, is the oldest known fossil representative of the living harvestmen and already had the characteristic traits of a small oval body and eight long legs. One of three small millipedes from the site, meanwhile, preserves evidence of the body pores that their living relatives use to exude noxious chemicals as a deterrent against predators.

'Following Stan Wood's initial discovery, the quarry was systematically taken apart by an international team of geologists and palaeontologists in an ambitious effort to get a wide-ranging picture of the life and environments of the early Carboniferous.'

The wider environment

During the early Carboniferous, East Kirkton was a volcanic region lying just 10 degrees north of the equator, with lush tropical vegetation dominated by extinct gymnosperm and pteridosperm plants. Although the major volcanoes in the area were dormant, underground volcanic heat produced hot springs and mineral rich waters, which sometimes supported abundant life in the lakes, but at other times became toxic and killed off many organisms. Most of the early tetrapods living in the area were still heavily dependent on water for reproduction – their unprotected eggs were laid, fertilized and developed in the water as amphibians do today. Inevitably, these eggs would have provided an ample food supply for the many fish that inhabited the lake and river waters.

Joggins Fossil Cliffs

48

DEFINITION A REMARKABLY PRESERVED CARBONIFEROUS FOREST ON THE COAST OF NOVA SCOTIA

DISCOVERY THE PETRIFIED TREES OF JOGGINS WERE FIRST NOTED IN 1852 BY GEOLOGIST WILLIAM DAWSON

KEY BREAKTHROUGH ONE OF THE JOGGINS FOSSILS, *HYLONOMUS*, WAS CONFIRMED IN 1963 AS THE EARLIEST KNOWN REPTILE

IMPORTANCE THE EVOLUTION OF REPTILES SAW ANIMALS FREED FROM THE NEED TO RETURN TO WATER FOR REPRODUCTION

The geological importance of the rock strata exposed at the Joggins Cliffs on the coast of Nova Scotia was first noted more than 150 years ago. Today, these deposits, originally laid down between 314 and 318 million years ago, are recognized as a UNESCO World Heritage Site.

In 1852, Canadian geologist William Dawson (1820–99) showed his British colleague Charles Lyell (1797–1875) the strata exposed in cliffs at Joggins, on the northern edge of Canada's Bay of Fundy. Lyell was especially intrigued by the Carboniferous coal-forming forests and the fossil trees whose sandstone casts, sometimes up to 6m (20ft) high, can still be seen standing in their original positions of growth. Looking closely at the tree stumps, the two geologists discovered a scattering of small fossil bones buried within them, and on closer examination Dawson realized they belonged to a small lizard-shaped creature that turned out to be an ancient amphibian. Seven years later, Dawson returned to search the same location for more fossils, and this time he discovered a fossil reptile, which he named *Hylonomus lyelli* in honour of his colleague. However it was not until 1963 that this 20cm (8in) creature was finally confirmed as the earliest known reptile.

The Joggins environment

Modern research has provided a detailed view of life in this rainforest swamp 314–318 million years ago. Today, Joggins lies at a latitude of 46 degrees North, but tracing back the movement of the tectonic plates reveals that, during late Carboniferous or 'Pennsylvanian' times, the site lay at just 12 degrees North. Poorly drained swamps and marshes surrounding an inland sea were home to rainforest vegetation dominated by giant tree-sized clubmosses up to 9m (30ft) tall, while the better-drained river banks and

OPPOSITE The Carboniferous coal deposits of North America and Europe were originally vast forests and swamps, formed when these landmasses lay in tropical latitudes. Their organic plant debris was so abundant that it accumulated and was buried and transformed by compression into the coal seams that fuelled the 19th-century Industrial Revolution.

floodplains were covered with dry coniferous scrub that was particularly susceptible to periodic wildfires. Growth of the swamp vegetation more or less kept pace with the rising sea levels and subsiding forest floor as the fallen and rotting vegetation was increasingly compressed into peat. Occasionally the whole region was inundated by seawater, killing off much of the plant and animal life, but each time the sea retreated, the organisms gradually returned and the ecosystem re-established itself. This cycle, which is characteristic of late Carboniferous coal-forming environments, was repeated some 14 times, depositing more than 900m (3,000ft) of 'coal measure' strata and 45 separate coal seams over the course of about a million years. Modern studies have linked this pattern of periodic deposits to the development of an ice age, with a cycle of alternating cool 'glacial' and warmer 'interglacial' climates.

Arthropods and others

Like their modern tropical counterparts, the coal swamps of Joggins were biodiversity hotspots, home to a wide range of evolving land animals. These include a dozen or more species of early tetrapods – extinct amphibian-like and reptile-like animals, none of which was more than 1m (40in) long. The true giants of the time, however, were the terrestrial arthropods, including the enormous herbivorous myriapod *Arthropleura* – a millipede-like creature that could grow to more than 2m (80in) long. Alongside *Arthropleura* and its relatives lived the first land snails, scorpions and whip spiders, while giant dragonflies darted through the vegetation. The cause for such 'gigantism' among Carboniferous arthropods is still a subject of some controversy, but one popular theory is that the creatures were able to thrive because of higher oxygen levels at the time.

'Among these fallen rocks are sandstone blocks covered in the footprints of small tetrapods, and close examination reveals that the feet of these animals had five slender, scale-covered toes, each ending in a sharp claw.'

So far, some 148 species have been identified from the Joggins Cliffs, but there is considerable potential for more discoveries and a stretch of beach some 15km (9 miles) long has been designated as a UNESCO World Heritage Site. To quote the organization's own description: 'The classic coastal section at Joggins, Nova Scotia, is of outstanding universal value. It contains an unrivalled fossil record preserved in its environmental context, which represents the finest example in the world of the terrestrial tropical environment and ecosystems of the Pennsylvanian "Coal Age" of the Earth's history.'

A giant step for reptiles

The most recent major discovery to emerge from the Bay of Fundy is a series of fossil footprints made by some unknown reptile that lived around 4 million years before *Hylonomus*. The prints were found in slightly older strata than those exposed at Joggins, but not far away on the remote Maringouin Peninsula. The well-known power of the bay's tides (which can vary by up to 17m/56ft) creates considerable erosion that is continually causing the coastal

cliffs to collapse and expose previously unseen rock strata. Among these fallen rocks are sandstone blocks covered in the footprints of small tetrapods, and close examination reveals that the feet of these animals had five slender, scale-covered toes, each ending in a sharp claw. The discovery prompted further searches at other localities in the area, and similar footprints have now been found on ancient sandy surfaces at Tynemouth Creek, near St Martin's in southern New Brunswick.

These latest prints are particularly interesting because the sandstones also preserve raindrop prints and deep mud cracks, showing that the creatures lived on dry tropical river plains that suffered seasonal drought with occasional downpours of rain. So the prints not only record the presence of the earliest reptiles, but also show that they were truly capable of living on dry land. Now the hunt is on for the remains of the animals that created them.

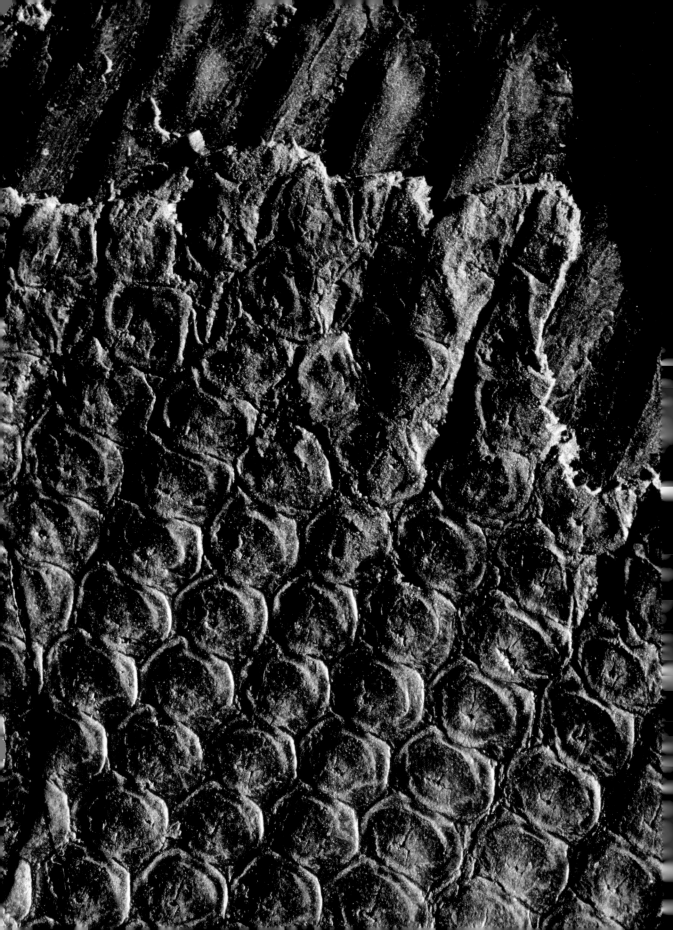

49

Coal forests and glaciations

DEFINITION COAL FORESTS SPREAD THROUGHOUT TROPICAL
LATITUDES IN CARBONIFEROUS TIMES AND SURVIVED REPEATED
SEA-LEVEL CHANGE

DISCOVERY THE WAXING AND WANING OF THE FORESTS IS REVEALED
BY COAL-BEARING CYCLOTHEM STRATA AROUND THE WORLD

KEY BREAKTHROUGH RECENT RESEARCH SHOWS HOW THE FOREST
PLANTS CHANGED RADICALLY BETWEEN COLD GLACIALS AND WARM
INTERGLACIALS

IMPORTANCE THE CARBONIFEROUS FORESTS LAID DOWN DEPOSITS
OF FOSSIL HYDROCARBON FUEL THAT ARE WIDELY USED TODAY

Late Carboniferous or 'Pennsylvanian' times saw the beginning of the
longest and most severe glaciation since the Snowball Earth period. But this
was also a time when Europe and America were covered in vast forests and
swamps, whose organic debris formed much of the world's coal.

Coal was the first of Earth's 'fossil fuels' to be exploited on a large scale,
and its use powered the 18th-century Industrial Revolution in Europe and
beyond. But coal mining also revealed a great deal about the conditions in
which coal forms – information that can help to build up a detailed picture
of the prehistoric environment.

Typically, coal-rich strata were laid down in cyclic patterns of sedimentation,
called cyclothems, usually a few metres thick. Each cyclothem involves a
sequence of terrestrial soils and coals, overlaid by shallow marine limestones,
muds and sands. As seen at the Joggins Fossil Cliffs of Nova Scotia (see
page 197), this inundation was normally followed by a return to terrestrial
conditions and recolonization of the land with swamps and forests. In places,
these cyclic patterns were repeated hundreds of times and accumulated
hundreds, even thousands of metres of coal-measure strata (although the
individual seams are rarely more than a few metres thick). Despite their
immense depth, these deposits have only been preserved where regional
tectonic activity caused them to subside into basins. Within the basins, the
coal measures were buried beneath younger sediments, and thus protected
from subsequent erosion before later uplift brought them back to the surface.

Since the 1930s, the underlying cause of these cyclic patterns in coal
formation has been linked to transitions between glacial and interglacial

OPPOSITE Transformed
into coal, the stem of
this late Carboniferous
tree is still covered
in the characteristic
scars of the leaf-
like structures that
developed from the
stem as it grew.

phases that created fluctuations in climate and sea level. However, it's only recently that scientists have tested this long-standing idea through a study of plant fossils associated with the different parts of the cyclothem.

According to new research carried out by Howard Falcon-Lang of Royal Holloway, University of London, and William DiMichele of Washington's Smithsonian Institution, typical coal forests and swamps were mostly established during the transition from glacial to interglacial conditions, when sea levels were low, and to a lesser extent during interglacial phases of rising sea level. Extensive peat deposits formed in these swamps as a result of changing sea levels and a climate that swung from seasonally dry to constantly humid conditions and back again. Frequent flooding with seawater covered the organic deposits with carbonate muds, protecting them from weathering, oxidation and erosion.

BELOW The process of compression and lithification normally destroys organic structure within coal deposits. Plant fossils are mostly preserved in the marginal deposits of coal seams, where other fine-grained sediment is interbedded with the organic layers.

Tropical coal-forming plants

The best-documented coal-measure ecosystems are the North American swamp forests, whose fossil plant communities are dominated by extinct groups such as cordaitaleans, lepidodendrids and marrattialean tree ferns. The cordaitaleans were woody trees, similar to conifers but with much larger, strap-shaped leaves. They ranged from shrubs to 30m (100ft) trees, carried their seeds and pollen in cones, and may have had stilt-like roots, allowing them to grow in watery environments like modern mangroves.

The lepidodendrids, also known as scale-trees, were among the largest and most common of the giant trees of the time, adapted to wet swampy environments and growing up to 30 m (100 ft) tall. The name 'scale-tree' is derived from the diamond-shaped, scale-like appearance of the scars left by the fleshy leaf bases on the bark.

The marrattialean tree ferns were much smaller than these giants, but were also the most common plants. They grew to a maximum height of 8m (26ft), and had a crown of long compound fern-like leaves up to 3m (10ft) long. They inhabited the wettest parts of the swamp forests, and had a buttressed appearance thanks to a mass of roots at the base of the trunk.

Glacial forests

The extensive Carboniferous glaciations locked up so much of Earth's water that sea levels fell and drainage of the low-lying swamp forests improved. Alongside the cooler, seasonally dry climate, this ensured that very different plants dominated the forests during the glacials. Although these glacial forests are only preserved at a few sites in North America, their fossils are predominately conifers and pteridosperms.

The conifers were not unlike their modern counterparts, and were probably ancestral to them. However, they included various extinct groups, such as the Voltziales – large trees that carried their pollen and seeds in cones developed at the tips of leafy branches, which had needle-like leaves.

'Typically, coal-rich strata were laid down in cyclic patterns of sedimentation, called cyclothems, usually a few metres thick. Each cyclothem involves a sequence of terrestrial soils and coals, overlaid by shallow marine limestones, muds and sands.'

The pteridosperms, or seed ferns, meanwhile, combined a woody conifer-like trunk with fern-like foliage and seed-producing reproductive structures. A diverse group, they ranged from low-growing prostrate forms, through vine-like plants to small trees whose 10m (33ft) trunks were formed from bundles of leaf bases and resinous wood. Fern-like fronds up to 3m (10ft) long grew from the crown, and seeds were up to 7cm (2.8in) long.

Surviving the glaciations

So if the coal forests changed their character and largely disappeared during each glacial phase, how did they then manage to reappear after each glaciation with more or less the same plants?

New research suggests that during the glacial phases, some coal-forest plants survived in geographically restricted refuges, from which they recolonized the tropical lowland swamps at the onset of each warm interglacial period. The 1994 discovery of the araucarian conifer *Wollemia nobilis* (the so-called 'Wollemi pine'), a living fossil that has survived increasing aridity for tens of millions of years in remote Australian valley sites, show how such species can unexpectedly cling on in small but viable populations.

50

Early Permian life

DEFINITION THE PERMIAN PERIOD SAW REPTILES SPLIT INTO SEVERAL DISTINCT GROUPS AND SPREAD INTO A VARIETY OF ENVIRONMENTS

DISCOVERY IN 1974, PALAEONTOLOGIST THOMAS MARTENS DISCOVERED A RARE SITE PRESERVING UPLAND PERMIAN FOSSILS AT BROMACKER IN EASTERN GERMANY

KEY BREAKTHROUGH THE BROMACKER FOSSILS PRESERVE UNIQUE CREATURES INCLUDING *EUDIBAMUS*, THE FIRST BIPEDAL REPTILE

IMPORTANCE THE PERMIAN REPTILES INCLUDED THE ANCESTORS OF MAJOR ANIMAL GROUPS SUCH AS THE DINOSAURS AND MAMMALS

The Permian Period, between 299 and 251 million years ago, marks a particularly important stage in Earth history and the evolution of life, as tetrapods diversified and spread beyond aquatic environments, evolving towards a more fully reptilian biology.

The Permian world saw the great supercontinent of Pangea (see page 146) at its greatest extent. Its presence dominated the global climate and the prospects for life – vast inland areas were isolated from the moderating influence of the oceans and developed into arid desert. The environment was hostile, but as plants found a foothold in drier environments, and insects followed them, it was inevitable that tetrapods would also adapt, moving away from their wetland origins into the drier uplands. This required the development of tougher protective skin structures and different skeletal postures.

But even if some of the pioneering land-living tetrapods, such as the seymouriamorphs, were well adapted as adults to the terrestrial way of life, they were still biological amphibians, forced to return to the water to breed. It was only with further crucial changes to the life cycle that they evolved from amphibians into reptiles, and the most important of these was the development of internal fertilization and self-contained, membrane-covered 'amniote' eggs capable of remaining viable when laid on dry land.

Yet despite the enormous advances made by life during the Permian, the period ended in the biggest extinction event in the history of life, with many of the land vertebrates wiped out and an even more dramatic death toll in the seas. Life on Earth would never be quite the same again, but the catastrophe provided new evolutionary opportunities for some life forms.

OPPOSITE True egg-laying (amniote) reptiles were well established by early Permian times. Equipped with tough, scaly and waterproof skins and free of dependence on water for reproduction, they diversified across the varied landscapes of the time. Being cold-blooded (ectothermic) they could also tolerate a considerable range of temperatures and environments, from tropic forests to arid deserts.

Permian biodiversity hotspots

Just as today, it was the tropical and equatorial landscapes of Pangea that were the greatest hotspots of biodiversity and evolution – especially low-lying, waterlogged regions such as lakes, river floodplains and swamps. Although the South Pole was still covered by a significant ice sheet, global temperatures were slightly above those of today. Sea level was also some 60m (200ft) higher than at present, although it dropped a dramatic 80m (270ft) at the end of the period. The heat, humidity and seasonal monsoon rains provided ideal conditions for plant life and cold-blooded vertebrates.

Early Permian strata from Texas reveal life some 295 million years ago around an enormous ancient river delta, much like today's Mississippi, which extended far into the ocean. There was an abundance of vaguely lizard-like and crocodile-like tetrapods, up to 3m (10ft) long, occupying the waters and shorelines of numerous lakes, rivers and swamps, where they fed on a plentiful supply of fish, including freshwater sharks and lungfish. Numerous amphibian groups were still diverse and abundant, but they experienced increasing competition on land from the diversifying reptiles.

The amphibians included aquatic animals with reduced limbs and long tails for swimming, such as the 60cm (24in) *Diplocaulus*, with its bizarre wing-

ABOVE This fossil from Moravia, in the Czech Republic, preserves an early Permian age larval tetrapod called *Discosauriscus austriacus*. This 'seymouriamorph' amphibian clearly shows signs of adaptation for an aquatic lifestyle, but many adult seymouriamorphs were terrestrial and thought to be close to the reptiles.

shaped head. The dominant amphibians, however, were crocodile-like temnospondyls, such as the 2m (80in) *Eryops*, with heavy muscular limbs and massive skeleton. This powerful top predator was an ambush hunter that preyed on smaller tetrapods on land, and on fish in the water.

Increasing competition came from extinct reptile groups, including pelycosaurs such as the 3m (10ft) predator *Dimetrodon* and the equally large *Edaphosaurus*, one of the first reptiles adapted to eat plants. Both had prominent sail-like structures rising from their backs, supported by bony extensions of their vertebrae and linked by a web of skin suffused with blood vessels. These sails may have acted as heat exchangers, helping the animals to warm up in the morning sun and cool down in the soaring midday heat. Early diapsid reptiles, more closely related to modern survivors, include the 60cm (24in) lizard-like *Araeoscelis*, which probably preyed on the increasingly abundant insect life. The lush vegetation included giant horsetails, ferns, seed ferns, conifers on the drier margins of the oxbow lakes, and other conifers and seed ferns on higher ground.

'The environment was hostile, but as plants found a foothold in drier environments, and insects followed them, it was inevitable that tetrapods would also adapt, moving away from their wetland origins into the drier uplands.'

An upland environment in Germany

As a rule, upland environments rarely survive in the geological record – they are dominated by erosion and worn down to sea level in the long term. However, in 1974 German palaeontologist Thomas Martens discovered a unique early Permian site (around 290 million years old) at Bromacker, near Tambach-Dietharz in what was then East Germany. Originally, this tropical environment (lying at 20 degrees North) was an upland floodplain, criss-crossed by small rivers, which developed ponds and swamps in the wet season. When these disappeared during dry periods, many of the aquatic and some terrestrial animals perished.

Excavated by US and German scientists since the 1980s, the Bromacker fossils include well-preserved footprints and articulated tetrapod skeletons, including the 60cm (24in) predatory amphibian *Seymouria*, the 2m (80in) reptile-like herbivore *Diadectes* and the sailbacked *Dimetrodon*. *Seymouria* is one of the best-known Permian tetrapods, and was thought to be a terrestrial reptile-like animal with massive limbs that held its body well off the ground. However, recent discoveries of closely related fossils preserve larval stages, revealing that the seymouriamorphs reproduced in the water, with external fertilization of naked eggs and free-swimming, tadpole-like larvae.

Bromacker also has some unique species that were probably especially adapted to its upland environment. One of these, a 25cm (10in) plant eater called *Eudibamus*, is the first known bipedal reptile. It had long slender hind legs, shorter forelimbs and a long tail that acted as a counterbalance when it ran upright, presumably to escape its predators.

51 The Great Dying

DEFINITION THE END OF THE PERMIAN PERIOD WAS MARKED BY THE
BIGGEST MASS EXTINCTION IN EARTH'S HISTORY

DISCOVERY A MAJOR CHANGE IN THE PATTERN OF LIFE AT AROUND
THIS TIME WAS CLEAR FROM THE EARLY 19TH CENTURY

KEY BREAKTHROUGH MASSIVE VOLCANIC ERUPTIONS IN SIBERIA
COINCIDED WITH THE EXTINCTION AND MAY HAVE BEEN ONE OF
ITS MAJOR CAUSES

IMPORTANCE THE GREAT DYING RESHAPED THE WORLD AND LEFT
THE WAY OPEN FOR THE EVOLUTION OF MESOZOIC LIFE

Around 251 million years ago life on Earth suffered its greatest ever setback, when the most devastating extinction in history wiped out between 80 and 95 percent of marine species and some 70 percent of land-living family groups. We still do not know exactly what caused this catastrophe.

Throughout the 19th century, fossil discoveries showed that, in the broadest terms, the Palaeozoic Era of 'ancient life' was marked out by marine species including extinct groups of trilobites, graptolites, nautiloid cephalopods, corals, echinoderms and jawless fish. Land environments, meanwhile, were dominated by primitive plant groups such as clubmosses and horsetails, along with many different extinct kinds of early amphibian and reptilian tetrapods. By contrast, the succeeding Mesozoic Era saw the rise of many new marine groups of molluscs, arthropods, fish and reptiles. Plant life on land was also transformed with the rise of groups such as conifers, ginkgos and cycads, and a whole new range of reptiles.

Remarkably, though, the enormous change between these two eras was not a gradual one – instead it happened in the geological blink of an eye at the end of Earth's Permian Period. The entire Palaeozoic world, which had evolved over nearly 300 million years, came to an abrupt end. Evidence for this end-Permian extinction is buried in the rock strata and fossil remains from the time – but political issues have prevented some of the most important sites from being examined in detail until relatively recently.

Although the break between Palaeozoic and Mesozoic life was generally recognized, the full extent of the end-Permian extinction event did not emerge until the middle of the 20th century. It only came to light when

OPPOSITE The end of Permian times was marked by the biggest extinction event in the history of life, which also marked the end of the Palaeozoic biota and its replacement by emerging Mesozoic life forms both in the sea and on land.

geologists attempted to accurately define the boundary between the Permian and the succeeding Triassic Period. The best sequences of end-Permian marine strata are preserved in the rock records of South China, the southern European Alps, Greenland and various central Asian localities from Iran east to Kashmir. These reveal that the extinction was not just a single event but had two precursors – one around 270 million years ago and the other 10 million years later. Each of these extinctions had its most dramatic effect upon life in shallow tropical waters.

Extinction in the oceans

Most of our data about the scale of the extinction comes from organisms with carbonate shells, which are more likely to be preserved in the fossil record. Large collections of shell fossils gathered from strata straddling the Permian/Triassic boundary reveal that several major groups were already in slow decline during late Permian times, but others became extinct far more abruptly. For example, two major coral groups, the tabulate and rugose corals, were entirely wiped out, destroying much of the fabric of Palaeozoic reefs. (Modern corals have since evolved from a separate group of soft corals.) Other reef organisms did survive, though with reduced diversity and numbers. For instance, the bryozoans, or 'moss animals', declined from 128 to 28 genera in late Permian times. Similarly, bivalves were reduced from 50 to just nine genera in the late Permian, and most other molluscs were also badly affected. Of 14 known genera of Permian sponges, only four survived into Triassic times, and it was some time before the group recovered.

RIGHT An infrared satellite image shows a small part of the vast Siberian Traps – plateau basalts that erupted around 250 million years ago at the end of Permian times. Vegetation that now grows on top of the ancient volcanic rock is shown pink and green depending upon its water content. The eruption of the Siberian Traps has been implicated as a cause of the end-Permian extinction.

Terrestrial victims and survivors

Our understanding of the Permian/Triassic extinction on land comes from just two regions, the Russian Urals and South Africa's Karroo Basin, which happen to have continuous fossil records across the event. These fossils show that life on land was apparently far less affected by the Great Dying, and that changes from Palaeozoic to Mesozoic plants and animals were far more gradual. The clubmosses and horsetails, water-loving plants that dominated the Carboniferous and Permian swamp forests, were reduced in numbers and diversity, and also shrunk in size from giant trees to shrub-sized plants, but only a few plant groups, such as the cordaitaleans and glossopterids, went completely extinct, along with eight orders of insects. In the aftermath, there was a rise in the diversity and numbers of ferns, conifers and seed ferns, which were more tolerant of the drier conditions.

As with water-loving plants, numerous amphibian groups that depended on water for reproduction had already suffered badly as their habitats dried up throughout the Permian. The changing environment provided better opportunities for amniote reptiles, which flourished throughout Permian times and diverged into two main groups, the sauropsids and synapsids (primarily distinguished by their skull structure). Both of these went on to make further major evolutionary developments. The sauropsids produced one group, the archosauriforms, that would achieve particular importance in the late Permian and again in Mesozoic times when they gave rise to the dinosaurs. The synapsids, meanwhile, gave rise to sail-backed 'pelycosaurs' such as *Dimetrodon*, and later to the increasingly mammal-like therapsids – active animals that included both carnivorous predators, such as the 1m (40in) long *Lycaenops*, and plant eaters such as the 5m (17ft) *Moschops*. Two-thirds of the therapsids were wiped out in the end-Permian extinction, but a few survived into the Mesozoic (see page 213) and became the ancestors of modern mammals.

'Fossils gathered from strata straddling the Permian/Triassic boundary reveal that several major groups were already in slow decline during late Permian times, but others became extinct far more abruptly.'

Possible causes

So far, no single cause for the end-Permian extinction has emerged, but the outpouring of the Siberian Plateau basalts across some 7 million square km (2.7 million square miles) of what is now northern Asia may have been a major contributing factor (see page 109). These vast eruptions released huge volumes of ash and greenhouse gases into the atmosphere, causing dramatic climate change. Other possible culprits include a massive release of methane from ocean-floor sediments causing widespread ocean anoxia and destruction of sealife (see page 381), and a rapid drop of some 25m (82ft) in sea level exposing much of the shallow continental shelf, killing most of the tropical reefs and their diverse life. The impact of a large asteroid or comet, similar to the event that marked the end of the Mesozoic (see page 261), has also been suggested as a possible trigger, but no geological evidence has been found to support it.

52

Triassic survivors

DEFINITION SPECIES THAT SURVIVED THE END-PERMIAN EXTINCTION RAPIDLY DIVERSIFIED AND SPREAD ACROSS THE PLANET

DISCOVERY FOSSILS AT GRAPHITE PEAK, ANTARCTICA, PRESERVE ONE OF THE BEST RECORDS OF LIFE IN THE EARLY TRIASSIC

KEY BREAKTHROUGH *THRINAXODON*, FOUND AT GRAPHITE PEAK, IS ONE OF THE FIRST CYNODONTS – EARLY RELATIVES OF MAMMALS

IMPORTANCE THESE EARLY TRIASSIC FOSSILS OFFER A RARE GLIMPSE OF THE WORLD THAT EXISTED BETWEEN THE GREAT DYING AND THE AGE OF DINOSAURS

How does life recover after a global catastrophe? The geological record of early Triassic rock strata and their fossil content provide some idea of what happened after the end-Permian mass extinction, and perhaps surprisingly some of the best information comes from Antarctica.

More than two-thirds of all land-living tetrapod families were wiped out in a global extinction 251 million years ago, which marked the end of not only the Permian Period, but the entire Palaeozoic Era of Earth's history. While it might seem virtually impossible for life to recover from such a major catastrophe, the biology of adaptation and reproduction ensured that it did. Extinction events tend to have the greatest impact on species that are already rare – they may be completely wiped out or so reduced in numbers that they cannot recover. Common species, on the other hand, may already owe their abundance to their reproductive strategies (such as the production of numerous offspring).

With our natural human tendency to be 'mammal-centric', it's easy to forget that amphibians and egg-laying reptiles, which were the dominant vertebrates of Permian and Triassic times, typically produce large numbers of offspring in the form of eggs. Under normal conditions, predation and high rates of mortality in the larval stages ensure that few of these infants make it through to adulthood, but with reduced competition, and especially with the removal of rare top predators, more offspring will tend to survive. As a result, the so-called 'disaster species' that make it through an extinction event can rebound in numbers quite quickly. A familiar modern demonstration of this phenomenon are land plants such as ferns, whose rapid reproduction and hardy nature allows them to quickly repopulate

OPPOSITE In the aftermath of natural disasters such as wildfires and volcanic ash clouds, ferns are often among the first life to recolonize a landscape. A classic example of 'disaster taxa', they have an ancient history extending back to mid-Devonian times, and reproduce by means of spores rather than flowers and seeds.

53

Rise of the dinosaurs

DEFINITION THE EARLIEST FOSSIL EVIDENCE FOR DINOSAURS AND
THEIR CLOSE RELATIVES COMES FROM THE MIDDLE TRIASSIC

DISCOVERY *ASILISAURUS*, DISCOVERED IN TANZANIA, IS THE OLDEST
DINOSAUR RELATIVE FOUND SO FAR

KEY BREAKTHROUGH NEW FOOTPRINTS DISCOVERED IN POLAND MAY
COME FROM EVEN EARLIER DINOSAUR RELATIVES, AND PUSH THE
ORIGIN OF THIS GROUP BACK INTO THE VERY EARLY TRIASSIC

IMPORTANCE THE DINOSAURS WERE TO BECOME THE DOMINANT
LAND ANIMALS THROUGHOUT THE MESOZOIC ERA

The oldest dinosaur relative yet discovered – a four-legged omnivore called *Asilisaurus* – was discovered in middle Triassic strata from southern Tanzania in 2010. But what exactly were the dinosaurs, and how and when did they evolve?

The dinosaurs, which came to dominate life on land during the Mesozoic Era and diversified into a huge variety of forms and lifestyles, were all members of a much larger 'crown' group of reptiles known as the archosaurs, which include the extinct pterosaurs, the living crocodiles and an avian line that encompasses the dinosaurs and their descendants the living birds (together known as the ornithodires). The evolutionary split between the crocodile group and the ornithodires had already occurred by mid-Triassic times, and at first the crocodile group were dominant. It was another 35 million years, into the early Jurassic, before the dinosaurs took over.

The early evolution of the dinosaurs is poorly documented, and only a few incomplete fossil remains survive to tell the story, such as the mid-Triassic *Lagerpeton*, a metre-sized 'dinosauromorph' with long slender hind limbs found in Argentina. A few specialized forms, such as *Silesaurus* from late-Triassic strata in Poland, are better known. Flourishing around 230 million years ago, this animal grew to around 2.3m (7.6ft) long and was a fast-moving plant eater. It is known from the remains of some 20 individuals, and similar fossils have been reported from New Mexico (see page 222), but the relationship between these animals and the true dinosaurs is problematic and so they are generally referred to as dinosauriforms. However, a new find suggests that *Silesaurus* was more closely related to the dinosaurs than *Lagerpeton* was.

OPPOSITE The forests of early Mesozoic times were very different from those of the Carboniferous; the dominant large plants were cycads, ginkgos and conifers, along with surviving ferns and sphenopsids such as the horsetails. This new vegetation formed the basis of the food chain for new types of reptile including the first dinosaurs.

'Ancient ancestor lizard'

Recently discovered by a team of US, German and South African researchers working in Tanzania, the partial skeleton of the new dinosauriform *Asilisaurus kongwe* (from Swahili words meaning 'ancient ancestor') has features that confirm it is related to *Silesaurus*, and the two genera have now been placed together in a new group called the Silesauridae. They share a number of features with the dinosaurs, and are currently considered to be the most closely related 'sister group' of the true dinosaurs. Bones from at least 14 animals were discovered in a single Tanzanian deposit, allowing the reconstruction of a near-complete skeleton that shows *Asilisaurus* had long slender limbs and a quadrupedal stance adapted for fast running. It also had leaf-shaped teeth adapted for a plant-eating lifestyle and a beak-like front to its lower jaw.

> 'The emerging dinosaurs may have been able to take advantage of abandoned ecological niches and the disappearance of many of the top predators, just as some other "disaster taxa" certainly did.'

Analysis suggests that the silesaurids, along with the first true dinosaurs (the ornithischians and sauropodomorphs), evolved along separate lines from a common ancestor that was carnivorous. As the oldest known ornithodire, *Asilisaurus* demonstrates that the silesaurids and the dinosaur lineage are of equal age, even though actual dinosaur body fossils have not yet been found in mid-Triassic strata. Furthermore, the variety of archosaur remains from Tanzania and new evidence from Poland (see below) seem to be telling us that the archosaurs went through a major diversification in Middle or even early Triassic times – much earlier than previously thought. During the early and middle Triassic, most of Earth's landmasses were grouped together in the supercontinent of Pangea that stretched from pole to pole (see page 146), so in geographic terms, a connection between animals in Poland and Tanzania is plausible.

Footprints from Poland

The 2010 discovery of fossil footprints from 250 million-year-old rocks in Poland's Holy Cross Mountains was heralded by headlines announcing 'Dinosaur origins pushed further back in time', and these prints do indeed seem to preserve the earliest known evidence for dinosaur-like reptiles in the early Triassic. The small prints are only a few centimetres long, suggesting that their maker, named *Prorotodactylus*, was about the size of a domestic cat, walked on all four legs and weighed no more than 1kg (2.2lb).

The Polish site preserves a large number of trackways, with the dinosaur prints forming just 2 or 3 percent of the total, so these primitive dinosaurs seem to have been rare animals. The other prints belong to amphibians, lizards and rauisuchians, predatory relatives of the crocodiles.

One of the most intriguing aspects of the find is that it pushes the origin of the dinosaurs back a further 15–20 million years, bringing it within 2 million years of the greatest mass extinction in the history of life, the end-Permian event, when some 90 percent of all life had been wiped out (see page 209). So was the emergence of this new reptile group linked to the mass extinction itself? According to this hypothesis, the emerging dinosaurs may have been able to take advantage of abandoned ecological niches and the disappearance of many of the top predators, just as some other 'disaster taxa' certainly did (see page 213). The logical conclusion is that without the extinction event, there might have been no dinosaurs – or at the very least, they might have not have risen to become the dominant land animals of the next 180 million years.

But can we take a discovery like this at face value? The problem here is essentially concerned with the nature of the trace fossil evidence – no bones of the putative *Prorotodactylus* have yet been found, and the footprints are merely impressions left on soft sand and mud as the animal walked over the ground. Well-preserved prints like these can indeed replicate the shape and form of an animal's foot and give some indication of anatomical features such as toes, but even so, the claim that these prints were made by an ancestral dinosaur or close relative is open to interpretation, and debate is sure to continue unless more definitive evidence appears.

LEFT 250 million-year-old footprints from early Triassic times, recently discovered in Poland, predate the earliest known skeletal remains of dinosaurs by some 15–20 million years. They are claimed to represent the oldest dinosaur-related fossils.

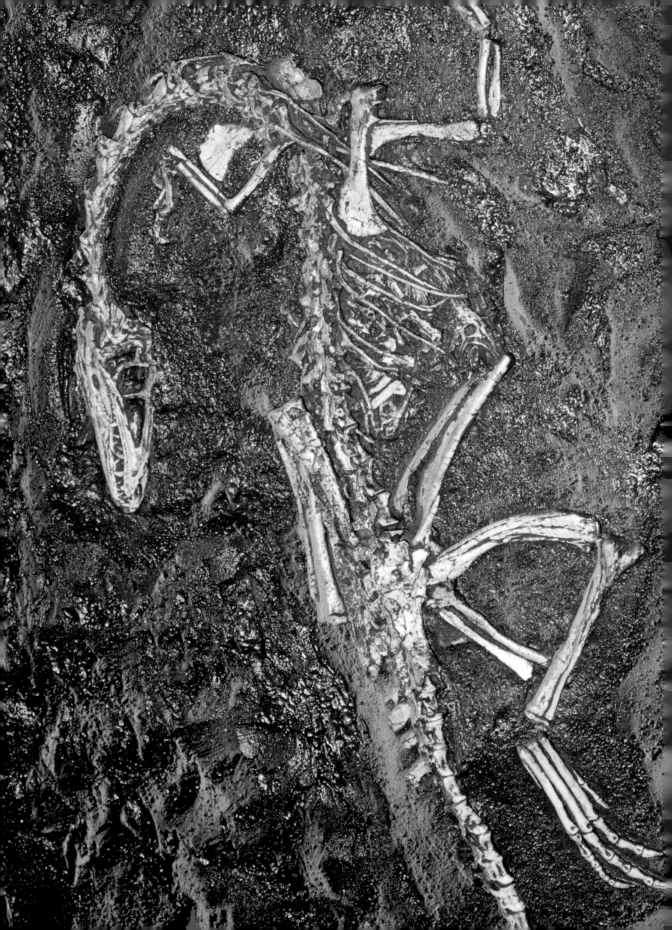

Late Triassic life

DEFINITION THE EARLIEST TRUE DINOSAUR FOSSILS COME FROM
LATE TRIASSIC TIMES WHEN THE GROUP WAS BEGINNING TO
DIVERSIFY AND RISE TO PROMINENCE

DISCOVERY NEW MEXICO'S GHOST RANCH SITE WAS DISCOVERED IN
1947, AND ARGENTINA'S VALLEY OF THE MOON IN THE 1950S

KEY BREAKTHROUGH *HERRERASAURUS* AND *EORAPTOR* SHOW EARLY
SPECIALIZATION AMONG MEAT-EATING DINOSAURS

IMPORTANCE THESE FOSSILS COME CLOSEST TO THE COMMON
DINOSAUR ANCESTOR FROM WHICH ALL OTHERS EVOLVED

Late Triassic times saw some significant changes in the land-living reptiles, with the first appearance of truly dinosaur-like animals. However, at first these creatures were very much in the minority – the dominant animals were still the plant-eating rhynchosaurs and the predatory cynodonts.

Few places preserve a fossil record of the transition between the mid-Triassic fauna of mixed reptiles and the dinosaur-dominated world of the late Triassic and beyond. Perhaps the two most famous sites capturing this intriguing phase in the history of life are the so-called 'Valley of the Moon' in the San Juan Province of Argentina, and Ghost Ranch in the US state of New Mexico.

The fossil riches of Argentina's San Juan and La Rioja provinces were first excavated in the late 1950s by Argentine palaeontologist Dr Osvaldo Reig (1929–92), within a remote region named the 'Valley of the Moon' on account of its barren, rocky terrain. Around 227 million years ago, this landscape was a broad floodplain with active volcanoes nearby. The numerous lakes and rivers were surrounded by giant 40m (130ft) tall conifers, such as *Protojuniperoxylon*, ferns, such as *Cladophlebis*, and horsetails. Lying at a latitude of 40 degrees South, the climate was strongly seasonal. Its diverse vertebrate fauna was dominated by herbivorous rhynchosaurs such as *Hyperodapedon* (whose hind feet had large claws for digging up edible roots and tubers), but also included pig-like dicynodonts such as *Ischigualastia*, the carnivorous cynodont *Exaeretodon* and the 7m (23ft) *Saurosuchus*, a carnivorous archosaur that was among the top predators of the time. Other plant eaters included the small 50cm (20in) long gomphodonts, and armoured aetosaurs, including the 3m (10ft) long *Stagonolepis*, also known from Elgin in Scotland.

OPPOSITE The 1947 discovery of hundreds of virtually complete skeletons of a small, lightly built, bipedal dinosaur-like animal called *Coelophysis*, found at Ghost Ranch, New Mexico, was one of the great fossil discoveries of the 20th century.

55 Life in Triassic seas

DEFINITION THE TRIASSIC PERIOD SAW REPTILES RETURN TO THE SEA
AS FEARSOME MARINE PREDATORS

DISCOVERY FINE FOSSILS WERE DISCOVERED BY 19TH-CENTURY
MINERS AT MONTE SAN GIORGIO ON THE SWISS/ITALIAN BORDER

KEY BREAKTHROUGH RECENT STUDIES INDICATE THAT A MARINE
EXTINCTION AT THE END OF THE TRIASSIC WAS LINKED TO
VOLCANIC ERUPTIONS AND THE OPENING OF THE ATLANTIC

IMPORTANCE SOME OF THESE REPTILES SURVIVED TO GIVE RISE TO
EVEN LARGER PREDATORS IN JURASSIC AND CRETACEOUS SEAS

Life in the oceans of the Triassic was dominated by reptiles, such as nothosaurs and placodonts, which had evolved from land-living ancestors. However, around 199 million years ago, the end of the Triassic was marked by another major extinction, and marine life suffered particularly badly.

Probably our best window into the marine life of Triassic times comes from the remarkable fossils of Monte San Giorgio, which today lies high in the European Alps on the border of Switzerland and Italy. These mid-Triassic (233 million-year-old) shales have been worked since the late 19th century for their chemical properties, especially for extraction of bitumen. As a by-product of the quarrying, hundreds of complete skeletons of marine reptiles and thousands of fish and molluscs have been found, and today many of these can be seen in the Natural History Museum of Milan and the Palaeontological Institute and Museum of the University of Zurich. According to a UNESCO World Heritage Site citation in 2003, these fossils provide 'the single best known record of marine life in the Triassic Period'. They owe their excellent preservation to oxygen-poor waters that periodically killed off the animals, and also prevented their remains from being scavenged.

The strata of Monte San Giorgio were originally laid down in a bay at the western end of the tropical Tethys Ocean (see page 146), as the supercontinent of Pangea began to fragment. The lack of normal currents in this area allowed the seawater to stratify into distinct layers, with little or no 'turnover' between them. As a result, the deeper waters became deficient in oxygen and toxic to normal marine creatures. There seems to have been little or no life on the seabed, except for a few bivalves that could tolerate the

OPPOSITE The ammonoids are a major group of extinct Mesozoic marine cephalopods with characteristic coiled shells, distantly related to the living squid and pearly nautilus. They arose in Palaeozoic times and developed many forms with distinctive features that can easily be recognized.

low oxygen levels. Otherwise, most animals were free swimmers, ultimately dependent on surface-dwelling plankton for their food. The plankton was consumed by small fish and cephalopods, which in turn were eaten by bigger fish and the marine reptiles. Some 30 species of primitive bony fish have been found at the site, many of which disappeared in the end-Triassic extinction. At the top of the food chain were various large reptile predators such as dolphin-like ichthyosaurs (three genera), seal-like nothosaurs (four genera) and more specialist forms, such as the placodonts (two genera) and the bizarre *Tanystropheus*. Additionally, there were five shark species, four of which were small animals with specialized teeth for crushing cephalopod shells.

The nothosaurs

The most common reptiles were a group of diapsids known as the nothosaurs, which arose from the smaller, aquatic lizard-like pachypleurosaurs and evolved into the larger, entirely marine plesiosaurs (which replaced them completely at the end of Triassic times). Typically, the nothosaurs had a long body, neck and tail, with paddle-like limbs and webbed feet for swimming. Their elongated heads were flattened and adapted for catching fish with numerous outward-pointing sharp teeth. *Ceresiosaurus* was a 3m (10ft) long nothosaur that may have been amphibious and laid its eggs on land. Other representatives included *Paranothosaurus*, *Lariosaurus* and the tiny 30cm (12in) long *Neusticosaurus* – the most common of all, with several hundred known fossil skeletons.

'*Tanystropheus* is one of the most bizarre reptiles known, with a 3m (10ft) neck as long as the rest of its body and tail put together.'

The ichthyosaurs first evolved from some, as yet unknown, terrestrial ancestor around 245 million years ago in mid-Triassic times, and became extinct in the late Cretaceous, around 90 million years ago. Primitive ichthyosaurs, including the 3.5m (11.5ft) long *Shastasaurus* from Monte San Giorgio, looked more like finned lizards with long flexible tails, than their dolphin-like descendants, such as the 1.2m (4ft) long *Mixosaurus*. The latter is one of the best-preserved and commonest reptiles found at Monte San Giorgio – stomach areas within some specimens still retain indigestible 'hooks' from the tentacles of their cephalopod prey, while others preserve embryos within the body cavity, showing that, despite being reptiles, they were fully adapted for an entirely aquatic life and gave birth to live young.

The placodonts, meanwhile, were another group of sauropterygian ('lizard-flipper') reptiles that evolved in early Triassic times from unknown ancestors. The earliest, such as Monte San Giorgio's 1.5m (5ft) long *Paraplacodus*, were lizard-like in form, but were already specialized for eating shellfish with protruding front teeth and flattened cheek teeth for crushing shells. Later species evolved bony plates on their backs to protect them from predators such as sharks – and in some animals the evolution of the armour plate converged to produce remarkably turtle-like forms.

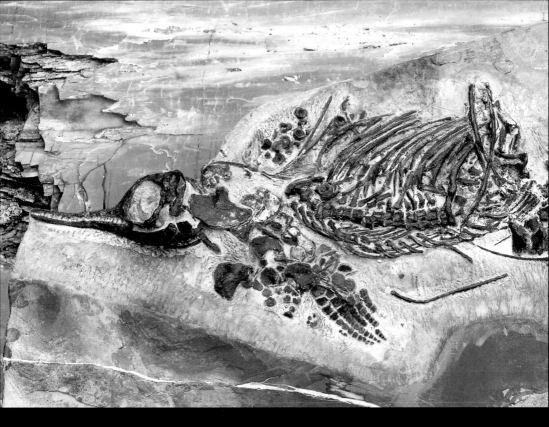

Perhaps the strangest of all these Triassic swimmers, however, is *Tanystropheus*. Classified as a prolacertiform archosauromorph, it is one of the most bizarre reptiles known, with a 3m (10ft) neck as long as the rest of its body and tail put together. Its head, meanwhile, had a long narrow snout with sharp interlocking teeth for catching slippery fish. A new specimen, found in 2006, preserves scaly skin impressions and indications of bulky muscles at the base of the tail, which shifted the animal's centre of gravity backwards for stability as its neck swung swung from side to side.

The end-Triassic extinction

The exact cause of the extinction that brought an end to this diverse marine life is not clear, but (as with the end-Permian extinction), it coincides with a massive eruption of continental flood basalts – the Central Atlantic Magmatic Province, formed between 205 and 191 million years ago (see page 109). Triggered by the break-up of the Pangean supercontinent, the eruption spread lava across an area of some 11 million square km (4.2 million square miles), now split between Europe, Africa and the Americas. Such a massive outpouring of lava, one of the largest known, may well have altered the chemistry of the atmosphere and oceans – it has been estimated that around 22 percent of marine families and about half of all genera became extinct.

ABOVE More than 10,000 specimens have been recovered from around Monte San Giorgio since the strata were first quarried for their bitumen in the 19th century. They include clams, ammonoids, fish and many kinds of marine reptiles, such as *Mixosaurus*, shown here.

Over the last decade an astonishing series of fossil discoveries from Daohugou, in the Nincheng County of Chinese Inner Mongolia, has revealed a quite unexpected population of early mammals living alongside the dinosaurs in mid-Jurassic times, around 164 million years ago.

According to long-held theories backed up by a wealth of fossil evidence, land-based life in Mesozoic times was dominated by the so-called 'ruling reptiles' – not only dinosaurs, but also other archosaurs such as crocodiles and flying pterosaurs. Fossil mammals have been found in Mesozoic rocks since the early 19th century, but most seemed to be small, rat- or cat-sized insect eaters. But now, thanks to the remarkable new Mongolian fossils, we are discovering that early mammals were already surprisingly diverse.

The Daohugou environment

The mid-Jurassic strata found at Daohugou were originally freshwater lake and river deposits in which the remains of various aquatic and terrestrial animals became buried. Fortunately, many of their soft tissues, such as hair and skin, were also preserved in outline and by mineral replacement, giving a remarkably complete picture of life in the region.

Daohugou lay in quite high latitudes, around 60 degrees North, but since there were no polar ice caps at the time, the climate would still have been temperate. The landscape around the lake was well forested, with plants including gymnosperms, conifers, cycads, clubmosses, horsetails and ferns.

Animal fossils from the site range from spiders and flies to amphibians and lizards, feathered dinosaurs, pterosaurs and the primitive mammals.

OPPOSITE The parachute-like membrane used by the 44 living species of flying squirrel for gliding from tree to tree was thought to have been a relatively recent mammal adaptation. However, the recent discovery of a primitive, mid-Jurassic age mammal from Daohugou in Mongolia shows that it has some remarkably ancient precursors.

The fossil preservation is so good that even the wings of flies and other delicate arthropod tissues have been fossilized. In 2010, the discovery of *Eoplectreurys*, the earliest known haplogyne spider, extended the history of this small family, today confined to the tropical Americas, back some 120 million years. Amphibians from the site, meanwhile, include *Chunerpeton*, the earliest known salamander, growing to 18cm (7in) long. Some 200 virtually complete specimens of its 'tadpoles' have also been preserved.

Early feathers?

However, one of the most extraordinary and frustrating finds at Daohugou is the partial skeleton of an animal called *Pedopenna* ('foot feather'), unveiled in 2005. All that remains of this 'maniraptoran' theropod dinosaur are the 12cm (5in) long foot bones, whose bird-like structure suggests affinities to both the troodontid and dromaeosaurid groups. Fascinatingly, the foot bones are fringed with impressions of long symmetrical feathers and an overlay of shorter ones. If the dating of the Daohugou site is correct (and there is some controversy about this), then *Pedopenna*, which must have been about 1m (40in) long in life, is the oldest known feathered dinosaur, predating even

RIGHT Another of Daohugou's fossil surprises has been the discovery of *Castorocauda*, a 50cm (20in) long semi-aquatic mammal that combines features from the otter, platypus and beaver. These ancient fossils have overturned the idea that early mammals were all small rodent-like animals.

the 'first bird' *Archaeopteryx* by about 15 million years (see page 253). In 2008, more feathered dinosaur remains were discovered at the site – this time from a small maniraptoran called *Epidexipteryx*, which displayed long straight ornamental feathers almost as long as its 25cm (10in) body.

The mammals of Daohugou

Even more important from the evolutionary point of view, however, are Daohugou's mammalian remains. Perhaps most superficially familiar is *Pseudotribos*, preserved as a near-complete skeleton with traces of its original fur. This small animal, 10cm (4in) long, had strong limbs and a robust pelvis and shoulder girdle, possibly for mole-like burrowing. It retained a primitive sprawling posture, and is classified as 'basal mammaliaform'. Its molar (cheek) teeth have a form familiar from true mammals, known as tribosphenic (with three distinct peaks on the biting surface), but they seem to have functioned in a different way, placing this animal in a sister group to modern mammals and revealing that there was more than one evolutionary pathway to combine the slicing and crushing functions of mammal cheek teeth. Other primitive features of the animal, such as its constricted waist, suggest an affinity to monotremes such as the living platypus and echidna.

'Fossil mammals seemed to be small, rat- or cat-sized insect-eaters. But now, thanks to the remarkable new Mongolian fossils, we're discovering that early mammals were in fact surprisingly diverse.'

Discovered in 2006, *Volaticotherium* is even more of a surprise. This small animal has a 3.5cm (1.4in) long skull, a body just 14cm (5.5in) long and an estimated weight of around 70g (2.4oz). It differs from other known Mesozoic mammals in having a sizeable furry skin fold, or patagium, that extended from the body to the wrists and ankles of its elongated limbs and onto the tail. Unique among mammals of this age, the membrane resembles that of the living flying squirrels (*Glaucomys*) and shows that *Volaticotherium*, as its name implies, was capable of gliding flight.

Such behaviour requires considerable evolutionary adaptation in order to develop an 'aerofoil' large enough to support the animal's weight and generate lift. *Volaticotherium* seems to have been a tree-living, foraging mammal that could launch itself into the air to glide between trees. However, it is unlikely to have had enough control of its flight to capture insect prey while gliding, as insectivorous bats, for example, do today. Like modern gliding mammals and many if not most Mesozoic mammals, it was also probably nocturnal.

One final mammalian surprise is *Castorocauda*, a 50cm (20in) long semi-aquatic mammal with a mixture of platypus, otter and beaver features. Weighing around 0.8kg (1.8lb), it had a 20cm (8in) flat, scaly tail, webbed feet and seal-like teeth for catching fish and invertebrates. *Castorocauda*'s specialized fish-eating adaptations confirm the suspicion that Mesozoic mammals were far more diverse than previously thought, pushing back the fossil record of the mammalian conquest of water by over 100 million years.

57

Africa's giant dinosaurs

DEFINITION A GROUP OF FOSSILS FROM TENDAGURU, TANZANIA THAT REVEALS SOME OF THE EARLIEST GIANT DINOSAURS

DISCOVERY THE TENDAGURU SITE WAS DISCOVERED BY GERMAN MINING ENGINEER BERNHARD SATTLER IN 1906

KEY BREAKTHROUGH AT 24M (80FT), *GIRAFFATITAN* WAS ONE OF THE TALLEST AND HEAVIEST DINOSAURS KNOWN

IMPORTANCE THE WIDE RANGE OF FOSSILS FOUND AT TENDAGURU PROVIDE A UNIQUE VIEW OF A LATE JURASSIC ENVIRONMENT

Between 1907 and 1931, more than 200 tonnes of late Jurassic dinosaur remains were excavated in tropical East Africa and shipped to Europe. With some bones more than 2m (6.6ft) long, the transportation of these huge fossils was one of the greatest feats in the history of palaeontology.

In August 1937, thousands of Berliners lined up to see the skeleton of a new giant African dinosaur, called *Brachiosaurus*, unveiled in the swastika-bedecked main hall of the Museum für Naturkunde. At 11.9m (39ft) high and 22.6m (74ft) long, it was the largest mounted skeleton of a land animal in the world.

The Tendaguru fossils from Tanzania reveal a picture of life around a subtropical river plain 152 million years ago. The site was separated from the sea by lagoons and shallow bays, with sandbars, coral reefs, barrier islands and a strongly seasonal climate. The land-living animals included plentiful plant-eating stegosaurs, huge long-necked sauropods and abundant ornithopod iguanodonts. The coexistence of so many herbivores shows not only that plant food was abundant, but also that the various species were exploiting different types of foliage across the varied and complex environment within which they lived. Needless to say, the large numbers of herbivores attracted meat-eating predators, in the form of theropods such as *Ceratiosaurus* and *Elaphrosaurus*.

All these dinosaurs were accompanied by a variety of other vertebrates including lizards, pterosaurs and a small primitive mammal relative called *Tendagurutherium* that is only known from a few teeth. The marine fossils include primitive holostean bony fish, a shark, cephalopod belemnites, corals, sea lilies and clams.

OPPOSITE The first *Brachiosaurus* fossils were discovered in Colorado in 1900. With thigh bones more than 2m (80in) long, it was clearly a very large animal. In the following decade, the recovery of a near-complete skeleton from German East Africa (Tanzania) showed just how large it really was.

A rare land environment

The Tendaguru deposits offer a good demonstration of how and why remains of land-living animals are preserved at some localities and not others. Most landscapes are essentially dominated by erosion, with their ground-up rocks and the organic remains contained within them steadily transported towards the sea, so that even if fossils are recovered, information about their original environment and articulated layout is lost. For terrestrial deposits to survive intact, special geological circumstances are needed, and Tendaguru provides an excellent example of one of such situation.

As the supercontinent of Gondwana (see page 145) rifted apart from late Carboniferous through to early Jurassic times, the landscape of East Africa became stretched and formed a series of fault-bounded valleys and basins – depressions where terrestrial sediments accumulated, eventually building into deposits several kilometres deep.

During the middle Jurassic, these basins were repeatedly flooded and drained as sea levels rose and fell, and shallow marine sediments were deposited along the continental shelf that became the present-day East African coastline. Then in the late Jurassic, between 155.7 and 130 million years ago, the newly opening Indian Ocean flooded into the large Mandawa Basin three times, producing a layered sequence of strata some 140m (460ft) thick, with three dinosaur-bearing 'horizons' separated by marine sediments.

Abundant fossil pollen in the Tendaguru strata shows that the forests were dominated by various conifers along with the ginkgo, gleicheniacean ferns and bryophytes. This was a time of greenhouse conditions, with south-easterly trade winds delivering heavy rain from December to February. However, there may also have been seasonal droughts.

The dinosaurs

One of the most common dinosaurs at the site was the armoured stegosaur *Kentrosaurus*, which grew to 5m (17ft) long, with a bulky 1.5-tonne body and forelimbs shorter than the hind ones. It would have been slow moving in life, and probably used the long bony spines on its back for protection. It may also have been able to rear up on hind limbs to reach high plant foliage – though not for long.

'The coexistence of so many herbivores shows not only that plant food was abundant, but also that the various species were exploiting different types of foliage across the varied and complex environment within which they lived.'

The most famous of the Tendaguru species, however, is *Giraffatitan*, a 24m (80ft) giant that weighed up to 50 tonnes. Originally classed as *Brachiosaurus*, this giant has recently been renamed following the discovery that it belongs in a separate genus from the *Brachiosaurus* of North America. It was the most common sauropod at the site, ranging over both the coastal plains and the drier interiors. The long forelimbs, toes and high wrist lifted the shoulder some 4.3m (14ft) high, and the neck was a further 9m (30ft) long. *Giraffatitan*'s skull had a long snout with chisel-shaped teeth, and nostrils positioned high on the head, which may also have acted as a sound-resonating chamber.

Dryosaurus, meanwhile, was a 4m (13ft) iguanodont – an agile animal weighing 100kg (220lb), capable of running at 33–43 km/h (20–27mph). Fossil hatchlings show that it developed rapidly, moving from four-footed to bipedal locomotion in its first five months. The discovery of some 30 closely associated individuals may indicate social behaviour.

The main predator at Tendaguru was *Elaphrosaurus*, a 6m (20ft), slender theropod whose long neck and legs suggest that it was a fast runner. Of the non-dinosaur vertebrates, meanwhile, *Rhamphorhynchus* is one of the most important. The group to which this long-tailed pterosaur belonged first evolved in the late Triassic and were still flourishing in Jurassic times. A skilful flier with a 1.75m (70in) wingspan, it had long, pointed and slightly curved teeth that it used to catch fish.

58

Australia's polar dinosaurs

DEFINITION A GROUP OF DINOSAURS FROM AUSTRALIA THAT WOULD
HAVE LIVED IN POLAR LATITUDES DURING THE EARLY CRETACEOUS

DISCOVERY THE FIRST AUSTRALIAN DINOSAUR FOSSIL WAS FOUND IN
EASTERN VICTORIA IN 1903

KEY BREAKTHROUGH THE RICH DEPOSITS OF 'DINOSAUR COVE' WERE
DISCOVERED IN 1984 AND EXCAVATED USING MINING TECHNIQUES

IMPORTANCE THE EXISTENCE OF DINOSAURS IN POLAR LATITUDES
WITH LONG, DARK AND COLD WINTERS SHOWS THEY WERE FAR
MORE ADAPTABLE THAN MODERN REPTILES

The discovery of dinosaurs that lived in high latitudes, well within the Antarctic Circle, has revealed that some of these reptiles were able to adapt to a life in which winters were long, dark and cold. But how did they survive in such a hostile environment?

The first dinosaur fossil to be recognized in Australia was discovered on the Cape Patterson coastline of eastern Victoria in 1903 by William Ferguson (1882–1950), a geologist prospecting for coal. Known as the Cape Patterson Claw' it was a single 5cm (2in) toe bone from a large theropod dinosaur.

At the time, the discovery of Australian dinosaur remains was noteworthy, but not particularly unexpected – it was only in the 1960s, with the growing awareness of plate tectonics and the realization that Australia had spent much of the Mesozoic Era in south polar latitudes, that the find became controversial – how could cold-blooded reptiles, which today are confined to the planet's warm temperate and tropical regions, have lived so far south? In 1979, Australian geologists returned to Victoria's early Cretaceous sandstones in search of answers – and what they found led to a radical rethink of the dinosaurs' ability to cope with life in harsh environments.

Dinosaur Cove

One cliff-side site has proved particularly rewarding. Nicknamed 'Dinosaur Cove', it preserves numerous fossils from a single ancient river deposit. Recovering these fossils was no simple task – it required tunnelling into the side of the cliff itself – but some 80 dinosaur bones were recovered during the first 16 days of mining in 1984, and over the following decades several thousand more fossils have been found. Another site of similar age was

OPPOSITE Only warm-blooded birds and mammals are capable of withstanding the freezing temperatures of today's polar regions, but in early Cretaceous times dinosaurs and amphibians apparently thrived close to the poles. They were helped by significantly warmer temperatures at the time, but still had to contend with up to three months of midwinter darkness.

discovered near the seaside town of Inverloch, and the total range of fossils now found includes lungfish, amphibians, turtles, mammals and a bird, all of which lived some 110 million years ago. There is a surprising diversity of small dinosaurs, dominated by plant-eating hypsilophodontids, along with the fragmentary remains of a coelurosaur and a possible oviraptorosaur. The most surprising recent discovery has been a small tyrannosaurid, the first of these theropod predators to be found in Gondwana.

Leaellynasaura was a 3m (10ft) bipedal hypsilophodontid herbivore, and is known from two partial skeletons that show a long flexible tail and a skull with enlarged eye sockets and optic lobes, suggesting that the animal had larger eyes adapted for the low light conditions of the polar Antarctic. *Atlascopcosaurus*, a slightly smaller relative, is only known from even more fragmentary material, but as small generalized plant eaters, it's reasonable to assume that these were among the most widespread animals.

Described in 1993, *Timimus* is a 3.5m (12ft) coelurosaur theropod with unusually slender legs. Although only a couple of thigh bones and some vertebrae have been found, the bone structure of this plant eater reveals cyclic growth patterns – possible evidence that this dinosaur spent part of each year in hibernation.

Even more recent has been the 2010 identification of a small tyrannosaur. Only a single hip bone is preserved, but it carries unmistakable features. Although small in size (about 3m/10ft long), this animal had some surprisingly advanced characteristics, suggesting that small tyrannosaurids were far more cosmopolitan than previously thought, and only later gave rise to the giant northern hemisphere predators of late Cretaceous times.

A polar environment

Most of the Dinosaur Cove bones came from a 'bonebed conglomerate', formed where disarticulated skeletal remains washed downriver accumulated in a relatively still area (like pebbles in a streambed hollow). Palaeomagnetic studies of contemporary rocks (see page 62) conclusively show that Australia at the time was still joined to Antarctica and lay at a high polar latitude of 75 degrees South.

Here, the midwinter darkness would have lasted about three months, and mean annual temperatures were between –6 and +5°C (21 and 41°F). The discovery of dinosaurs in an environment so totally unsuitable for modern reptiles has been used to support the argument that they were actually warm-blooded (see page 251) like modern mammals.

At first glance, it's also tempting to ask what these polar dinosaurs ate. However, the early Cretaceous climate was significantly warmer than today, so even though the environment was highly seasonal, there was no polar ice cap and the landscapes were well wooded, with large conifers, araucariacean pines, ginkgos, cycadophytes and ferns and horsetails in the understorey. Most of these plants were evergreen, so would have been available as food the year round, but how the dinosaurs coped with the long dark days is unknown.

Amphibians and mammals

What's more, the polar dinosaurs are not the only surprising animals to be found this far south. The last known amphibian temnospondyl, *Koolasuchus*, probably survived longer in these regions than anywhere else in the world, partly because of the geographical isolation, but also because of its tolerance of water temperatures that were too cold for the crocodiles that had already pushed temnospondyls into extinction elsewhere. *Koolasuchus* was a spectacularly large amphibian, some 4m (13ft) in length with a broad and flat head some 50cm (20in) long and jaws armed with 100 or so teeth, some of which were themselves up to 10cm (4in) long. With eyes on top of its skull, it was probably an ambush predator that lay hidden on the riverbed awaiting its prey. *Koolasuchus*, like living salamanders, must have been tolerant of cold water (as cold as –2° C/28.5°F in the case of some living salamanders) and may even have hibernated in soft riverbed muds.

A few small teeth and fragments of jawbone betray the presence of at least three distinct mammal genera, named *Bishops*, *Ausktribosphenos* and *Teinolophus*. *Bishops* is the most interesting of these – a small shrew-like animal with a primitive jaw structure that puts it in a recently established group of Gondwanan mammals called australosphenids – a group that also included the early ancestors of today's most 'primitive' mammals, the monotremes.

'Although only a couple of thigh bones and some vertebrae have been found, the bone structure of *Timimus* reveals cyclic growth patterns – possible evidence that this dinosaur spent part of each year in hibernation.'

59

India's journey north

DEFINITION THE SURPRISINGLY RAPID MOVEMENT OF THE INDIAN PLATE SINCE LATE CRETACEOUS TIMES, RESULTING IN A COLLISION WITH ASIA AND THE FORMATION OF THE HIMALAYAS

DISCOVERY THE PRECISE RATE OF INDIA'S MOVEMENT CAN BE CALCULATED FROM PALAEOMAGNETIC DATA

KEY BREAKTHROUGH ACCORDING TO A NEW THEORY, THE SPEED OF INDIA'S MOVEMENT IS DUE TO ITS RELATIVELY THIN CRUST

IMPORTANCE THE CREATION OF THE HIMALAYAS AND TIBETAN PLATEAU HAS HAD A MAJOR EFFECT ON BOTH REGIONAL AND GLOBAL CLIMATE

Around 50 million years ago, one of the biggest collisions in Earth's history created the spectacular Himalayan mountains. Yet the impetus for this event began some 80 million years before, when India broke away from Australia and Antarctica, and began to race north at a rate of 70mm (2.8in) per year.

Around 130 million years ago during the early Cretaceous, the world map looked very different from today, with Africa, India, Australia and Antarctica all clustered into the southern hemisphere supercontinent of Gondwana. Then, the opening of a new ocean tore India away from East Africa and sent it on a 5,600km (3,500-mile) journey north across the equator.

New measurements of India's trajectory show that for some 80 million years, from 130 until 50 million years ago, the massive continent travelled north at an astonishing 70mm (2.8in) a year – sometimes perhaps even 200mm (8in) per year. This might seem remarkably slow by human standards but for a huge continent-sized raft of Earth's crust it is extremely fast – normally plates move at no more than 40mm (1.6in) per year.

A recent analysis, by Prakash Kumar of India's National Geophysical Research Institute, provides a possible explanation, suggesting that India's uniquely fast progress may be due to the relative thinness of its ancient crustal rocks. With a maximum depth of 80km (50 miles), they are only half as deep as the continental crusts in Antarctica and Australia (while southern Africa's ancient crust is a remarkable 280km/175 miles thick). The thinness of the Indian Plate is thought to have reduced its drag through the asthenosphere (see page 57) and allowed it to 'race' away compared with the slower motion of the other, more deeply rooted continents.

OPPOSITE Seen from space, the vast Himalayan mountain chain is revealed to consist of roughly parallel ridges and valleys (running from top left to bottom right in this image) that reflect the underlying geological structure of the impact zone between the Indian and Asian plates.

India meets Asia

As India moved north, the newborn Indian Ocean opened behind it and the ancient Tethys Ocean to the north (see page 146) closed up as its sea-floor crust was driven down – subducted beneath the continental crust of Asia. Finally, some 50 million years ago in early Cenozoic times, India itself crashed into Asia. The collision slowed India's progress to 5mm (0.2in) per year, but was unable to stop it completely – instead, the Indian Plate ploughed on northwards, crumpling, folding and shearing the continental margins so that they were pushed in all directions. The effect can be simply modelled by pushing a book placed on a tablecloth over a slippery surface and watching how the cloth is folded around the book.

By early Miocene times (around 23 million years ago), the crumpled rocks had been pushed up some 9km (5.5 miles) above sea level to form the Himalayas, while deep in the mountain roots below, the pressures and temperatures became so great (up to 750°C/1,380°F) that the rocks were radically altered by the process known as metamorphism. New minerals, such as mica, garnet and sillimanite, were created as the sedimentary rocks were transformed into metamorphic rocks such as schist and gneiss.

'New measurements of India's trajectory show that it travelled north at an astonishing 70mm (2.8in) a year – sometimes perhaps even 200mm (8in) per year.'

Deep below ground along the margin between the colliding plates, some of the rocks were melted and recrystallized to form blocks of igneous granite. These granites, originally formed at depths of more than 10km (7 miles), have today risen up to form some of the highest Himalayan peaks, including Nuptse in Nepal, Changabang and Shivling in India, and Kanchenjunga on the border between the two. Everest, the highest peak of all, has a base made of metamorphic schists and gneisses, penetrated by granite dykes (sheet-like bodies created where the molten rock spread into them through a network of cracks). But for all this activity at its roots, the mountain's upper slopes were surprisingly unscathed – they are made from fossil-bearing Ordovician limestones originally laid down below sea level, and climbers are often surprised to come across marine fossils as they struggle towards Everest's 8,848m (29,029ft) summit.

Despite the immensity of the Himalayas, not all of the convergent motion of India and Asia could be accommodated by their formation, and so a large-scale structure called an 'overthrust' developed. The lower layers of the Indian Plate detached from the folded rocks on its surface and were pushed beneath Tibet as part of the process that caused the uplift of that mighty plateau (see page 81). To either side of the Himalayan mountain front, meanwhile, the rocks were pushed aside to form the flanking ranges of Afghanistan in the west and Burma in the east. It is estimated that the overall collision has 'shortened' the Earth's crust in this region by more than 1,000km (620 miles) so far – and the process is far from over.

Wearing away

The Himalayas continue to rise upwards at a rate of around 10mm (0.4in) per year, but the erosion created by their uplift has worn away an estimated 40km (25 miles) of rock from the rising mountains over the last 40 million years. Much of the debris has been transported to lower altitudes, filling in the Ganges and Brahmaputra valleys and becoming folded into the frontal ranges of the Himalayas.

At its peak around 16–19 million years ago, chemical weathering of the silicate rocks and burial of organic carbon conspired to remove huge amounts of carbon dioxide from Earth's atmosphere, possibly contributing to a prolonged period of global cooling and the growth of the Antarctic and Greenland ice sheets (see page 293). Today, the debris is still carried to the sea by the Indus, Ganges and Brahmaputra rivers, creating fan-shaped deltas that extend for tens of kilometres into the ocean. The Indus alone can deposit 1 million cubic m (35 million cubic ft) of sediment onto its floodplain annually, and carries 175 times this much material into the sea each year.

ABOVE Between 130 and 50 million years ago, the Indian Plate moved northwards until it converged on the Asian continent. The collision of the two plates not only created the great crumpled range of the Himalayas, but also uplifted the vast Tibetan Plateau. The combined effects have radically altered the climates and environments of southern Asia.

Changing the climate

Elevation of the Himalayas around 23 million years ago also helped to generate the dominant Asian weather system known as the monsoon. Moisture-laden winds from the Indian Ocean are drawn towards the Himalayas by high pressure over the Tibetan Plateau in summer. As they are forced upwards by the mountains, the water vapour carried by these winds precipitates to create heavy seasonal rains between June and September. This rainfall has radically changed the regional climate and the surrounding environment, allowing intensive agriculture and the growing of essential crops such as cotton, rice and oilseed, but also creating an ever-present threat of catastrophic flooding.

The evolution of flowering plants

DEFINITION ANGIOSPERMS OR FLOWERING PLANTS EVOLVED TO
DOMINANCE FROM EARLY CRETACEOUS TIMES

DISCOVERY *ARCHAEFRUCTUS*, THE EARLIEST KNOWN ANGIOSPERM,
WAS DISCOVERED AT LIAONING, CHINA, IN THE LATE 1990S

KEY BREAKTHROUGH NEW FOSSILS SUGGEST THAT THE FIRST
ANGIOSPERMS BECAME SUCCESSFUL BECAUSE OF THEIR EFFICIENT
SYSTEMS FOR TRANSPORTING WATER AND NUTRIENTS

IMPORTANCE FLOWERING PLANTS OR ANGIOSPERMS ARE THE
DOMINANT PLANTS ON EARTH TODAY, AND VITAL FOOD SOURCES

With more than a quarter of a million species, flowering plants are found across Earth's land surface and provide the planet's largest food supply. Charles Darwin described their origin and evolution as an 'abominable mystery' and even today, it is still a matter of considerable debate.

Today, the flowering plants, technically known as angiosperms, dominate many but not all of Earth's landscapes. Their 254,000 or more species, ranging from water lilies to magnolias, grasses to oak trees, make the angiosperms the most successful group of modern plants, but this was not always so. Their rise appears to have accompanied the advance of the placental mammals, modern birds and certain insect groups, all of which are essential components of life on land today. What's more, the relatively recent evolution of the grasses has had a major impact upon mammalian life – especially that of humans, since grasses and cereals provide the bulk of all our food.

However, the fossil record of the earliest angiosperms is patchy and frustrating. Their delicate tissues do not preserve easily, and they have a tendency to fall apart so that their distinguishing features are rarely fossilized together. Analysis of the genetic differences between major groups of living angiosperms provides scientists with a way to estimate the length of time since they diverged from their ancient common ancestry – the so-called 'molecular clock' (see page 130) – but such studies have produced conflicting data. Some date the first 'basal' angiosperms to the very beginning of the Cretaceous, but others put their origins in the Jurassic or even as far back as Triassic times. However, angiosperm fossils are certainly rare before 120 million years ago, well into the early Cretaceous, and their apparently

OPPOSITE *Leefructus*
is a newly discovered,
beautifully preserved
early Cretaceous
fossil angiosperm
from China, between
123 and 126 million
years old. Its features
include clusters
of simply veined
'trilobate' leaves,
and a fruit with five
carpels enclosed in a
receptacle. Its recent
description by a Sino-
American team led
by Ge Sun and David
Dilcher reinforces
evidence for the early
diversification of the
eudicot flowering
plants that dominate
today's flora.

sudden appearance explains why Charles Darwin considered them one of the major problems for his theory of evolution.

Angiosperm features

The name 'angiosperm' is derived from Greek and means 'seed vessel'. It was first coined as long ago as 1690 by German physician and botanist Paul Hermann (1646–95). Defining angiosperm features include the possession of flowers as reproductive organs, an enclosed covering (carpel or pericarp) around the female ovules that often develops into a fruit, and the possession of a nutritive tissue called the endosperm, which provides food for the developing plant embryo and sometimes the seedling.

Biologically, the angiosperms and the more ancient gymnosperms ('naked seeds') are the only plant groups to have evolved seeds. This has allowed them to become completely independent of damp or watery environments, and to colonize drier landscapes. It seems likely that the angiosperms evolved from gymnosperms in early Mesozoic times and subsequently became distinguished from them by a number of newer features. One problem faced by botanists is that most of the typical angiosperm features also appear in various gymnosperms – they are not individually diagnostic, but it is their presence alongside one another that defines the angiosperms. Unfortunately, the fossil plant record rarely preserves the whole association of these features, making it hard to identify the earliest angiosperms.

What is more, the various reproductive structures of flowering plants appear in a sequence throughout the plant's life cycle. Flowers form to attract pollinators, then disintegrate after fertilization of the ovule by pollen. The ovule forms a seed, while the carpel often grows into a fruit, to attract fruit-eating animals that can help disperse the seeds. Even a perfect fossil plant can only preserve a single stage in this reproductive cycle, and of course the processes of decay and fossilization generally involve disintegration, separation of parts and their transportation before burial. Tree trunks may remain in their original location, with branches, woody tissues and heavier fruits nearby, but leaves may be blown further afield, and pollen is even further dispersed. Roots and flowers, meanwhile, are very rarely preserved at all.

First fossil traces

At present, the oldest known fossil angiosperm is *Archaefructus*, found in 128 million-year-old early Cretaceous strata at western Liaoning, China, in the late 1990s. Identified as an aquatic herbaceous plant, it has dissected leaves and elongated reproductive structures with carpel-like clusters at the end. The whole reproductive structure was at first interpreted as a bisexual 'flower' (containing ovules but also stamens to release pollen). More recently, however, it has been reinterpreted as a single-sexed flower. Since bisexual flowers are generally assumed to be an essential characteristic of the earliest 'basal' flowering plants, and single-sexed flowers are a later development of a group called the eudicots, *Archaefructus* may now have lost its possible place as a basal angiosperm. Meanwhile, another angiosperm-like fossil from the same strata, *Synocarpus*, is thought to represent another eudicot with fruits produced by unisexual flowers.

'The fossil record of the earliest angiosperms is patchy and frustrating. Their delicate tissues do not preserve easily, and they have a tendency to fall apart so that their distinguishing features are rarely fossilized together.'

But to add another twist to the story, new *Archaefructus*-like fossils from the Chinese strata preserve the exceedingly rare association of a plant with roots and shoots. This new species, *Archaefructus eoflora*, has a reproductive structure with two carpels and a stamen – characteristics of a bisexual flower. So it seems that *Archaefructus* may after all have had the potential to evolve the wide range of flower types found in both modern angiosperms and their fossil representatives.

Angiosperm evolution today

Modern studies tend to picture the first flowering plants as opportunistic, fast-growing, weedy herbs and shrubs with thin leaves, growing in disturbed riverside habitats. Evolution of a more efficient transport system for carrying water from roots to leaves and increasing photosynthesis may have been a key to the success of these new plants, and the increasingly dense vein patterns found in early angiosperm leaves seem to support this idea.

However, there is still a major conflict between the appearance of angiosperms in the fossil record and the 'molecular clock' estimates of their origin, recently dated to around 215 million years ago. While this leaves a gap of 80 million years in the record of plant fossils, it does coincide rather suggestively with the rise of pollinating insects such as bees and flies.

61

Sauropod puzzles

DEFINITION SAUROPODS WERE FOUR-LEGGED, LONG-NECKED HERBIVORES, INCLUDING THE LARGEST DINOSAURS OF ALL

DISCOVERY SAUROPOD FOSSILS HAVE BEEN FOUND AROUND THE WORLD SINCE THE 19TH CENTURY, AND OUR IMAGE OF THEM HAS CHANGED MANY TIMES

KEY BREAKTHROUGH RECENT ANALYSIS OF SAUROPOD NECK POSTURES HAS TRANSFORMED OUR VIEW OF THEM ONCE AGAIN

IMPORTANCE STUDIES OF SAUROPOD METABOLISM SEEM TO UNDERMINE SUGGESTIONS THAT DINOSAURS WERE WARM-BLOODED

Of all the iconic dinosaurs, the giant sauropods are perhaps the most spectacular. The largest were pushing the biological limits for land-living animals whose bodies are not supported by water, yet they obtained their energy from plant food. So what do we know about their biology?

The sauropod dinosaurs reached their peak in terms of abundance and diversity during the late Jurassic from 'prosauropod' ancestors, but continued to flourish into late Cretaceous times. Originating late in the Triassic, they quickly evolved a number of successful lineages that dispersed globally thanks to the close association of the continents during this period. The diplodocids and brachiosaurs migrated widely, while the dicraeosaurids and titanosaurids were specific to the southern landmass of Gondwana. The similarities between the African dinosaurs of Tendaguru (see page 233) and those of the Morrison Formation in the western United States were thought to indicate the presence of a 'land bridge' through Europe that remained after the direct connection between Africa and America was severed. However, recent comparison between the Morrison and Tendaguru brachiosaurs has shown that they were not quite as similar as was once thought – and indeed the African brachiosaurs have now been reassigned to a new genus, *Giraffatitan*. As a result, it seems likely that these late Jurassic sauropods were probably both descended from an earlier, globally dispersed fauna.

Legs and locomotion

The sauropods include the largest ever land animals, with body weights of up to 50 tonnes – five times heavier than the largest elephants. Essentially quadrupedal, they had long necks with small heads, balanced by long tails. The distribution of their body weight and the structure of their legs differed

OPPOSITE High in the Bolivian Andes, one of the world's longest set of dinosaur tracks are exposed on an ancient rock surface of Cretaceous age, now tilted to a high angle by tectonic movements. The footprints are thought to have been made by a large titanosaur sauropod.

from those of elephants – sauropod hind limbs were more massive and carried proportionally more weight than the forelimbs, which had relatively weak elbows, wrists and inflexible feet. As a result, the forelimbs contributed less to the animal's forward movement than the hind limbs, so a sauropod's gait would have been quite different from that of the elephant. Strong claws on the hind feet helped to generate forward movement, but maximum speeds could have been no more than a fast walk (though thanks to the length of the animal's stride, this would still have amounted to 18 km/h (11mph). More normal would have been a smooth slow walk at 3–4km/h (1.8–2.5mph), involving very little vertical movement of the body.

The posture problem

The great lengths of sauropod necks are clearly related to the animals' feeding strategies. In order to grow so large, they would have had to consume very large amounts of plant material, and a long neck could have given access to a greater volume and variety of plant food than that available to any other contemporary herbivore. The details of neck posture are still hotly debated, since they have the potential to radically alter feeding strategies – for example, early brachiosaur reconstructions positioned the neck almost vertically and sometimes showed animals rearing on their hind legs to feed even higher in the tree canopy, thus avoiding direct competition with their own young or smaller species. With the diversity of sauropods present at locations such as Tendaguru, such 'niche partitioning' could have been important. But was it actually possible?

In 2009, scientists from the University of Portsmouth in the UK and Western University of Health Sciences in California set out to answer this

BELOW The sauropod dinosaurs were the largest animals ever to have lived on land, and the function and disposition of their immensely long necks and tails has been a matter of considerable debate as has the speed at which they could move. Model studies and computer simulations such as these can help to resolve questions about function and mobility.

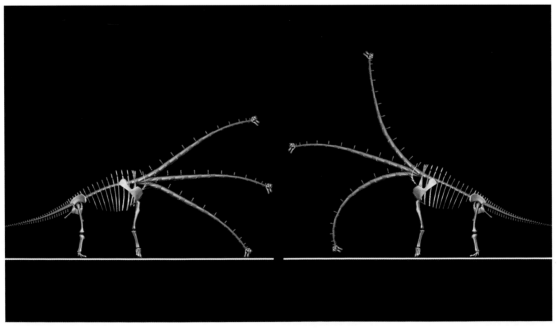

question through a detailed analysis of the neck structure and anatomy in giant sauropods such as *Brachiosaurus* and *Giraffatitan*. They concluded that the neck could not have been held vertically, but instead sloped upwards so that the head was 'only' about 6m (20ft) above the ground. Aside from structural evidence in the neck vertebrae themselves, there are supporting physiological arguments for this less ambitious posture. For instance, the length of the carotid artery from heart to brain is 9.8m (32ft). According to another recent study, this means that a hypothetical four-chambered heart weighing 386kg (850lb), capable of moving 17.4 litres (3.8 gallons) of blood with each stroke at around 14 strokes per minute, would have to pump blood with a pressure of roughly 1,000 millibars in order to reach a head that was held upright. This would have created very high pressure in the limbs and required adaptations such as pumps and valves in the veins, along with a very thick skin, to prevent abnormal accumulations of fluid.

'In order to grow so large, they would have had to consume very large amounts of plant material, and a long neck could have given access to a greater volume and variety of plant food than that available to any other contemporary herbivore.'

Feeding and metabolism

Teeth and jaw structure were also vital to feeding strategies – since brachiosaur teeth grew around the entire margin of the upper and lower jaw, they may have been capable of a vertical cutting bite, a suggestion supported by wear patterns found on some front teeth. The amount of food the sauropods needed to consume would have depended upon their metabolic rate, and crucially whether they were cold-blooded ectotherms like modern reptiles, or warm-blooded endotherms like mammals. This topic has been a subject for fierce debate, since a higher metabolic rate and warm-bloodedness would have implications for the behaviour of dinosaurs beyond the sauropods (see page 237). However, it may be significant that the ectothermic situation is quite different for big and small reptiles. The relatively large volume of the sauropods would have helped them conserve heat and maintain a fairly constant body temperature *without* the faster metabolism and greater food requirements seen in mammals – a condition known as gigantothermy.

However, the question still remains of how such large animals could sustain themselves eating relatively low-value plant material, especially when their intake was limited by small jaws and skulls. According to one recent estimate, a 30-tonne warm-blooded brachiosaur would have required a daily food intake of 344,000 kilocalories. Based on the calorific value of modern cycads, conifers and ferns, this would have meant eating 360–500kg (800–1,200lb) of plant food in a foraging 'day' that probably lasted 12 to 14 hours. At a rate of 1–6 bites per minute, each mouthful would have had to include 100–600g (3.5–21oz) of edible matter. Altogether, this seems to be an unsustainable feeding rate. A cold-blooded gigantothermic metabolism would have required a far smaller and much more feasible intake, so the sheer practicalities of eating seem to be telling us that the largest dinosaurs, at least, were probably gigantothermic.

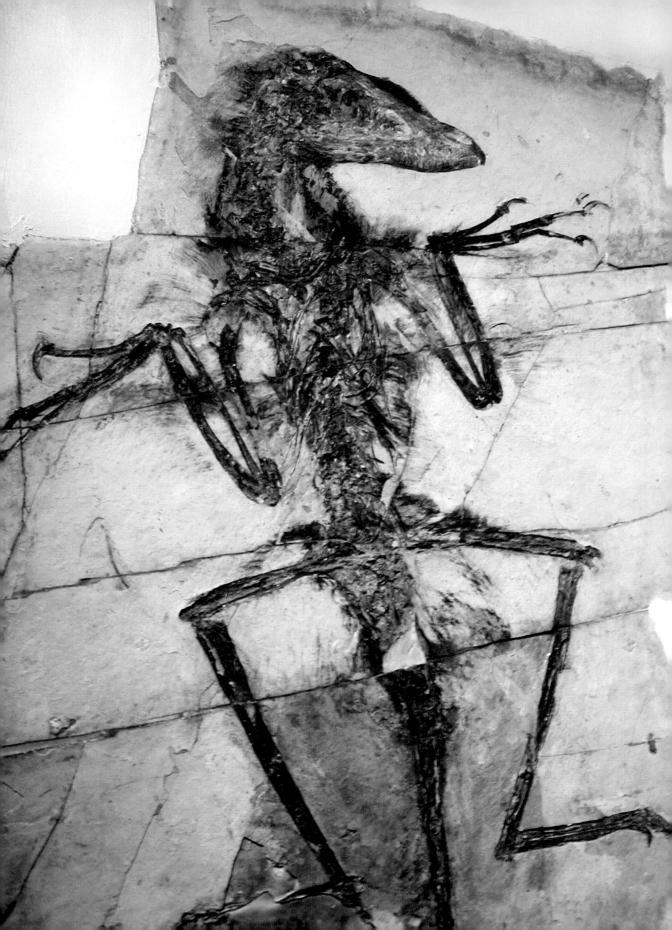

62

Linking birds and dinosaurs

DEFINITION BIRDS ARE NOW WIDELY RECOGNIZED AS A SPECIALIZED GROUP OF DINOSAURS THAT EVOLVED FEATHERS AND FLIGHT

DISCOVERY *ARCHAEOPTERYX*, A FOSSIL BIRD WITH SOME DISTINCTLY DINOSAUR-LIKE FEATURES, WAS FOUND IN GERMANY IN 1861

KEY BREAKTHROUGH SINCE THE 1990S, A SERIES OF FEATHERED DINOSAUR FOSSILS HAVE EMERGED FROM LIAONING, CHINA

IMPORTANCE THE DISCOVERY THAT MANY DINOSAURS HAD FEATHERS HAS TRANSFORMED OUR IMAGE OF THESE ANCIENT REPTILES

The realization that birds are dinosaurs is no longer particularly newsworthy, but the details of how feathers and flight developed, and how birds evolved from one small family of feathered theropod dinosaurs are still being worked out, and new discoveries still have the capacity to surprise.

Over the last two decades or so, a spectacular series of fossils preserving soft tissues have been uncovered in Asia, and especially from 124 million year old early Cretaceous strata at Liaoning, China. These fossils include a variety of theropod dinosaurs with fossilized feathers and more primitive feather-like structures. They range across most 'coelurosaur' theropod groups, from the tyrannosaurids and compsognathids through oviraptorosaurs, such as *Caudipteryx*, to very bird-like dromaeosaurids such as *Microraptor*. In 2010, Spanish palaeontologist Francisco Ortega announced possible evidence for feather-like structures in *Concavenator*, a member of a more primitive theropod dinosaur group called carcharodontosaurids. What's more, simple tubular filament structures have also been found in the skin of ornithischian 'bird-hipped' dinosaurs such as *Tianyulong* (a bipedal heterodontosaurid) and the small ceratopsian *Psittacosaurus*. Ironically considering their name, ornithischian dinosaurs are only very remotely related to the 'lizard-hipped' saurischians that are the true ancestors of birds – so it now seems that feather-like structures were a much more common feature among all dinosaurs, and birds simply inherited and adapted feathers from their dinosaur ancestors.

'Ancient wing'

In 1860, German quarrymen working the Jurassic limestones of Solnhofen uncovered a beautifully preserved fossil feather. Just 5cm (2in) long, its asymmetric structure showed that it came from the wing of an animal that

OPPOSITE Over recent decades, numerous small dinosaurs with a variety of skin structures ranging from hair-like fibres to well-developed feathers, including the spectacular *Sinornithosaurus* shown here, have been recovered from Cretaceous strata in China. These discoveries have revealed much detail about the evolution of feathers and birds.

could fly. The following year, an entire prehistoric bird was discovered and purchased by a local physician, who in turn sold it and several other fossils to anatomist Richard Owen (1804–92) and the British Museum for the considerable sum of £700 ($3,400 at the exchange rate of the time).

> 'Did flight develop from the ground upwards, with increasingly well-muscled and feathered 'arm-wings' that became strong enough for take-off and flapping flight, or did it evolve from tree-down gliders that then developed the anatomy required to flap their wings?'

Owen recognized that the fossil, soon named as *Archaeopteryx* (meaning 'ancient wing'), showed a very interesting mixture of primitive reptile features, such as a long bony tail and toothed beak, and more evolved, bird-like features, such as forearms adapted as wings with flight feathers. However, it was Thomas Henry Huxley (1825–95), an outspoken champion of Darwinian evolution, who realized that *Archaeopteryx* provided a wonderful example of the evolutionary transition between two major groups of animals. Huxley pointed out that the skeletal structure of *Archaeopteryx* was very similar to that of a small bipedal theropod dinosaur called *Compsognathus*, which had also been found at Solnhofen in 1861.

Feathered dinosaurs

But the world of palaeontology was still amazed more than a hundred years later, when the first feathered dinosaur from Liaoning was revealed in the mid-1990s. Initial studies of the 68cm (27in) long *Sinosauropteryx*, a close relative of *Compsognathus*, suggested that some small theropod dinosaurs had a body covering of 'fuzzy' down that was probably for insulation. In 2010, however, new analysis revealed that the filaments were pigmented with alternating dark and light bands, especially along the tail, suggesting a use in display.

Soon after, the discovery of further specimens from Mongolia and China, such as the 1m (40in) long oviraptorosaur *Caudipteryx*, revealed that other theropod dinosaurs had true primary feathers that were symmetrical and not associated with flight. Finds like this provided the first good evidence that feathers evolved before flight, and probably began as an adaptation for insulating the bodies of these small and active theropods. Additionally, they could have been used for sexual display, and perhaps as camouflage for the females.

Since then a whole array of fossil feather-like structures have been found on both adult and a handful of juvenile dinosaurs – such as a recently discovered (2009) fossil of the oviraptorosaur *Similicaudipteryx*. These reveal that even non-avian feathered dinosaurs went through a radical change in the structure of their feathers associated with moulting. Down-covered juveniles developed into adults with a variety of feather forms, as seen in modern birds. However, the dinosaurs show an even greater variety of feather types, including filaments and long ribbon-like structures with plumed tips that are not found in modern birds, suggesting that some developmental stages have been lost in feather evolution.

Powered flight

While recent fossil finds prove conclusively that 'feathers came first', they also add to the evidence in a long-running debate about the evolution of powered flight. Did flight develop from the ground upwards, with increasingly well-muscled and feathered 'arm-wings' that became strong enough for take-off and flapping flight, or did it evolve from tree-down gliders that then developed the anatomy required to flap their wings? The presence of wing-like feathers on both the arms and legs of the tiny maniraptoran *Microraptor* shows that at least one group of small dinosaurs became efficient gliders, but it is important to remember that the vast majority of known feathered dinosaurs are Cretaceous in age – they post-date the early evolution of the 'avialian maniraptoran' (bird) dinosaur branch and the appearance of flapping flight seen in animals such as *Archaeopteryx*. Nevertheless, they can still show how these remarkable developments might have happened.

In 2009, however, two feather-covered specimens of a crow-sized and fast-running troodontid theropod, called *Anchiornis*, were found in 155 million-year-old Late Jurassic strata from China. This is the first feathered troodontid that predates *Archaeopteryx,* and confirms that such feathered dinosaurs did exist before the iconic 'first bird'. Surprisingly, *Anchiornis* has remarkably bird-like aerodynamic feathers on its feet and legs to match those on its arms and tail, but its arm feathers form a rounded wing that was clearly not adapted for flight.

ABOVE The question of whether powered flight evolved from the ground up with the flapping of muscular wings or by gliding down from trees has exercised palaeobiologists for decades. Certainly, the first birds with powered flight, such as *Archaeopteryx* shown here, had evolved by late Jurassic times.

63

Dinosaur eggs and babies

DEFINITION DINOSAUR EGGS CAN REVEAL SURPRISING DETAILS ABOUT THEIR BREEDING HABITS AND YOUNG

DISCOVERY IN 1997, PALAEONTOLOGISTS WORKING IN ARGENTINA STUMBLED ACROSS AN ENORMOUS SAUROPOD NESTING GROUND AT NEUQUÉN, PATAGONIA

KEY BREAKTHROUGH DISTRIBUTION OF EGG BROODS REVEALS THAT SAUROPOD GROUPS RETURNED TO NESTING SITES YEAR ON YEAR

IMPORTANCE SITES SUCH AS NEUQUÉN REVEAL RARE EVIDENCE ABOUT THE SOCIAL HABITS OF DINOSAURS

As land-living reptiles, dinosaur young hatched from eggs. But questions over how dinosaurs nested and whether they cared for their offspring have intrigued palaeontologists for decades. A remarkable nesting ground in Patagonia is now helping to answer some of these questions.

In late Cretaceous times, more than 80 million years ago, the Auca Mahuevo region of the Neuquén Basin in northwestern Patagonia, Argentina, was a wide river floodplain, cut off from the Pacific Ocean by the growing Andean mountain chain with numerous active volcanoes to the west. Rapid weathering and erosion of the rising mountains resulted in huge volumes of sediment being carried east by rivers flowing into the Neuquén Basin. Combined with rapid cycles of climate change, this caused periodic floods of silt-laden material across the region, laying down new layers of clay that buried and 'sealed' the landscape. Today, these sedimentary strata have been re-exposed at the surface and are revealing their secrets.

In 1997, a team from the American Museum of Natural History, led by Luis Chiappe and Lowell Dingus, arrived at Auca Mahuevo in search of fossilized birds and dinosaurs. However, the team soon realized that the fragments of 'rock' scattered across the desert site were eggshells and even complete eggs and nests within a giant fossilized hatchery, laid bare by modern weathering. What was more, the eggs came not from birds, but from sauropod dinosaurs.

Since this initial discovery, thousands of clutches of sauropod eggs have been mapped out, revealing a site that covers several square kilometres of the Anacleto Formation rocks. Furthermore, the nesting site can be traced down to depths of some 85m (280ft) below the surface. Within the strata,

OPPOSITE Dinosaur eggs have been recovered from Jurassic and Cretaceous strata around the world, and have been given scientific names based on their shell structure and form. Relatively few can be firmly linked to the dinosaurs that laid them, and even when they still contain fossilized embryos it can be difficult to link the embryo to its parent.

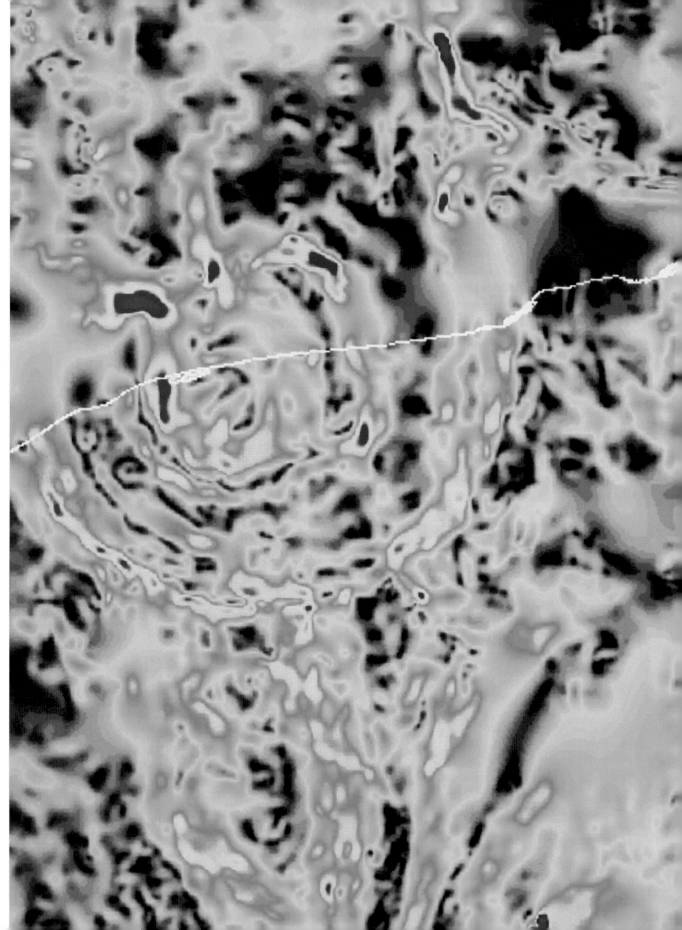

The end of the Mesozoic

The Mesozoic Era ended, quite literally, with a bang – the blast caused when an 11km (7-mile) wide rock from outer space crashed into the Gulf of Mexico 65.5 million years ago. The impact coincided with a major extinction event in which some 30 percent of all living organisms disappeared.

The last geological period of the Mesozoic Era, the Cretaceous, was marked by the final break-up of the supercontinent of Pangea and the evolutionary divergence of life on land. The Pangean break-up produced extended shorelines and vastly increased the area of continental shelves available to diversifying marine life. For the first time since the end-Permian extinction event, life regained and even overtook the high point of Palaeozoic biodiversity. Many of today's familiar animals and plants were already becoming increasingly important – on land, there were flowering plants, insects, birds and early mammals, while in the seas and oceans, bony fish, cartilaginous fish and turtles all flourished and diversified.

But, just as happened several times before in the history of life, a catastrophic setback awaited, in the form of a mass extinction event that wiped out the non-avian dinosaurs, pterosaurs, many marine reptiles and the ammonites. Since its discovery in the 1980s, the Gulf of Mexico impact event has been widely blamed for the extinction, but there may be more to the story.

Iridium from Gubbio

In the late 1970s, American geologist Walter Alvarez sampled a 2cm (0.8in) layer of clay from seabed deposits that mark the Cretaceous/Cenozoic boundary (traditionally known to geologists as the K/T boundary) in strata below the historic town of Gubbio in northern Italy. Alvarez asked his father,

OPPOSITE The concentric structures of the Chicxulub impact crater buried beneath the Yucatán Peninsula and Gulf of Mexico are revealed in this false-colour image from a gravity survey of the region (the white line marks the coast). The entire structure is today filled in by more than 1km (0.6 miles) of younger Cenozoic sediments.

Nobel Prize-winning physicist Luis Alvarez (1911–88), whether there was any way to measure how long the clay had taken to be deposited, and Luis suggested looking for the geochemical signature of platinum-group elements accumulated from tiny 'micrometeorites' that normally rain down on Earth at a known and fairly steady rate.

When the Gubbio clay was analysed, it did indeed prove to contain an anomalous amount of a platinum-group element – iridium, which is extremely rare in Earth's crust, but is more abundant in comets and asteroids. Concentrations of iridium in the clay were around 30 times greater than the background level (though still only measurable in parts per billion).

Announcing their discovery in 1980, the Alvarez team argued that the only reasonable explanation was that the iridium had been introduced to Earth's atmosphere by a very large extraterrestrial asteroid, which they estimated to be around 10km (6 miles) in diameter. As supporting evidence, they cited the presence of quartz crystals and tiny glass spherules that were also found in the boundary clay, and which they believed were formed by heating and compression during the impact. They went on to suggest that the impact was responsible for the extinction of the dinosaurs.

BELOW A dark clay layer marks the boundary between the uppermost limestones of the Cretaceous Period and the lowermost sediments of the younger Paleogene period. This clay layer was deposited in very different seabed conditions to the surrounding sediment and its geochemical analysis revealed the iridium anomaly that first revealed the end-Cretaceous impact.

At first, the far-reaching Alvarez hypothesis split the scientific community. One of the major objections from the doubters was the lack of an identifiable impact crater with the appropriate age. But although the Alvarez team did not yet know it, the crater had already been found in 1978, when an airborne magnetic survey identified a huge structure hidden deep beneath Mexico's Yucatán Peninsula. It was not until 1990 that Canadian geologist Alan Hildebrand learned of the crater's existence and dated it to the very end of the Mesozoic.

The impact and its effects

The Chicxulub impact is now estimated to have released energy equivalent to 100 million million tonnes of TNT, blasting a hole around 12km (7.5 miles) deep and around 100km (62 miles) wide, rimmed by 8km (5-mile) mountains that immediately collapsed to form a series of outer rings up to 180km (110 miles) across. Around 50,000 cubic km (12,000 cubic miles) of pulverized limestone were blasted through the atmosphere as dust, droplets of molten rock, shocked quartz and other mineral fragments.

A searing blast of heat swept across the Americas, igniting forests and incinerating all living things. Soot from the wildfires blackened the skies, blocked out sunlight and cooled the atmosphere. The shock generated enormous earthquakes and giant tsunami waves. Ten hours after impact, the waves gained height as they approached the eastern seaboard of the Americas and swept far inland, drowning any survivors and blanketing the landscape with a layer of seabed sediment. The intense heat also released hundreds of millions of tonnes of carbon and sulphur dioxides from seabed sediments into the atmosphere. The sulphur dioxide mixed with water droplets to generate widespread acid rain, while the carbon dioxide, as a greenhouse gas, would have led to long-term climate warming after the initial crash in temperatures.

'The Alvarez team argued that the iridium had been introduced to Earth's atmosphere by a very large extraterrestrial asteroid, which they estimated was around 10km (6 miles) in diameter.'

The Deccan Traps

But it may be a mistake to lay the entire blame for the extinction at the door of the Chicxulub impact. Some scientists have argued that many of the great Mesozoic animal groups were already in long-term decline by this time, and the impact also happens to coincide with a vast eruption of plateau basalts in the Deccan region of India – perhaps the largest volcanic eruption in Earth history, which expelled some 8 million cubic km (1.9 million cubic miles) of basalt lavas covering some 1.5 million square km (580,000 square miles), piling up to more than 2km (1.2 miles) deep in places. Vast amounts of carbon dioxide were released into the atmosphere, warming global climates, acidifying ocean waters and probably making a significant contribution to the end-Cretaceous extinction, just as the similar Siberian Trap eruptions of the end-Permian had probably helped to bring an end to the Palaeozoic.

65

The Cenozoic recovery

DEFINITION FOLLOWING THE EXTINCTION OF THE DINOSAURS,
MAMMALS, BIRDS AND OTHER REPTILES DIVERSIFIED RAPIDLY

DISCOVERY DISCOVERED IN 1901, CRAZY MOUNTAIN BASIN IS ONE
OF THE RICHEST SOURCES OF EARLY CENOZOIC MAMMAL FOSSILS

KEY BREAKTHROUGH GENETIC STUDIES SINCE THE 1970S HAVE
CLARIFIED THE MAJOR SUBDIVISIONS IN MAMMAL ANCESTRY THAT
ORIGINATED AROUND THIS TIME

IMPORTANCE THESE SMALL EARLY ANIMALS GAVE RISE TO THE
ENTIRE RANGE OF PRESENT-DAY MAMMAL LIFE

Mammals today range in size from tiny shrews to giant whales, with an enormous diversity of form and habit. Yet their fossil record shows that all this variety has developed in little more than 60 million years from the rat-sized animals that survived the end-Cretaceous extinction.

The extinction of the dinosaurs 65.5 million years ago saw global landscapes emptied of large herbivores and carnivores, creating huge opportunities for the surviving terrestrial animals, none of which were anything like the size of the large dinosaurs. Survivors included non-dinosaur reptiles such as lizards and crocodiles, and the avian line of dinosaurs we know better as birds. Most significant for the future, however, were the mammals, descendants of the late Permian therapsids that had been eclipsed by the archosaurs during the Triassic Period (see page 213) and gone on to develop into more familiar mammalian forms in the shadow of the dinosaurs (see page 229).

The exact origins of these early Cenozoic mammals and the way in which they diversified have both proved difficult to resolve – it seems there were successive waves of mammal origination and extinction as plants and environments recovered and climates changed. The timing of these changes seems occasionally chaotic and lacking in structure, but this probably reflects the different responses of plants and animals – especially during the rapid warming of early Paleogene times.

Plant pioneers

Proximity to the Chicxulub impact site (see page 263) meant that the plant life of the Americas was particularly badly hit by the end-Cretaceous extinction. The immediate aftermath actually saw an *increase* in plant diversity, but that

OPPOSITE The Americas suffered the greatest impact from the Chicxulub event at the end of the Cretaceous. Wildfires devastated landscapes, and the landscape took more than 10 million years to fully recover.

collapsed back to just 25 known species around 55.8 million years ago, before recovering again during the early Eocene, around 50 million years ago. Plant numbers doubled as global mean annual temperatures rose some 4°C (7°F) to 26°C (79°F), and extensive tropical-type swamp forests developed (see page 273). It was against this background of fluctuating plant cover that the early mammals were struggling to survive.

The major difficulty with discovering the details of these early mammals, is that the fossil evidence is so fragmentary. Most of them were rodent-sized creatures, with few larger than a beaver, and their remains do not fossilize easily in forest environments where soils are regularly disturbed by a whole host of other organisms, accelerating their decay. Only the most resistant parts, such as teeth, are usually preserved, and these make up the bulk of the fossil record for small mammals. Fortunately for palaeontologists, however, mammal teeth are sufficiently complex and distinctive enough to distinguish their different species and relationships.

The North American record

Some of the best evidence for the changing terrestrial life of early Paleogene times comes from western North America, where the fossil-rich rocks of Montana and Alberta include sites such as Crazy Mountain Basin, discovered in 1901 by palaeontologist Albert Silberling (1883–1951) and a group of Princeton students.

'The exact origins of these early Cenozoic mammals have proved difficult to resolve – it seems there were successive waves of mammal origination and extinction as plants and environments recovered and climates changed.'

Three different groups of very early Paleogene mammals have been found here: primitive 'multituberculates', similar to animals that had first evolved in late Jurassic times; the somewhat more advanced marsupials, which had originated in the early Cretaceous; and finally the more advanced placentals, which had also originated in early Cretaceous times around 125 million years ago. The proportions of the three groups changed over time, with the marsupials initially declining while the multituberculates flourished. The placentals did even better, with an explosive radiation into some 88 genera belonging to 29 families. But there were also high rates of extinction and turnover, with more than half of these genera apparently extinct by 61 million years ago. Despite these setbacks, successive waves of placental mammals gradually replaced the multituberculates.

Examples of these early Paleogene mammals include *Ptilodus*, a squirrel-sized multituberculate some 50cm (20in) long, with a prehensile tail, long limbs and a reversible foot that allowed it to both climb and descend trees head-first. With a broad, short-snouted skull, laterally placed eyes, serrated cheek teeth and chisel-shaped incisors, it probably had rodent-like habits. The pig-sized *Conoryctes*, meanwhile, which survived from the Palaeocene to the Eocene, belongs to the primitive placental taeniodonts, a North American

group of short-limbed specialized feeders with front teeth developed into tusks, and claws on their feet, perhaps for digging out their plant food.

Plesiadapis was another basal placental that spread across North America and Europe via Greenland before the North Atlantic Ocean opened up. A relatively large, lemur-like animal, it weighed up to 5kg (11lb). With mobile limbs and strongly curved claws, it may have been a tree climber, or perhaps a ground dweller like the living marsupial possum. The teeth were rodent-like, with large incisors and reduced canines, indicating that it was probably omnivorous or fruit eating. At one time, the plesiadapiformes were considered to be primitive primates, but they are now thought more likely to have been a non-primate sister group to our early ancestors.

Modern mammals appear

Modern mammal groups first appeared at the beginning of Eocene times, around 55 million years ago. While isolated continents, such as Australia, acquired unique mammalian faunas, the northern continents were home to a group of placentals known as the Boreoeutheria. Since the 1970s, molecular analysis of living placentals has shown a fundamental split between this group and other mammal families such as the Afrotheria (elephants and their relatives, plus aardvarks, tenrecs and golden moles) and the Xenarthra (armadillos, sloths and anteaters). The Boreoeutheria are themselves subdivided into Laurasiatheria (insectivores, bats, pangolins, carnivorans, odd- and even-toed ungulates and whales) and Euarchontoglires (primates, tree shrews, rodents and rabbits, and a number of extinct Paleogene groups such as leptictids, taeniodonts and pantodonts).

66

Primate beginnings

DEFINITION PRIMATES ARE THE MAMMAL GROUP TO WHICH WE HUMANS BELONG – AN ORDER THAT INCLUDES LEMURS, MONKEYS, APES AND THEIR FOSSIL RELATIVES

DISCOVERY NEW FINDS OF DIVERSE ANTHROPOID PRIMATES FROM NORTH AFRICA AND ASIA THROW LIGHT ON THIS GROUP'S ORIGINS

KEY BREAKTHROUGH NEW STUDIES SUGGEST THAT MONKEY-LIKE ANTHROPOID PRIMATES FIRST EVOLVED IN ASIA

IMPORTANCE THESE RECENT DISCOVERIES OVERTURN THE TRADITIONAL VIEW THAT PRIMATES FIRST APPEARED IN AFRICA

Humans, apes and monkeys are grouped together as the 'anthropoid primates' and it's inevitable that we have a vested interest in the origins and evolution of our own ancestors. Yet attempts to clarify early anthropoid evolution have proved remarkably difficult.

Part of the problem in pinning down our early ancestors comes from the sometimes contradictory evidence about when and where the first primates evolved, and the difficulty of pinning down the taxonomic identity of some of the extinct groups. Fossil primates hit the headlines in a big way in 2009, thanks to the promotion of a stunning fossil nicknamed 'Ida'. This 47 million-year-old creature, formally named *Darwinius masillae*, was discovered in a near-perfect state of preservation at Messel, Germany. As the time it was described as an important link to the anthropoid line of primates, but its classification remains controversial.

According to 'molecular clock' estimates (based on the amount of genetic difference between living anthropoids and other primates), the major primate groups probably split from one another in the late Cretaceous, but the fossil record seems to support an origin in more recent Paleogene times. The fossils also offer conflicting evidence about the original site of anthropoid origins, pointing to both North Africa and Asia.

New discoveries from Libya

Paradoxically, however, the recent discovery of 39 million-year-old mid-Eocene fossils from Libya in North Africa actually seems to support the idea that early anthropoids came from Asia and 'colonized' Africa, rather than evolving there. Excavated by a team led by Jean-Jacques Jaeger

OPPOSITE The 58cm (23in) fossil of *Darwinius masillae* was originally discovered in 1983 in lake deposits at Messel in Germany, dated to around 47 million years old. Although promoted in 2009 as a link to the anthropoid primates, it is certainly a remarkably preserved primitive adapiform. More detailed study and analysis of the skeleton should help clarify its evolutionary relationships.

ABOVE This reconstruction of a scene on the Libyan coast 39 million years ago groups newly discovered early anthropoids *Afrotarsius* (top left), *Karanesia* (top right), *Biretia* (bottom left) and *Talahpithecus* (bottom right) in their coastal woodland habitat.

of the University of Poitiers, the most surprising thing about the Libyan fossils is their diversity. They include three tiny species, named *Afrotarsius libycus*, *Biretia piveteaui* and *Talahpithecus parvus*, each of which is a representative of a distinct anthropoid family (the Afrotarsiidae, Parapithecidae and Oligopithecidae respectively). They may well have coexisted, and none weighed more than 500g (18 oz). Since it would have taken a long time for the first basal anthropoids to develop into such distinct families, it seems their ancestry must lie even further back, in early Eocene times.

Jaeger's site, at the Dur At-Talah escarpment in southern Libya, preserves coastal deposits from a tidal shoreline. Several separate fossil 'horizons' were laid down at different times, some containing the remains of land-living animals. Apart from the anthropoids, fossils found at the site include some rodents and extinct proboscideans (distant relatives of the modern elephants), such as *Moeritherium*. This amphibious plant eater grew up to 3m (10ft) long, stood some 70cm (38in) tall and led a similar lifestyle to the living hippos.

The African background

There are two possible explanations for the presence of diverse anthropoids in North Africa at this time. One possibility is that there could be a significant gap in the African fossil record, so the African ancestors of these fossils have not been preserved or simply have not yet been found. However, this seems unlikely, as a number of other fossil-rich Eocene sites in North Africa have been thoroughly excavated in recent decades, without turning up any evidence for older anthropoids that might be ancestral to the Libyan animals.

At present, the oldest African anthropoids date from 40 million-year-old strata in Algeria – just a million years before the Libyan fossils and too close in time to be plausible ancestors. Another Algerian fossil, found in 1992, was named *Algeripithecus* and dated at around 50 million years old. At first it was thought to be an early anthropoid, but it is in fact a strepsirhine, part of a group of primates represented today by the lemurs of Madagascar, the galagos of central Africa and the lorises of southern Asia.

Out of Asia?

The more plausible explanation for the sudden appearance of diverse anthropoids in North Africa, however, is that they arrived from elsewhere, and discoveries made in Myanmar (Burma) and China over the last few decades tend to support this theory. They show there was a greater diversity of basal anthropoids present in Asia during Eocene times, including the 45 million-year-old *Eosimias* (meaning 'ancestral ape'), which may be one of the earliest anthropoids of all.

Eosimias was discovered in China in the mid-1990s, and since then fossils belonging to the eosimiid family have been found elsewhere in China and in Myanmar. Most are simply isolated teeth, resulting in considerable debate over their anthropoid status. But some *Eosimias* foot bones discovered in China in 2000 preserve specialized features that are restricted to the anthropoids. These verify that *Eosimias* was indeed an anthropoid, while its mixture of primitive and advanced features point to its ancestry among the 'haplorhine' primates (a group that includes both the modern anthropoids and the tarsiers).

Like most early anthropoids, *Eosimias* was very small, weighing no more than 100g (3.5oz), and probably looked like a tarsier with small spoon-shaped incisors, enlarged canines and a distinctive jawbone. Its ankle structure points to a monkey-like foot posture, with soles that faced downwards rather than inwards. It probably walked along the tops of branches on all fours (rather like the present-day South American squirrel monkey) instead of clinging to vertical surfaces and leaping from tree to tree.

A new find from Myanmar

In 2009, a new fossil primate called *Ganlea*, from 38 million-year-old strata in Myanmar, was classified as part of another extinct group of primitive primates, known as the amphipithecids, whose relationships have been a matter of debate. Were they anthropoids, or more closely related to the extinct, lemur-like adapiforms? Like *Eosimias*, this new primate has the greatly enlarged and heavily worn canine teeth of a fruit eater that used its canines to pry open tough tropical fruits to reach the nutritious seeds inside. This kind of specialist seed eating is found in living anthropoid saki monkeys from the Amazon Basin, but has not been seen in the adapiforms, supporting the idea that *Ganlea* and its relatives were anthropoids. The continuing discovery of a wide range of Eocene fossil anthropoids in Asia supports the idea that they originated and diversified here, before several already distinct groups colonized Africa during mid-Eocene times.

'Paradoxically, however, the recent discovery of 39 million-year-old mid-Eocene fossils from Libya in North Africa actually seems to support the idea that early anthropoids came from Asia and "colonized" Africa.'

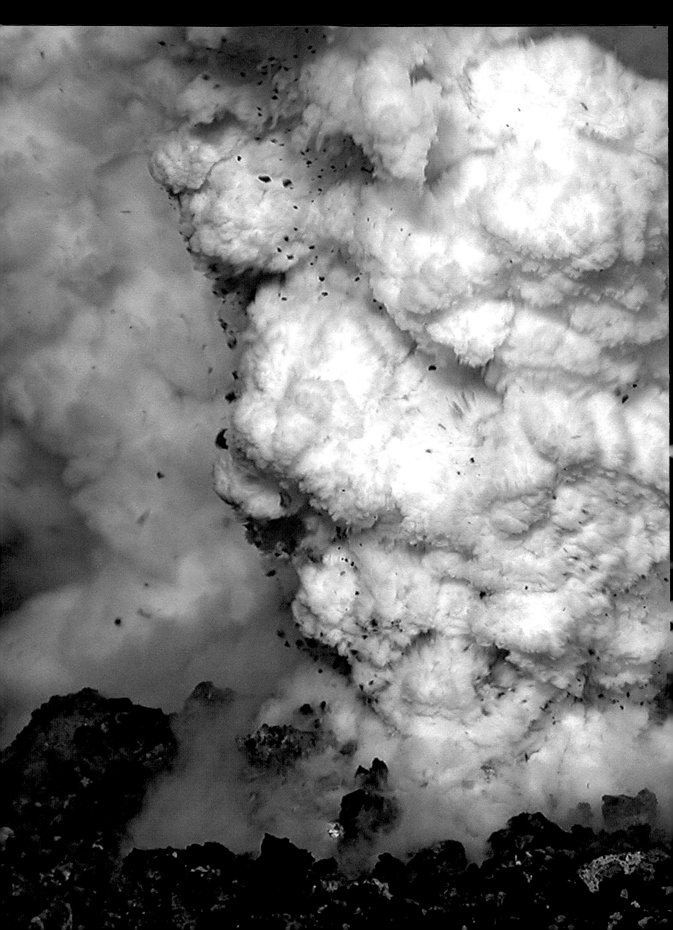

67

PETM – The Great Warming

DEFINITION A PERIOD OF EXTREME GLOBAL WARMING THAT PEAKED AROUND 56 MILLION YEARS AGO

DISCOVERY RECENT ANALYSIS OF MAMMALIAN TEETH SUPPORTS THE IDEA OF A SHARP RISE IN TEMPERATURE

KEY BREAKTHROUGH THE TOOTH STUDY SUGGESTS THAT TEMPERATURE RISES COULD HAVE BEEN TRIGGERED BY A METHANE RELEASE LINKED TO THE OPENING OF THE ATLANTIC

IMPORTANCE THE PETM CREATED WIDESPREAD CLIMATE CHANGE THAT INFLUENCED ALL LIFE ON EARTH AT THE TIME

Around 55.8 million years ago, the temperature of ocean surface waters rose sharply by 4–9°C (7–16°F), causing rapid climate change and a profound reorganization of the world's mammal faunas. If such an event were to be repeated, it could be even more catastrophic than current global warming.

The marine and continental rock record both preserve evidence for a dramatic rise in global temperatures some 55.8 million years ago, marking the end of Palaeocene times and the beginning of the Eocene. Known as the Palaeocene-Eocene Thermal Maximum or PETM, it has been linked to the release of a huge amount of carbon into the atmosphere, producing a significant greenhouse effect. At such a remote date, this global climate crisis had nothing to do with ice ages or artificial influences, so what could have caused it?

Methane and its source

Geochemical analysis of carbonate rocks formed at the time shows an abrupt shift in the ratios of different carbon isotopes (atoms of the same element with different masses), indicating that huge amounts of isotopically 'light' carbon were pumped into the atmosphere – an event known as a carbon isotope excursion (CIE).

Various mechanisms have been suggested to explain this sudden rise in atmospheric carbon, but there may be no single explanation. According to some estimates, at least 1,500 billion tonnes of carbon were released in a relatively short time frame – far too much to have been 'outgassed' directly by any volcanic eruptions of the time. However, it's possible that volcanism associated with the opening up of the North Atlantic could have erupted into

OPPOSITE Several independent lines of evidence, from fossils to geochemical traces, show that Earth experienced a major phase of global warming, known as the Palaeocene-Eocene Thermal Maximum, around 55.8 million years ago. The main cause seems to have been the release of huge amounts of carbon into the atmosphere, some of which could have come from volcanic eruptions.

68

The Great Rift Valley

DEFINITION A MAJOR AREA OF CONTINENTAL EXTENSION RUNNING
THROUGH EAST AFRICA

DISCOVERY US GEOLOGISTS IDENTIFIED A TECTONIC 'TRIPLE
JUNCTION' AT THE HEART OF THE RIFT VALLEY IN 1968

KEY BREAKTHROUGH SUBSIDENCE ASSOCIATED WITH THE VALLEY'S
FORMATION ALLOWS SEDIMENTS TO ACCUMULATE HERE,
PRESERVING FOSSIL REMAINS

IMPORTANCE THE RIFT VALLEY IS OF PARTICULAR SIGNIFICANCE TO
HUMANS, SINCE IT WAS IN THIS REGION THAT OUR SPECIES AND
ITS CLOSE RELATIVES FIRST EVOLVED

The 'Cradle of Humankind' in East Africa's Great Rift Valley owes its existence to a combination of geological, environmental and evolutionary processes that began in Ethiopia more than 30 million years ago. The evolution of humanity was just one part of this ongoing geological drama.

From Mozambique, at 17 degrees South, the Great Rift Valley extends for 6,000km (3,700 miles) across the equator through Ethiopia, the Red Sea and on up through Jordan to the Dead Sea at 33 degrees North. Its rivers, lakes and grasslands support abundant and diverse life, and the history of this region is particularly interesting to us humans. Over the last 5 million years or so, this 'African Eden' has seen the evolution of successive waves of the hominid family, culminating in our own *Homo sapiens* species. Yet the landscape itself owes its existence to underlying geological processes that have been at work for much longer, over more than 30 million years.

Rifting processes

Geologically, the Great Rift Valley is one of the few regions of Earth where a continental plate is being actively torn apart by the processes of plate tectonics (see page 41). Extension of the lithospheric rocks has stretched and thinned Earth's outer layer through the formation of large parallel faults that descend through the whole thickness of crust, breaking it into fault blocks. As the blocks slip downwards to fill the slowly growing space, they form a valley floor in which new sediments accumulate. On either side, the valley is flanked by uplifted blocks, with mountains that are volcanic in places.

Such continental rifting can be driven by two different mechanisms, one passive and the other active, and both are seen in the development of the

OPPOSITE Lake Victoria lies in the middle of an elevated geological structure known as the Kenyan Dome, around which the Great Rift Valley diverges. In this satellite image, the western branch is clearly demarcated by Lake Tanganyika (bottom left) while the eastern branch is less obvious.

Great Rift. Initially, some 70 million years ago, the active upwelling of hot material in the mantle beneath Ethiopia expanded the crust into a large dome some 200–500km (125–310 miles) across. With an average elevation of more than 1,500m (5,000ft), the Ethiopian Highlands still form the largest uplifted plateau in Africa. Stretching and fracturing at the top of the dome led to the formation of massive and extensive fissures, from which about 1 million cubic km (200,000 cubic miles) of flood basalts erupted (see page 109). In late Paleogene times, around 30 million years ago, these fissures also gave rise to some large explosive volcanoes.

The Ethiopian triple junction

Another result of the doming was the formation of a 'triple junction', a tectonic feature first proposed in 1968 by US geophysicists W. Jason Morgan, Dan Mackenzie and Tanya Atwater, and associated with three divergent plate boundaries at roughly 120 degrees to one another. In the case of the Great Rift, the junction is centred on Ethiopia's Afar Triangle, and is formed by the separation of the Arabian, African and Indian plates. During the late Paleogene, the junction gave rise to three rift valleys – one running southwards into East Africa, another trending north and the third roughly eastwards. Later flooding with ocean water transformed the northern and eastern branches into the Red Sea and Gulf of Aden respectively.

The southern rift, meanwhile, continued to extend southwards, filling with major lakes and rivers and creating further volcanic activity from early Miocene times around 20 million years ago. A series of long, narrow tectonic basins developed, sinking by up to 6km (3.7 miles) relative to their surroundings. Each gradually filled with sedimentary deposits several kilometres thick, which record the valley's history of changing climate and tectonic and volcanic activity, as well as preserving fossil remains of animals and plants. The Great Rift Valley became an evolutionary hotspot for many life forms, from fish in its lakes to primates and hominids in its forests. It also provided a major corridor for the migration of animals out of Africa.

Olduvai Gorge

Historically, the most famous hominid-related site in the valley is Tanzania's Olduvai Gorge, close to the huge Ngorongoro volcanic caldera. Now a World Heritage Site, the gorge preserves a series of lakeside sediments and volcanic lavas laid down through the last 2 million years across an area of some 250 square km (100 square miles). Excavation of these strata since the 1950s, mostly by the Leakey family and their numerous collaborators, have produced a fossil bounty of at least 10,000 human-related artefacts and an even greater number of vertebrate fossil remains. These included, in 1959, *Australopithecus boisei*, the first australopithecine hominid found outside of South Africa (see page 286), and a year later, the earliest human species, *Homo habilis*, alongside its younger relative *Homo erectus*.

'During the late Paleogene, the junction gave rise to three rift valleys – one running southwards into East Africa, another trending north and the third roughly eastwards.'

The remains seem to be concentrated within a relatively small area of the gorge and it was generally thought that animals and hominids were drawn here by drinkable lake waters. However, Africa became increasingly arid from around 7 million years ago as global climates see-sawed towards the Quaternary Ice Age (see page 293). Rivers became seasonal and, along with some small lakes, intermittently dried up, while other lakes became increasingly alkaline and undrinkable for most animals. A recent detailed study has confirmed that the Olduvai lake suffered in this way, and periodically dried out altogether, raising the question of why the animals were attracted to the area.

Fortunately, the discovery of carbonate 'tufa' mineral deposits from freshwater springs, and clays from wetlands full of freshwater fossil organisms, has provided an answer – there were clearly other sources of fresh water in the vicinity. As the ice ages set in, resources of permanent fresh water would have become increasingly important as lifelines for migratory animals. Rainwater falling on the Ngorongoro Highlands 30km (19 miles) to the east formed a groundwater supply that flowed westwards beneath the flanks of the highland volcanoes towards the lake basin before emerging as a series of springs and wetlands.

These are the first spring deposits associated with human remains to be identified in East Africa – and they provide a unique window onto the precarious lifestyle of our early hominid relatives.

69 Changing climates of the past

DEFINITION PREHISTORIC VARIATIONS IN TEMPERATURE, RAINFALL AND ATMOSPHERIC GASES CAN BE STUDIED IN MANY WAYS

DISCOVERY MEASUREMENT OF THE ANCIENT ATMOSPHERE FROM ICE CORES AND OTHER DEPOSITS WAS PIONEERED IN THE 1950s

KEY BREAKTHROUGH POLAR SCIENTISTS HAVE EXTRACTED DATA FROM ICE CORES GOING BACK 800,000 YEARS

IMPORTANCE CHARTING PAST CLIMATES HELPS US TO UNDERSTAND THE RANGE OF INFLUENCES ON EARTH'S ATMOSPHERE, PAST AND FUTURE

The geological record shows that Earth's climate has changed in the past due to natural processes associated with atmospheric gases including carbon dioxide. This record of climate change not only reveals the past history of our planet, but also offers lessons about its future development.

Earth's climate has always changed and always will – the rock record tells us so. Several past ice ages have dramatically lowered sea levels, with severe impacts upon life in the oceans and on land. Conversely, there have been ice-free periods when global sea levels were much higher than at present and extensive forests and animal communities developed at the poles despite the long dark winters (see page 237). Changes between different global climates can occur over a wide range of timescales, ranging from the abrupt (a matter of decades) to the more gradual (millions of years). The fossil record shows that life as a whole has generally adapted to these changes, but it's also inevitably that such changes can have severe effects on certain species.

Records of past climate change

Scientists study past climate with a variety of techniques, frequently inferring climate conditions from indicators such as fossil types and rock features. Pollen in river, lake and land sediments, and the shells of tiny organisms called foraminiferans in seabed sediments, both offer hints of the prevailing conditions in which they were deposited. Ancient glacial scarring and petrified deserts, meanwhile, can reveal the comings and goings of ancient ice ages and parched global supercontinents.

Since the onset of the recent Quaternary Ice Age, we also have the benefit of trapped samples of precipitation and atmospheric gases preserved in glacial

OPPOSITE The growth of organisms with skeletal structures such as mineral shells or organic stems tends to reflect the changing environmental conditions and especially the prevailing climate over its lifespan. Tree rings record not only the age of the plant, but also how relatively beneficial or detrimental the climate was for the growth of the particular species. Narrow growth bands reflect minimal growth and difficult conditions.

ice. The importance of these ice samples, created in layers by the annual process of summer melting and winter re-freezing, was first recognized by Danish scientist Willi Dansgaard (1922–2011) in the early 1950s, and further developed by French glaciologist Claude Lorius and Swiss climatologist Hans Oeschger (1927–98). Today, drilled 'core samples' from layered ice sheets in Greenland have revealed information about climate and atmosphere going back 120,000 years, while records from Antarctica have been extended across some 800,000 years. Even finer details, recording decadal and annual changes to weather and climate can be preserved in the 'growth rings' of cave stalactites, corals and trees.

All these measures reveal that past climate change has been linked to changing levels of 'greenhouse gases' in the atmosphere – especially carbon dioxide. Greenhouse gases in the atmosphere warm Earth's surface by allowing a wide range of radiations from the Sun through to reach the ground and sea, but then preventing the infrared (heat) radiation those surfaces radiate from escaping back into space. Increasing carbon dioxide has repeatedly warmed global climates, melting ice caps and glaciers, causing ocean waters to warm and expand, and raising sea levels to flood low-lying coastal regions. At times, oxygen levels in ocean waters have fallen, acidity has increased and there have been significant extinctions in marine life. The enlarged surface area of the oceans during such warm periods also increases the evaporation of water into the atmosphere, changing global rainfall patterns with significant impact on the wider environment and the life it supports.

'Ominously, studies of ice and sediment cores reveal that within the last glacial of the Quaternary Ice Age, the climate swung from warm to cold over just a few decades, and remained cold for centuries before gradually warming up.'

Changes in the Cenozoic

Our understanding of global climate change is most complete for the past 55 million years or so – the period since early Eocene times. The story begins with a phase of unusually high temperatures, with global averages 6–7°C (11–13°F) above those of today, and polar temperatures still warmer at 10–20°C (18–36°F) above the current norm, ensuring a world with no polar ice. Geochemical evidence shows that this period – the Paleocene-Eocene Thermal Maximum (PETM, see page 273) – was accompanied by a major release of carbon into the atmosphere and ocean waters. Atmospheric carbon dioxide levels were already high, but this boost drastically increased global temperatures so much that it took around 100,000 years for the climate to recover.

The subsequent fall in temperature was far from even, but by late Eocene times, around 34 million years ago, Antarctic glaciers had developed and coalesced into an ice sheet for the first time. It took a lot longer, until just 2.6 million years ago, for a large Arctic ice sheet to develop, initiating the Quaternary Ice Age with its drastic effect on global environments and life (see page 293). At low latitudes, changes to rainfall and sea level impacted on the evolution and dispersal of our own human species.

Oscillating ice age climates

The climate within the Quaternary Ice Age was often much colder than it is today, but also slightly warmer at times, oscillating between cold glacials and warm interglacials. Ominously, studies of ice and sediment cores reveal that within the last glacial (100,000–11,500 years ago), the climate swung from warm to cold over just a few decades, and remained cold for centuries before gradually warming up.

When ice sheets were at their maximum, they locked up so much water that sea levels fell 120m (390ft) below present levels. Huge areas of shallow seabed became dry land, opening routes for migration. Asia was connected to North America across the Bering Strait, the British Isles became part of mainland Europe and the Southeast Asian archipelago bridged much of the route to Australia. During interglacials, meanwhile, temperatures rose up to 5°C (9°F) above today's average and sea levels rose up to 9m (30ft) above present levels, isolating island promontories and flooding low-lying coasts.

Today, we are living in an interglacial warm phase that geologists call the Holocene, blessed with an unusually stable climate that developed around 11,500 years ago. However, this stability has given us a false sense of climate security – and there is mounting evidence to show that this stable phase has now come to an end. It is no longer a matter of whether climate change will happen, but when it will happen, and how drastic it will be.

ABOVE The retrieval of ice cores from holes drilled through the polar ice sheets has provided a wealth of information about year on year climate change through the last several hundred thousand years of the recent Quaternary Ice Age.

The australopithecine clan

70

DEFINITION EARLY APE-LIKE HOMINIDS ARE THE IMMEDIATE
ANCESTORS OF HUMANS AFTER OUR ANCESTORS SPLIT FROM THOSE
OF CHIMPS AROUND 7 MILLION YEARS AGO

DISCOVERY THE FIRST AUSTRALOPITHECINE TO BE DISCOVERED WAS
FOUND IN SOUTH AFRICA IN 1925 BY RAYMOND DART

KEY BREAKTHROUGH RECENT FINDS OF *ORRORIN* AND *SAHELANTHROPUS*
HELP ILLUMINATE THE OLDEST BRANCHES OF OUR FAMILY TREE

IMPORTANCE AUSTRALOPITHECINE FOSSILS REVEAL WHEN HUMAN
CHARACTERISTICS SUCH AS UPRIGHT WALKING FIRST EVOLVED

Over the past 6 million years, the human family has evolved from ape-like beings into modern humans. The ape-like side of this divide was dominated by an extinct group known as the australopithecines, or 'southern apes', and their immediate ancestors, living between 6 and 4 million years ago.

In 1871, Charles Darwin predicted that it was 'somewhat more probable that our early progenitors lived on the African continent than elsewhere', since this was where 'the gorilla and chimpanzees … man's nearest allies' currently lived. However, it was to be more than 50 years before the fossil evidence for this evolutionary connection was first found, and a further 20 years before it was generally accepted.

Then, in the 1980s, strong independent support for the connection came from a genetic analysis of living humans from around the world (see page 298). Today, we know that the evolutionary history of our human family extends back more than 6 million years to a time when we shared a common ancestor with the chimps. And that ancestor, from which all humans and the other 20 or more extinct members of the wider human family are descended lived, as Darwin predicted, in Africa.

Historical discoveries

The search for human ancestry was hampered for decades by religious and cultural strictures – Darwin himself was severely criticized for even suggesting in his 1858 book *On the Origin of Species* that light might be thrown 'upon the origin of man and his history'. Even though the first extinct human-related species, *Homo neanderthalensis*, had been found in Germany in 1856, its significance was not generally acknowledged until the 1880s.

OPPOSITE The recovery of virtually complete fossil skulls from our remote australopithecine relatives allows scientists to reconstruct their possible appearance in life, as seen in this reconstruction of *Australopithecus afarensis*. While the facial musculature and skin structure can be modelled with considerable confidence, the extent of facial hair, size and shape of the ears and nasal opening are much more speculative.

However, it was not until 1925 that the deep ancestry of the human family started to come to light, with the South African discovery of *Australopithecus africanus* by Raymond Dart (1893–1988). The significance of this 1m (40in) tall, upright-walking and ape-like species was not fully appreciated until after the Second World War, when (along with a number of other similar fossils) it was grouped together in a genus known as the australopithecines. Most of the South African fossils had very ape-like skulls, but some were more heavy-boned and gorilla-like than others. As a result, they are now divided into two groups – the lighter-boned *Australopithecus* and the separate but closely related *Paranthropus*.

Australopithecines from Olduvai

Over several decades from the late 1940s, the search for our human ancestry in Africa was dominated by Kenyan-born English anthropologist Louis Leakey (1903–72) and his family. They mostly concentrated their searches in East Africa and the Great Rift Valley (see page 277), and particularly at Olduvai Gorge and around Lake Turkana. Despite finding another australopithecine (*Paranthropus boisei*) at Olduvai in 1959, the Leakeys mostly concentrated on searching for immediate ancestors of the genus *Homo*. What was more, when the oldest levels at Olduvai were radiometrically dated from volcanic rocks to around 1.8 million years old, it became clear that the australopithecines had arisen some considerable time before this.

'Today, we know that the evolutionary history of our human family extends back more than 6 million years to a time when we shared a common ancestor with the chimps.'

Of course, it took some time to locate fossil-bearing strata of the right age, and as a result it was only in 1974 that US palaeontologists Don Johanson and Tom Gray discovered the 3.3 million-year-old remains of a key australopithecine species, *Australopithecus afarensis*, far to the north of Olduvai at Hadar in Ethiopia. Just 1m (40in) tall, 'Lucy', as the near-complete skeleton was nicknamed, was an upright-walking but still very ape-like being, with a small (400ml/13.5fl oz) brain.

Nevertheless, Lucy was relatively advanced compared to living chimps, suggesting that the common ancestor of chimps and humans must have been quite a bit older. By this time, however, advances in evolutionary genetics had led US molecular biologists Allan Wilson (1934–91) and Vincent Sarich to suggest that the two lineages diverged around 5 million years ago.

Even older

The search for older strata and fossils continued without success until the 1990s, when some very fragmentary remains were discovered in the Afar region of Ethiopia, by a team led by Tim White of the University of California at Berkeley. Named *Ardipithecus*, they were some 4 million years old. Continued field work eventually turned up further remains, which took years to excavate and prepare. Meanwhile in 2000, a French team working in the Tugen Hills of Kenya, led by Martin Pickford and Brigitte Senut of

France's National Museum of Natural History, found even older, but still fragmentary, skeletal remains. Named *Orrorin* and dated to around 6 million years old, they included a thigh bone whose form indicated that, even at this early time, these small ape-like beings were walking upright. The following year another French team, led by Michel Brunet, found a remarkably intact skull in the Djurab Desert of Chad, which seems to be slightly older at just over 6 million years. Brunet claimed that this new creature, named *Sahelanthropus*, had a mosaic of ape and more advanced features and that it too walked upright. Other experts argued that it was in fact a fossil chimp.

Ardipithecus ramidus

Finally, in 2009, the lengthy reconstruction of Tim White's crushed specimen 'Ardi' was completed. Although 'only' 4.4 million years old, the reconstruction of *Ardipithecus ramidus* gives a whole new insight into our early ape-like ancestors. Ardi stood about 1.2m (4ft) high, with a body weight of around 50kg (110lb) and a brain size of 300–350ml (10–12fl oz). With very long arms, relatively short legs and feet with long opposable toes, it was still well adapted for climbing trees. Its hands were more flexible than those of apes, perhaps allowing it to walk along the top of branches and reach out for fruit. Surprisingly, the feet were also adapted for walking upright, though in a different and far less efficient way to humans – clearly *Ardipithecus* needed to be able to climb trees to obtain food, and to walk across open ground from tree to tree. The remains of some 35 similar beings shows that males and females were of similar size, perhaps reflecting a degree of social cooperation greater than that seen in the living great apes. The early evolution and divergence of our ape-like relatives was clearly a lot more complex than we previously thought.

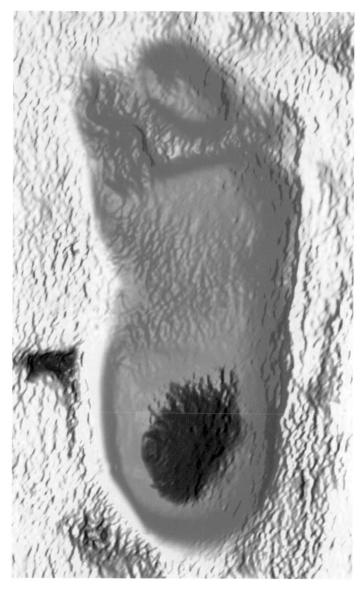

ABOVE Well-preserved human-related footprints found at Laetoli in Tanzania, and dated to between 3.8 and 3.5 million years old, were most likely made by the australopithecine species *Australopithecus afarensis*.

71

The real 'Great Flood'

DEFINITION A GARGANTUAN FLOOD CAUSED BY THE DRYING OUT
AND SUBSEQUENT REFILLING OF THE MEDITERRANEAN

DISCOVERY SEA-FLOOR ROCKS STUDIED IN THE 1970S REVEALED
THAT THE MEDITERRANEAN DRIED OUT 5.6 MILLION YEARS AGO

KEY BREAKTHROUGH NEW MODELS OF THE FLOOD THAT REFILLED
THE SEA REVEAL ITS ENORMOUS SCALE

IMPORTANCE THE 'MESSINIAN SALINITY CRISIS' CREATED AN
IMPORTANT LAND BRIDGE FOR MIGRATING SPECIES

Around 5.6 million years ago, the Mediterranean Sea almost dried up as tectonic forces closed its connection to the Atlantic. Waters evaporated and sea levels fell – but just 300,000 years later, the dam burst and the Atlantic waters poured in, producing the biggest flood in Earth's history.

Although the general outlines of this remarkable story have been known since the 1970s, new information suggests that the reflooding of the Mediterranean Basin was an even more catastrophic event than previously thought. It now appears that perhaps as much as 90 percent of all the incoming water flooded into the basin in less than two years. On a geological timescale, this was virtually instantaneous, and sea levels would have risen at an astonishing rate of up to 10m (33ft) per day.

The Messinian salinity crisis
In the 1970s, deep-sea drilling operations in the Mediterranean recovered sedimentary rocks and fossil remains which revealed that, some 5.6 million years ago in late Miocene times, areas of the seabed had been dry land, with salty lakes and wooded vegetation. This brief episode in the Mediterranean's long history is known as the 'Messinian salinity crisis', named after Messina in Sicily, one of several sites where salt deposits from these lakes are still exposed today. (Others are in southern Spain and northeast Libya). The existence of these new vast new landscapes created an important 'land bridge' from Africa into Europe, allowing an interchange of species that included our own remote human ancestors.

This dramatic drying-out of the Mediterranean, it soon became clear, was caused by the closure of the Straits of Gibraltar, the narrow bottleneck that

OPPOSITE Ancient layers of salt laid down around 5.6 million years ago in late Miocene times record a time when the entire Mediterranean evaporated away into a few huge salt lakes.

linked the Mediterranean with the wider Atlantic Ocean, as tectonic forces drove Africa north into the Eurasian Plate. Once the sea had been isolated and could no longer be replenished by colder Atlantic waters, evaporation did the rest, and sea levels dropped steadily over the next 300,000 years.

Around 5.5 million years ago, as global climates continued to cool towards the Quaternary Ice Age (see page 293), more fresh water from increasing river drainage around the basin created large brackish lakes. Finally, the rock record showed the reappearance of marine sediments 5.33 million years ago, indicating that Atlantic waters had returned in an event known as the Zanclean Flood (named after the pre-Roman name for Messina). The entire basin soon refilled, severing the land bridge between Africa and Europe with a considerable impact on the dispersion of our human ancestors into Europe and their survival in the ensuing ice age (see page 305).

'The water flowed in at a rate of 100 million cubic m (3.5 billion cubic ft) per second – roughly a thousand times greater than the present-day flow of the Amazon.'

Modelling the flood dynamics

Ever since it was discovered, the exact nature of this massive flood event has been a subject for speculation and fantasy, conjuring images of a Niagara-like torrent pouring across the Straits of Gibraltar. In 2009, however, a team led by Daniel Garcia-Castellanos of the Research Council of Spain, added some much-needed detail, modelling the rate at which water flowed back into the Mediterranean by using observations of the way that active river erosion can transform landscapes by cutting down through solid rock. Remarkably, they were able to learn a great deal about the nature of this huge flood from studying the development and eventual deterioration of much smaller mountain lakes.

A variety of other sources, such as sediment cores from deep-sea boreholes and geophysical data from ground and satellite measurements, have also provided information about the geography of the Mediterranean Basin during the late Miocene, along with a detailed timescale for the flooding. These showed that sea levels fell in the Mediterranean by as much as 1,000m (3,300ft) around the mouth of the Rhone and up to 2,500m (8,200ft) in the Nile Delta, perhaps creating the biggest and deepest sub-sea-level depression in Earth's history.

The Zanclean deluge

From all this data it appears that the initial reflooding began relatively slowly – probably initiated by tectonic subsidence of the Camarinal Sill – a tilted sheet of volcanic rock near Gibraltar that had hitherto been substantial enough to hold back the waters of the Atlantic. Over thousands of years, bursts of Atlantic water poured over and down the sill, which formed a gently sloping rock ramp, inclined at 1–4 degrees, rather like a weir or dam overflow. The ramp was steadily worn down by erosion, creating river-like incisions that deepened by perhaps as much as 50cm (20in) per day. These eventually

ABOVE A computer generated model shows the refilling of the Mediterranean Basin 5.33 million years ago, as Atlantic waters cascaded over the Camarinal Sill and through the Straits of Gibraltar.

gouged out a valley through which a deluge of flood-waters poured at speeds of up to 40m (130ft) per second. The water flowed in at a rate of 100 million cubic m (3.5 billion cubic ft) per second – roughly a thousand times greater than the present-day flow of the Amazon. Calculations suggest that 90 percent of the Mediterranean Basin was refilled within a period of between a few months and two years.

In fact, these figures seemed so dramatic that the scientists thought that there must be something wrong with either the data, the model or their calculations. After finding no obvious errors, they decided to search for physical evidence of the predicted erosion channel beneath the Straits of Gibraltar, now filled in by sediment. Fortunately, another series of boreholes had been drilled during testing for a proposed transport tunnel, and this data, along with numerous seismic profiles, revealed a U-shaped channel 200km (125 miles) long, up to 11km (7 miles) wide and 250–650m (820–2,130ft) deep. The boreholes confirm that the channel is infilled with sediment and rock debris eroded from its wall in the flood.

Human history might not repeat itself, but geological history can. Tectonic forces are still driving North Africa towards Europe, and it is possible that in the future, the Straits of Gibraltar will once again seal off the Mediterranean.

72

The Quaternary Ice Age

DEFINITION A COLD PHASE BEGINNING 2.75 MILLION YEARS AGO
IN WHICH EARTH'S CLIMATE HAS ALTERNATED BETWEEN CHILLY
GLACIALS AND WARMER INTERGLACIALS

DISCOVERY EVIDENCE FOR THE ICE AGE WAS FIRST NOTICED BY
LOUIS AGASSIZ IN THE 1830S

KEY BREAKTHROUGH JAMES CROLL AND MILUTIN MILANKOVITCH
LINKED THE GLACIAL CYCLES TO CHANGES IN EARTH'S ORBIT

IMPORTANCE THE ICE AGE HAS SHAPED THE EVOLUTION AND SPREAD
OF HUMANITY THROUGHOUT THE HISTORY OF OUR SPECIES

Around 120,000 years ago, in the latter part of the Quaternary Ice Age, hippos wallowed in the River Cam in Cambridgeshire, England. This is hardly our traditional picture of an ice age, so how did they get there, and how did they survive?

The existence of glaciers and ice sheets is very much part of our modern picture of the Earth, but the geological record shows that, throughout our planet's history, this situation has been the exception rather than the norm. Global climate is relatively warm at the moment, almost certainly as a result of human activity, but the current warm phase is no hotter than several other 'interglacial' phases that have periodically occurred in the last 2.75 million years of the Quaternary Ice Age.

Until the 19th century, the very idea of a world engulfed in ice was unthinkable. The truth was only revealed through the studies of Louis Agassiz (1807–73), a Swiss geologist who proposed in 1837 that the glaciers of the European Alps had once extended over a much larger region, and who went on to identify many telltale glacial features in landscapes that are now normally ice-free. Today, the geological record suggests there have been seven major ice ages throughout Earth's 4.5 billion-year history, three of which have occurred in the last 542 million years.

What causes an ice age?

Normally, global climates and the atmosphere/ocean system, warmed by heat from the Sun and protected by greenhouse gases, have been enough to prevent the formation and persistence of snow and ice at high polar latitudes. In order to overcome the two factors that generally keep Earth warm, it

OPPOSITE It was only in the late 1830s that European scientists realized that the snow caps and glaciers of the European Alps are just the high-altitude remnants of a much more extensive phenomenon from the recent past – the Quaternary Ice Age.

seems that the underlying mechanism behind ice ages must control solar 'insolation' and reduce greenhouse gases for long enough to cool the planet.

The current period of glaciation began some 35 million years ago with the appearance of glaciers on Antarctica. However, this formation of persistent ice required a long prehistory of cooling, especially in ocean waters, from a warm peak around 55 million years ago. Within this long, slow descent, there were marked phases of rapid and severe cooling around 15 million years ago, and 2.75 million years ago, just before the beginning of Quaternary times.

Indirect or proxy measures of temperature change in deep-ocean waters make use of oxygen isotopes preserved in the shells of marine micro-organisms called foraminiferans. Water evaporating from the ocean surface tends to contain more of the lighter oxygen isotopes than that which is left behind, and if the evaporated water becomes locked into ice sheets instead of returning to the sea, the isotopic content of the seawater, used by the foraminiferans to make their shells, becomes heavier.

Studies confirm that the cycles of change became magnified from the start of the Quaternary Period, as high-latitude ice sheets periodically expanded and contracted between long phases of increasing cold and short phases of warmth (sometimes warmer than today). During some of the cold phases, the ice sheets reached as far south as present-day New York and Moscow.

'The current period of glaciation began some 35 million years ago with the appearance of glaciers on Antarctica. However, it required a long prehistory of cooling from a warm peak around 55 million years ago.'

Cycles of glaciation

This dramatic pattern of oscillation has three distinct periods. Up until around 2.75 million years ago, there were frequent small swings in climate every 23,000 years, and glaciers began to appear in high latitudes and at high altitudes in temperate regions. Early in the Quaternary, the period increased to 41,000 years and the fluctuations grew in intensity. Glaciation became more extensive and had a greater effect on environment and life, as permafrost conditions extended well beyond the ice itself.

However, 900,000 years ago there was a further step-change in frequency, to a 100,000-year cycle. With this significantly longer period, glacial phases became more profound and far-reaching. Glaciers coalesced into massive ice sheets that spread to lower temperate latitudes. Around their fringes lay permafrost wastelands, especially in the continental interiors of Asia and North America, where few trees could survive. Sea levels dropped by more than 100m (330ft) as ocean water became locked up in the ice sheets, and newly exposed areas of continental shelf connected previously isolated landmasses, linking northeast Africa to the Arabian Peninsula, Southeast Asia to the Indonesian archipelago, Asia to North America and the British Isles to mainland Europe. The impact on life in general, and our human ancestors in particular, was devastating.

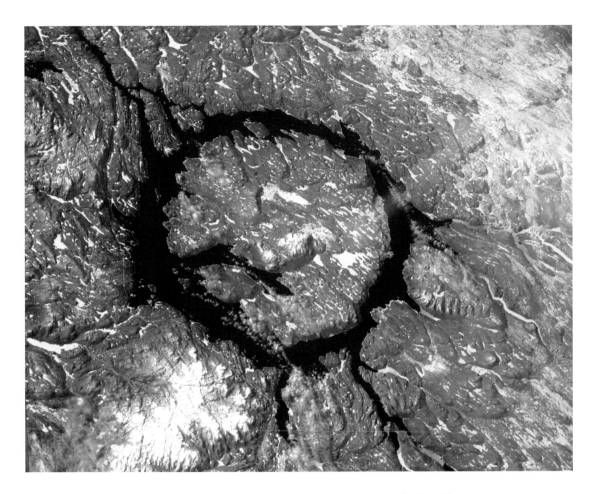

The intervals of these three major cycles are linked to changes in Earth's orbit – astronomical patterns first recognized by Scottish scientist James Croll (1821–90) in the mid-19th century and developed by Serbian mathematician Milutin Milankovitch (1879–1958) in the 1920s. The 23,000-year cycle is connected to changes in the angle of Earth's axis of rotation, the 41,000-year cycle is linked to the tilt of the axis relative to the Sun, and the 100,000-year cycle relates to the changing shape of Earth's orbit around the Sun.

The geological record of climate change (see page 281) has verified the existence of these 'Milankovitch cycles', but has also revealed many briefer periods of intense change. These can occur very rapidly over hundreds or thousands of years and have been particularly important for plant and animal life. There are many other factors at play affecting the pattern of global climate on both very long and very short timescales: plate movements can affect the global circulation of warm and cold water; intense volcanic activity can pump dust, ash and greenhouse gases into the atmosphere to initially cool, and later warm, climates; and last but not least, human industrial activity can promote global warming by re-releasing carbon dioxide and methane locked up as hydrocarbons in the geological past.

ABOVE The ice sheet that once covered most of Canada created a vast glacier-scarred terrain of flattened wetlands. They also eroded the weakened rim of the Manicouagan impact crater in Quebec, creating a unique circular lake filled with meltwater that is about 70km (45 miles) wide.

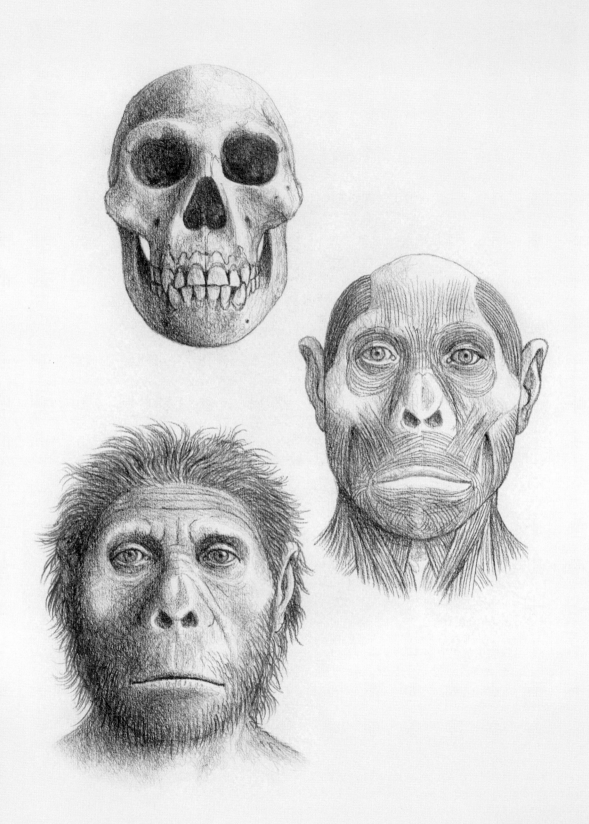

73

Genus *Homo*

DEFINITION THE GENUS THAT INCLUDES MODERN HUMANS AND OUR
IMMEDIATE ANCESTORS, EVOLVING AROUND 2 MILLION YEARS AGO

DISCOVERY THE FIRST FOSSIL HUMAN RELATIVES, THE
NEANDERTHALS, WERE DISCOVERED IN GERMANY IN 1864

KEY BREAKTHROUGH GENETIC ANALYSIS IN 1987 SHOWED THAT ALL
LIVING HUMANS SHARE A COMMON ANCESTOR, 'MITOCHONDRIAL
EVE', WHO LIVED IN AFRICA 200,000 YEARS AGO

IMPORTANCE FOSSIL HUMANS PUT OUR OWN SPECIES IN AN
EVOLUTIONARY CONTEXT, WHILE GENETIC EVIDENCE SHOWS THAT
WE ARE ALL UNITED BY A RECENT COMMON ANCESTRY

Our immediate human family includes all species belonging to the genus
Homo – somewhere between 9 and 12. Experts disagree about the precise
number because extinct species can only be determined through preserved
fossil anatomy, not by the 'acid test' of mating and reproduction.

Our species, *Homo sapiens*, was first defined by the Swedish botanist and
taxonomist Carl Linnaeus in 1753, with the simple description 'know thyself'
(a Greek maxim borrowed from the Temple of Apollo at Delphi). Linnaeus
was the first to formally classify humans along with monkeys and apes into
a single zoological 'order', the Anthropomorpha (later changed to Primates).
However, Linnaean classification carried no evolutionary implications, and
it was not until the middle of the 19th century that the possibility of humans
having evolved, rather than being the result of instantaneous and divine
creation, was first countenanced.

The Asian connection

In 1864, the first extinct member of the human family was formally named
as *Homo neanderthalensis*, based on fragmentary remains found a few years
before near Düsseldorf in Germany. By the 1890s, however, questions were
being raised about whether humans originated in Africa, as Darwin had
suggested (see page 285), or in Asia.

Charismatic German biologist Ernst Haeckel (1834–1919) championed the
Asian origin theory so persuasively that a young Dutch anatomist, Eugène
Dubois (1858–1940), set off to Indonesia in 1887 in search of a 'missing link'
that would prove the story. Amazingly, in 1891 he discovered human-like
fossils at Trinil in Java, naming them *Homo erectus* since the leg bones clearly

OPPOSITE Originally
thought to be remains
of *Homo erectus*, fossils
found in 1.7 million-
year-old deposits at
Dmanisi, Georgia,
reveal features that are
in fact more primitive,
such as smaller brain
size. As a result they
have tentatively been
placed in a separate
species *Homo georgicus*,
as shown in this artist's
reconstruction.

belonged to an individual that could stand upright. But as so often happened, the significance of Dubois's find was not recognized by the experts of the time. And before that recognition came, the discovery was eclipsed by that of another Asian species, 'Peking Man' (*Sinanthropus pekinensis*), found in China and described by Canadian anatomist Davidson Black (1884–1934) in 1927.

The discovery of *Sinanthropus* seemed to reinforce an Asian origin for humans and a theory, known as the multiregional hypothesis, that modern humans in different parts of the world had evolved separately from an earlier human species. By the 1940s, however, 'Peking Man's' close similarity to Dubois's 'Java Man' was recognized. As a result, the species was subsumed into *Homo erectus*, which was finally acknowledged as an important human ancestor alongside *Homo neanderthalensis*.

We are all Africans

In 1960, however, Louis Leakey and his family turned up fragmented, 1.8 million-year-old hominid fossils at Olduvai Gorge in East Africa's Great Rift Valley (see page 277). These were clearly more advanced than the australopithecines previously found at the site (see page 286), and Leakey was convinced they belonged to the most ancient human species, which he named *Homo habilis* ('handy man'). However, in order to fit his new finds into the direct human family, he redefined the genus *Homo*, reducing its minimum 'watershed' brain size from 750ml (26fl oz) down to 650ml (23fl oz), much to the dismay of many other experts. Nevertheless, the find re-established the African ancestry of the human family, and from it arose the idea that there had been two moves out of Africa, firstly by *Homo erectus* around 1.8 million years ago and then much more recently by *Homo sapiens* around 70,000 years ago.

'Mitochondrial Eve lived around 200,000 years ago in Africa, and her descendants spread into Asia and beyond around 70,000 years ago.'

The African fossil record shows that *Homo sapiens* evolved around 200,000 years ago, but since *Homo erectus* seems to have died out around 500,000 years ago, there is an obvious evolutionary 'gap' for our direct ancestor. This is currently filled by *Homo heidelbergensis*, whose incomplete remains were first found in Germany in 1907, and have since been identified in Africa. Living between 800,000 and 200,000 years ago, it too seems to have migrated out of Africa, reaching Europe before it became isolated from its African cousins. According to current thinking, African *Homo heidelbergensis* then gave rise to *Homo sapiens*, while the European branch gave rise to the Neanderthals.

Mitochondrial Eve and beyond

In 1987, a team of scientists from the University of California at Berkeley conducted a revolutionary analysis of the global distribution of mitochondrial DNA in living humans (a type of DNA that is inherited from the mother alone, and whose slow mutation can therefore be used to trace the time since 'divergence' from a common ancestor – see page 130). They identified an

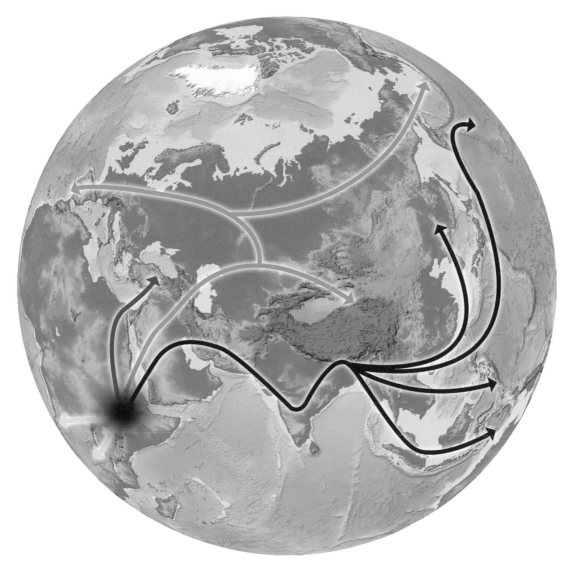

ancestor shared by all members of *Homo sapiens*, nicknamed 'mitochondrial Eve'. She lived around 200,000 years ago in Africa, and her descendants spread into Asia and beyond around 70,000 years ago. This discovery effectively put an end to the original multiregional hypothesis.

Today, however, the details of our immediate ancestry have been blurred by new finds in Europe, such as the *erectus*-like *Homo georgicus* (1.7 million years old) and the Spanish *Homo antecessor* (1.2 million–800,000 years ago), which may have been ancestral to *Homo heidelbergensis*. While the genetic evidence for the African ancestry of our species has been upheld, new evidence found in DNA recovered from Neanderthal and other skeletal remains (see page 305) indicates that our ancestors were interbreeding with the Neanderthals – though not frequently enough to generate a new species. What is not in doubt is that despite small genetic differences, modern humans around the world are still a single interbreeding species.

ABOVE Since the evolution of modern humans in Africa around 200,000 years ago, evidence suggests there have been at least three successive waves of humans beyond the continent. The first, around 100,000 years ago got no further than the Levant (blue). The second, from around 60,000 years ago, got to Asia and Australia (red), while a third wave moved north throughout Eurasia and into the Americas (orange) from around 48,000 years ago.

74

The first art

DEFINITION FINDS OF GRAPHIC AND SCULPTURAL SYMBOLS OF
CULTURAL SIGNIFICANCE FROM EUROPE AND BEYOND

DISCOVERY MAJOR RECENT FINDS INCLUDE THE CHAUVET CAVE IN
SOUTHERN FRANCE AND THE SWABIAN VENUS FROM GERMANY

KEY BREAKTHROUGH BLOMBOS CRAYONS FROM SOUTH AFRICA SHOW
THAT GRAPHIC SIGNS ORIGINATED MORE THAN 75,000 YEARS AGO,
PROBABLY BEFORE HUMANS SPREAD FROM AFRICA

IMPORTANCE THE DEVELOPMENT OF ART IS ONE OF THE FEW WAYS
IN WHICH WE CAN INVESTIGATE EARLY HUMAN CONSCIOUSNESS

One of the defining characteristics of our species is the capacity to communicate abstract and complex ideas in a variety of ways, from language, gestures and signs, to images. Such images can take a variety of physical forms, and some can be preserved over many thousands of years.

Since language and behaviour cannot be directly 'fossilized', it is often only through the presence of these prehistoric works of art that we can learn about the evolution of the human cultures. Over the last century or more, the discovery of such objects and rare sites has excited huge amounts of speculation, and recent discoveries have revolutionized our ideas about when and where the human 'habit of art' originated.

A Eurocentric view

For a long time, the apparent abundance of rock art from prehistoric caves in France, Spain and elsewhere seemed to suggest that *Homo sapiens* did not develop their artistic tendencies until they had reached Europe. We now know this happened around 35,000 years ago, at a time when Europe was home to a wide range of ice age animals including the mammoth, woolly rhino and giant deer, alongside a scattered population of the separate species *Homo neanderthalensis*. The Neanderthals' presence raised the interesting question of whether they produced any 'art' similar to that of the incoming modern humans – and if not, why not? But there was also a more pressing issue – when first discovered, none of the prehistoric artwork could be dated, and so the chronology of its development was unknown. One major clue came from the widespread depiction of the extinct ice age fauna, suggesting that the art was produced before those animals became extinct between 15,000 and 10,000 years ago (see page 329).

OPPOSITE Carved from a 6cm (2.4in) section of juvenile mammoth tusk around 35,000 years ago, the 'Venus of Hohle Fels' formed part of an extraordinary flowering of the Aurignacian culture in southern Germany.

Nevertheless, it was generally assumed that the first art had developed from crude and primitive two-dimensional manifestations to more sophisticated, coloured and occasionally three-dimensional forms. The natural chronology seemed to progress from crude outline drawings and engravings to archetypal 'cave paintings' such as the horses of Lascaux in France and the bison of Altamira in Spain.

Chauvet upsets the chronology

In 1994 a team of French cavers led by Jean-Marie Chauvet worked their way into a cave on the left bank of the river Ardèche in southern France, close to the famous natural bridge of the Pont d'Arc. Here they discovered one of the most important decorated caves in the world, and one of the first to be subject to modern scientific scrutiny and archaeological techniques from the moment of its discovery. The walls of the cave are covered with images ranging from mammoths, cave bears, lions and rhinoceros to horses and wild cattle – some outlined in red, others in black with some shading. There are also pigmented hand-prints and stencils, along with an engraved owl and horse. But apart from the wonderful preservation and quality of its art, the most shocking aspect of 'La Grotte Chauvet' is its age. The style seemed to place the cave in the 'Solutrean' culture, previously assumed to be no more than 22,000 years old, but minute samples of charcoal recovered from many of the images have been carbon-dated to between 36,000 and 37,000 years ago, placing the artwork firmly within the Aurignacian culture of the first modern humans to reach Europe. If correct, this early date completely upsets previous ideas about chronology, so it is not surprising that some have questioned its reliability.

'Over the last century or more, the discovery of such objects and rare sites has excited huge amounts of speculation, and recent discoveries have revolutionized our ideas about when and where the human "habit of art" originated.'

Swabian carvers

In the 1930s, excavations in the limestone caves of Swabian Jura in southern Germany uncovered the first of some 25 tiny, centimetre-sized ivory carvings. The beautifully sculpted figures included big cats, horses, mammoth, a bird and a larger 28cm (11in) human figure with a lion's head. Recent excavations at Hohle Fels cave near the town of Blaubeuren have turned up further remarkable finds, dated to around 35,000 years ago, that are also products of the Aurignacian culture. They include the oldest known carved human figure, made from a single piece of mammoth ivory some 6cm (2.4in) long, and the oldest known musical instrument.

The female figurine, known as the Venus of Hohle Fels, has greatly exaggerated sexual features and reduced arms, legs and head. The head itself is perforated – perhaps for hanging as a pendant. While we cannot make assumptions about the mindset of the people who made these objects, the enlarged breasts, thighs and belly would seem to be linked to fertility. The musical instrument, meanwhile, is a slender flute 21cm (8.4in) long, made from the hollow wingbone of a griffon vulture and perforated with

five finger holes. Altogether, the level of sophistication in imagery, form and manufacturing show that art and music were already an integral part of everyday life in Aurignacian society. These remarkable artefacts must be predated by a considerable history of earlier production, perhaps stretching back for millennia – but no signs of it have been discovered so far.

Art's African roots – Blombos Cave

However, in 1991, archaeologists discovered two small engraved pieces of ochre buried in 75,000 year-old deposits on the floor of a cave at Blombos on the southern coast of South Africa, alongside bone and stone tools and some 60 perforated shell beads. The 5cm (2in) ochre fragments have flattened sides, showing where they were ground to produce a blood-red powder that seems to have been widely used for burial ceremonies. The sides are also engraved into a lozenge-shaped geometric pattern. Although the significance of this pattern is unknown, it is clearly intentional and is therefore by far the oldest known abstract 'art'. Altogether, these artefacts show for the first time that symbolic and decorative culture did indeed develop while our species was still confined to its African homeland. The work may seem far removed from the Swabian carvings in terms of cognitive skill and sophistication, but at more than twice the age, the Blombos discoveries give the first hint that prehistoric art had deep African roots.

ABOVE A 5cm (2in) fragment of naturally occurring red ochre mineral was found in South Africa's Blombos Cave. This 75,000-year-old 'crayon' was ground down on its side to produce the distinctive blood red pigment powder before being engraved with its enigmatic diagonal cross-hatching – the oldest known graphic art.

75 Neanderthal mysteries

DEFINITION NEANDERTHALS WERE THE CLOSEST GENETIC COUSINS OF OUR OWN SPECIES *HOMO SAPIENS*

DISCOVERY RECENT STUDIES OF NEANDERTHAL REMAINS SUGGEST THEY WERE FAR MORE LIKE US THAN ONCE THOUGHT

KEY BREAKTHROUGH ANALYSIS OF NEANDERTHAL DNA REVEALS THAT THEY INTERBRED WITH EURASIAN HUMANS, AND WERE PROBABLY CAPABLE OF SPEECH

IMPORTANCE STUDIES OF THE NEANDERTHAL GENOME WILL HELP TO REVEAL WHAT FEATURES MAKE *HOMO SAPIENS* UNIQUE

In recent decades our picture of the Neanderthals has transformed from the traditional view of them as brutish and stupid, to a more benign image of a tragic people pushed into extinction, alongside other ice age megafauna, by the arrival of modern humans. But what is the reality?

The Neanderthals are our closest and best-known relatives, so it might seem somewhat paradoxical for us to hold them up to ridicule. However, part of the problem lies in the history of their discovery – they were the first extinct member of the human family to be recognized in the mid-19th century, but it took decades for them to become generally accepted, as there was a general reluctance to acknowledge the possibility of human evolution.

Early illustrators depicted Neanderthals as monstrous, ape-like brutes, but in fact they were physically very similar to us. Some experts have claimed that a Neanderthal dressed in modern clothing could easily merge into the crowd in any big city of the Western world, but closer inspection would reveal distinctive differences in a typically pale-skinned, possibly red-haired and powerfully built individual. Above a thick neck, the face would be large, dominated by a big nose and deep-set eyes below a very prominent brow and low receding forehead. Although the jaw was massive, the absence of a protruding chin would be noticeable unless hidden by a beard. The Neanderthal brain, meanwhile, was similar in size to our own, but may have been organized somewhat differently.

Neanderthal DNA

In 1997, scientists from Germany's Max Planck Institute, led by Swedish geneticist Svante Pääbo, succeeded in extracting mitochondrial DNA

OPPOSITE This large broken scallop shell is among many remarkable recent discoveries made by anthropologist João Zilhão of Bristol University at a Neanderthal site in southeast Spain. Dated to around 50,000 years old, its perforations may have originally allowed it to be worn as a personal ornament.

(mtDNA) from Neanderthal remains, confirming that the Neanderthals were indeed a distinct genetic species. Since then, technological advances have made it possible to reconstruct and sequence the entire Neanderthal genome, published in draft form in 2009. The following year, Pääbo's team reported evidence within the genome that some interbreeding may have occurred between Neanderthals and non-African modern humans (accounting for 1–4 percent of the mtDNA), while the more deeply rooted African population of *Homo sapiens* were unaffected.

Intriguingly, it also seems that the Neanderthals possessed the speech-related 'FOX-P2' gene in a similar form to modern humans, suggesting they were capable of some form of language. Fossil evidence shows that they certainly possessed some of the anatomical 'equipment' for speech, though the structure of the Neanderthal throat suggests that their voices were higher-pitched than those of modern humans.

Living through an ice age

BELOW Some of the last sites to be occupied around 27,000 years ago by the Neanderthals were coastal caves on Gibraltar from which they had access to a variety of marine food resources.

We know today that the Neanderthals evolved around 300,000 years ago, probably in Eurasia, and came to occupy a vast territory that stretched from western Europe to the Urals, reaching as far south as Gibraltar and the Middle East. They finally died out more than 27,000 years ago – though not before coexisting with modern humans in their Eurasian heartland for more than 10,000 years. *Homo sapiens* probably first encountered the Neanderthals as they first spread into the Middle East around 120,000 years ago, and then

again with the main dispersal out of Africa into Europe and Asia around 60,000 years ago.

The heyday of the Neanderthals coincided with an uncomfortably dynamic phase in Eurasian history – the later stages of the Quaternary Ice Age. As the climate swung repeatedly back and forth between warmer and colder phases, there were drastic environmental changes that would have impacted heavily on both plant and animal life, including the Neanderthals.

Some have claimed that the Neanderthal body was 'cold adapted' for ice age survival, but recent analysis and comparison with the anatomy of other ancient human species does not bear out this idea. Indeed, recent mapping of the distribution of Neanderthals and contemporary modern humans shows that when confronted with extreme glacial conditions, both retreated south out of central Europe. The last Neanderthals survived in Mediterranean coastal sites, such as Gibraltar, until around 27,000 years ago, but by this time their populations were so fragmented, isolated and reduced in numbers (like those of other ice age animals) that they could never recover, as new waves of modern humans from the Middle East and beyond moved in to replace them.

Neanderthal lifestyles

Archaeological evidence suggests that the Neanderthals were essentially mobile hunters, living in small, male-dominated groups. They hunted game using heavy wooden spears tipped with stone points and had a diet that ranged from shellfish and tortoises to small mammals, deer and horses. Healed fractures in the torsos of some skeletons show how tough this life was, but there is also evidence for altruism, with some crippled or elderly individuals clearly helped to survive. Analysis of their teeth suggests that they were predominantly meat eaters, but the 2010 discovery of plant remains embedded in plaque shows that their diet was more varied than previously thought. The remains of grains, legumes (including peas and beans) and dates, along with starch (perhaps from barley and water lilies), suggests they may even have cooked some of this plant food.

'Some experts have claimed that a Neanderthal dressed in modern clothing could easily merge into the crowd in any big city of the Western world.'

Very rarely, bone tools and ivory ornaments have been found in apparent association with Neanderthal remains (most notably at the Grotte du Renne near Arcy-sur-Cure in France), but recent studies have dated these to less than 21,000 years old, when the Neanderthals were already extinct. Certainly, other Neanderthal sites yield nothing like the abundance and variety of tools and art associated with *Homo sapiens*. Indeed, it seems likely that the modern human ability to develop more varied tools and weapons was crucial in allowing our ancestors to develop the varied diets, clothing and shelter that helped them to survive the ice age in greater numbers than their cousins.

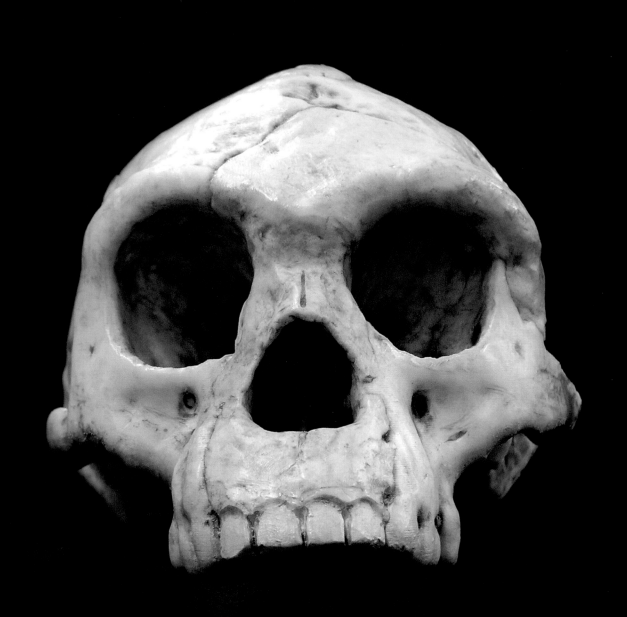

Unexpected cousins

DEFINITION NEW MEMBERS OF THE HUMAN FAMILY THAT EXISTED
UNTIL RELATIVELY RECENT TIMES

DISCOVERY THE FLORES 'HOBBITS' WERE DISCOVERED IN 2001,
AND THE DENISOVAN 'SPECIES X' WAS IDENTIFIED IN 2010

KEY BREAKTHROUGH BOTH ANATOMICAL AND MOLECULAR (DNA)
ANALYSIS REVEAL THE EXISTENCE OF NEW HUMAN SPECIES

IMPORTANCE THE PRESENCE OF OTHER HOMINIDS IN RECENT TIMES
PUTS OUR OWN SPECIES INTO CONTEXT AND RAISES QUESTIONS
ABOUT OUR INTERACTIONS WITH THEM

Until recently, the accumulated evidence of the fossil and archaeological record indicated that *Homo sapiens* has been the only human species on the planet since our last surviving Neanderthal relatives died out around 30,000 years ago. But in the last decade, all that has changed.

Today, we know that there were at least two other human species sharing the late Quaternary Ice Age world – the diminutive 'hobbits' of Southeast Asia (*Homo floresiensis*), and 'Species X' (the Denisovans) in Siberia.

The Flores 'hobbits'

In 2001, the discovery of the skeleton of a metre-high human-like individual, buried deep in cave-floor deposits on the Indonesian island of Flores, caused an international sensation. At first, Australian archaeologist Mike Morwood and his Indonesian–Australian team believed they had found the skeleton of a three-year-old child, but on closer examination, they realized this was something very different and very strange – the fossil remains of a dwarf human species. With a mixture of primitive and more advanced features, the diminutive and small-brained (c.400ml/13.5fl oz) female, named as *Homo floresiensis* in 2003, also had characteristics that confirmed she was an adult.

Altogether, the remains of some six individuals, associated with primitive stone tools and the bones of extinct animals such as the dwarf elephant *Stegodon*, have now been found at Liang Bua cave, dating to between 95,000 and 17,000 years ago. Initially, there were counter-claims that the dwarf skeletons were merely victims of some kind of pathological condition, and lengthy 'custody battles' over access to the remains and even the Liang Bua site itself. However, recent detailed analysis of the bones has revealed unique

OPPOSITE This partially reconstructed skull of the so-called 'Hobbit' *Homo floresiensis*, reveals a unique dwarfed human species that survived in isolation on the Indonesian island of Flores until perhaps 17,000 years ago. The species combines primitive and more advanced features of the genus *Homo*.

anatomical features, suggesting that far from being a population of dwarf *Homo sapiens*, the Flores people are a link back to early (pre-*Homo erectus*) members of the human family. It seems that the ancestors of these people became isolated on the island and gradually succumbed to 'island dwarfism' under pressure from limited food resources. How and when *Homo floresiensis* got to Flores in the first place, and whether they encountered the modern humans that first spread through the region more than 40,000 years ago, are questions that so far remain unanswered.

Ancient DNA

Since the 1980s, the study of evolution has been revolutionized through the science of DNA – the complex molecules that carry the genetic 'blueprint' of an organism in almost every cell of its body. Comparison of mitochondrial DNA (mtDNA – see page 130) allows taxonomists to estimate the amount of time that has passed since living species diverged from a common ancestor. Genetic comparisons also allow taxonomists to apply the 'acid test' of species separation – whether organisms can interbreed to produce viable offspring. Unfortunately, fossil DNA from extinct individuals is almost impossible to find and use, because the molecule is extremely fragile and normally starts to degrade within hours of death. So for extinct humans, as for the vast majority of fossil life, palaeontologists are forced to rely on anatomical features of the skeleton alone.

'At first, archaeologists believed they had found the skeleton of a three-year-old child, but on closer examination, they realized that this was something very different and very strange – the fossil remains of a dwarf human species.'

However, DNA fragments can occasionally be preserved in the right environment, and the development of techniques for the multiplication and 'sequencing' of small quantities of DNA has made it possible for the first time to recover genetic information from 'sub-fossil' remains. Desiccation is one of the most effective means of preservation, and the dried skins of extinct animals have proved a particularly fruitful source of DNA fragments – as early as 1984 the 150-year-old pelt of a quagga (a South African horse species driven to extinction in the 1870s) was sequenced to reveal its close evolutionary relationship to the zebra. Freeze-drying is the best preservative of all, however, and the frozen cadavers of ice age animals such as mammoths, found in high-latitude permafrost, can preserve a significant proportion of their original DNA (albeit in fragmented form). It is important to recognize, however, that the ability to extract and compare fragments of ancient DNA is a long way from fabricating the complete genetic sequences that would be needed to turn some scientists' dreams of recreating extinct species into reality.

When it comes to our human relatives, the best environments for preserving DNA are cool, dark cave interiors at high latitudes, where fossil remains are not much more than 50,000 years old. Sadly, the humid tropical conditions at the Flores site both today and in the past mean that it has not been possible

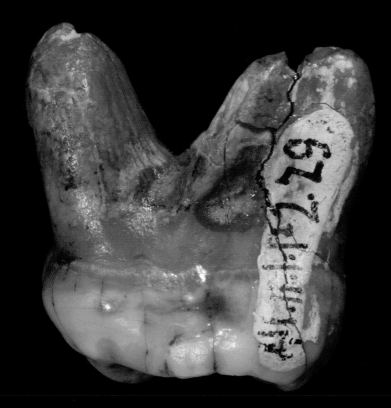

to recover DNA from the 'hobbit' remains. As a result, until recently the only hominid DNA recovered from a species other than our own has come from Neanderthals (see page 305).

The Denisovans

However, in 2008, archaeologists working at Denisova Cave in Siberia's Altai Mountains recovered a small bone from the little finger of a juvenile female. Radiocarbon dating of other remains found alongside the bone suggested this female lived around 40,000 years ago, and a team led by Johannes Krause and Svante Pääbo of the Max Planck Institute in Leipzig, Germany, were able to extract precious mtDNA from them. Differences from the mtDNA of both Neanderthals and modern humans indicated that this fingerbone came from a completely new human-related species, which last shared a common ancestor with us and the Neanderthals around 1 million years ago. A 2010 analysis of a tooth previously discovered at the same site confirmed this remarkable discovery.

With no anatomical features to describe, the species is so far unnamed, and they are known simply as 'the Denisovans' or 'Species X'. They predated our immediate ancestors *Homo heidelbergensis*, but could be related to the Spanish *Homo antecessor*, which lived between 1.2 million and 800,000 years ago. However, this is all speculation until more skeletal remains or further ancient DNA from other members of 'Species X' are recovered.

77

The Younger Dryas

DEFINITION A BRIEF BUT SHARP DECLINE IN THE NORTHERN
HEMISPHERE CLIMATE TOWARDS THE END OF THE LAST ICE AGE

DISCOVERY THIS 'BIG FREEZE' WAS IDENTIFIED FROM ICE CORE
STUDIES AND OTHER EVIDENCE IN THE 1950S

KEY BREAKTHROUGH STUDIES OF GLACIAL DEPOSITS IN NEW ZEALAND
HAVE CONFIRMED THAT THE YOUNGER DRYAS WAS COUNTERED BY
A PERIOD OF WARMING SOUTH OF THE EQUATOR

IMPORTANCE THE YOUNGER DRYAS PROVIDES A GOOD EXAMPLE OF
SUDDEN CLIMATE CHANGE AND EARTH'S COMPLEX RESPONSE TO IT

The Younger Dryas was a period of rapid cooling in the northern hemisphere from about 12,900 years ago to 11,700 years ago and is one of the best-known examples of rapid climate change. But while the north cooled, it seems the southern hemisphere was steadily warming.

The most recent glacial period in the Quaternary Ice Age began to loosen its chilling grip around 15,000 years ago, leading to a phase of general warming across the next 4,000 years as the huge ice sheets and glaciers that covered vast swathes of the northern continents melted and retreated. However, Arctic ice core records suggest this warming was interrupted by some distinctly cool phases that reversed the general trend. There is now increasing evidence that such 'cold snaps' were more common features during various deglaciation events than previously thought. Nevertheless, they are still worthy of note, since during one of the most recent, known as the Younger Dryas, climates in the northern hemisphere cooled rapidly by some 15°C (27°F) over just a few centuries. Needless to say, it is important to understand why this happened and what, if any, were the precursor events that might provide advance warning of a future recurrence.

Big chill

The Younger Dryas is named after a small tundra plant, *Dryas octopetala*, whose pollen is commonly found in northern hemisphere sediments laid down between 12,900 and 11,700 years ago. This cold snap seems to have been triggered by huge volumes of fresh, cold meltwaters flowing from the melting North American ice sheet into the North Atlantic. The influx damped down normal ocean circulation patterns – most notably the warm Gulf Stream flowing across the Atlantic towards Europe – and produced

OPPOSITE Lake Ohau, in New Zealand's Southern Alps, lies in a region that was subjected to intense glaciation during the Quaternary Ice Age. Mapping and analysis of its glacial deposits is helping determine the climate history of the southern hemisphere, especially during the northern hemisphere's 'Younger Dryas' cold snap.

cooler sea surface temperatures in the tropics. This in turn increased rainfall to the Amazon and weakened the intensity of the Asian monsoon still further afield. Regional temperatures fell by as much as 15°C (27°F) in Greenland and perhaps elsewhere. The event was catastrophic for many organisms that could not adapt to such a rapid change in climate and environments. But since most of the evidence for the Younger Dryas comes from the northern hemisphere, there are still questions about its global reach – most importantly, what impact did the Younger Dryas have in the southern hemisphere?

South of the equator

The southern continents certainly went through similar, 1,000-year-long reversals of warming at the end of the last glacial – for instance, there is good evidence for a cooling phase known as the Antarctic Cold Reversal that occurred between 14,500 and 12,900 years ago. After this, however, Antarctic temperatures and atmospheric carbon dioxide increased steadily to reach the present interglacial levels by around 11,500 years ago – at exactly the same time that the northern hemisphere was plunging into the Younger Dryas. This lack of synchronicity between Earth's hemispheres has perplexed scientists for some time.

New Zealand's island setting in the southwest Pacific, far from the influence of northern hemisphere ice sheets and North Atlantic deep-water circulation, make it an ideal place for testing the timing and duration of post-glacial climate change in isolation from other factors. The alpine mountains of

South Island preserve an excellent record of glacial advances and retreats but even here, the evidence for the exact timing of these events has proved somewhat uncertain – until recently.

In 2010, a team of scientists led by Michael Kaplan of Columbia University's Lamont-Doherty Earth Observatory published the results of a detailed study of glacial deposits called moraines at the head of a river in the Ben Ohau Range of the Southern Alps. Moraines are a jumble of rock and sediment scraped from the landscape by advancing glaciers, carried by them and then left behind when they retreat. Kaplan's team were able to date the moraines by measuring the level of an isotope called beryllium-10 (^{10}Be) in their surface layers. This 'cosmogenic' isotope is formed when soils are bombarded by high-energy particles entering the atmosphere from space. The amount of ^{10}Be present indicates how long the soil has been exposed at the surface, and therefore how long since the moraine was deposited by a retreating glacier.

The study confirmed that New Zealand's glaciers melted and retreated over a period of 1,500 years from 13,000 years ago, coinciding with the onset of the Younger Dryas cooling in the northern hemisphere. Evidently, there was a 'flip-flop' between post-glacial events in the two hemispheres, which matches trends in the temperature of the Southern Ocean seen in deep-sea sediment cores and changes in Antarctic climate and atmospheric carbon dioxide measured from ice cores.

What caused the flip-flops?

Scientists still aren't sure about the mechanism that caused northern and southern climates to slip 'out of sync' with one another, but two plausible explanations for this strange pattern rely on cold conditions in the North Atlantic making their effects felt much further afield than would normally be expected. As we saw, the melting of the North American ice sheets released huge amounts of cold fresh water into the North Atlantic, damping down the ocean's circulation and allowing sea ice to spread southwards. The exact path taken by these melting flood-waters has not yet been identified – although it is often assumed to have been through the St Lawrence Seaway, there is recent evidence that it may in fact have been further to the north, via the Arctic.

'During the Younger Dryas, climates in the northern hemisphere cooled rapidly by some 15°C (27°F) over just a few centuries.'

In one, 'wind-driven' theory, the weakening of the Atlantic Gulf Stream shifted the pattern of prevailing westerly winds southwards so that more heat was retained in the southern hemisphere. This increased upwelling in the Southern Ocean, which in turn released more carbon dioxide into the atmosphere from the ocean waters, warmed the southern hemisphere climate and triggered the retreat of the New Zealand glaciers. A rival 'ocean-driven' model suggests that the faltering Gulf Stream led to a global reconfiguration of ocean currents themselves, allowing warm air to simply 'pool' in the southern hemisphere and warm up the climate.

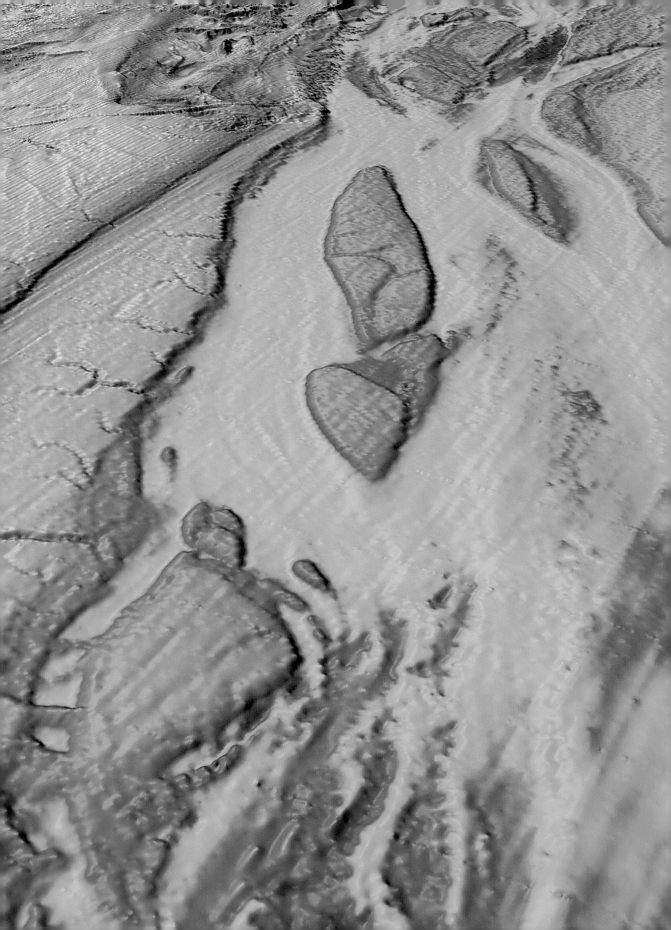

78

Megafloods

DEFINITION SUDDEN AND DRAMATIC FLOOD EVENTS TRIGGERED BY
CHANGES AT THE END OF THE LAST ICE AGE

DISCOVERY 'SCABLANDS' LINKED TO THE COLLAPSE OF GLACIAL
DAMS WERE IDENTIFIED IN NORTH AMERICA IN THE 1920S

KEY BREAKTHROUGH RECENT MAPPING OF THE ENGLISH CHANNEL
SUGGESTS IT WAS FORMED BY SIMILAR TORRENTIAL FLOODS

IMPORTANCE THESE HUGE FLOODS RESHAPED LANDSCAPES AND HAD
A DEVASTATING EFFECT ON ANYTHING THAT GOT IN THEIR WAY

Dramatic megafloods have occurred many times in Earth's history as natural dams formed by glaciers finally gave way, unleashing torrents of escaping meltwater to carve huge gashes in the landscape. Intriguing new evidence suggests that just such a flood separated Britain from continental Europe.

The biblical flood story is well known to anyone brought up in the Judeo-Christian tradition and was so firmly entrenched in the western mindset that some academic geologists were still trying to find geological evidence for it in the early decades of the 19th century. For instance, the English geologist and palaeontologist William Buckland (1784–1856) argued that there was evidence for the biblical flood in the landscapes and fossils of northern Europe. However, he changed his mind during the 1840s, when Swiss geologist Louis Agassiz reinterpreted these features as the result of a widespread ice age (see page 293).

In the 1920s, however, American geologist J. Harlen Bretz (1882–1981) argued that some peculiar erosion and deposition features in Washington State and Oregon were indeed the result of catastrophic flooding. He coined the term 'channelled scablands' to describe their form, but it took another 30 years before his idea gained acceptance. Since the 1950s, scientists have started to understand the details of how these features formed as glacial ice melted and their waters ran back to the oceans as catastrophic floods.

Glacial meltwaters

Each glacial phase of the Quaternary Ice Age locked up a significant proportion of the global water budget in snow and ice, which covered high-latitude landscapes, especially in North America, Greenland, northern

OPPOSITE This 3-D view of the English Channel seabed reveals details of the flood valley cut by meltwaters flooding from both Britain (top left) and mainland Europe (top right). The deep erosional scours in the bottom of the channel, coloured dark blue, are around 100m (330ft) deep.

ABOVE Palouse Falls in Washington State is the only major waterfall left along the path taken by floodwaters from glacial Lake Missoula between 15,000 and 13,000 years ago. The Palouse canyon was cut by the floodwaters through 150m (500ft) of lavas from the Columbia River Plateau Basalts, perhaps in a matter of days or weeks.

Europe, Russia and Antarctica. Since the end of the last glacial period, much of this ice has melted and found its way back into the oceans. Probably the most powerful of these events were the Missoula Floods that periodically burst from the glacial Lake Missoula, formed by the meltwaters of the Cordilleran ice sheet that covered much of North America. Its natural ice dam broke on some 40 occasions between 15,000 and 13,000 years ago, with the lake waters surging down the Columbia River Valley and flooding a large part of eastern Washington State.

Modern estimates suggest that Lake Missoula had a volume of more than 2,000 cubic km (500 cubic miles) and that the largest of the floods discharged 40–60 cubic km (10–15 cubic miles) of water per hour, at maximum speeds of around 35m per second (78mph). The immense power of the floodwater eroded some 200 cubic km (50 cubic miles) of rock and sediment from the landscape downstream, creating characteristic scablands with eroded channels, giant current ripples and boulders. Such enormous floods are now known as megafloods. To qualify for the name 'megaflood', a flood must have a flow rate of at least 1 million cubic metres (220 million gallons) of water per second.

The Alaskan megaflood

Scientists have recently identified signs of a similar glacial lake megaflood in Alaska. In 2010, Michael Wiedmer from the University of Washington and his colleagues reported evidence that at least four floods burst through breaches in the natural dams holding back the glacial Lake Atna around 17,000 years ago. This lake covered more than 9,100 square km (3,500 square miles) of what is now the Copper River Basin.

Wiedmer's team estimates that around 1,400 cubic km (340 cubic miles) of floodwater rushed over the landscape at around 3 million cubic m (660 million gallons) per second, creating enormous current ripples or sand waves. These huge dunes are not obvious because of their sheer size – today, they are masked by roads and buildings – but their 35m (115ft) height is an indication of the strength and size of the flood that formed them.

The isolation of Britain

Scientists since the mid-1970s have suggested that Britain's isolation from continental Europe might also have involved a catastrophic flood. In 2007, however, a team led by Sanjeev Gupta from London's Imperial College finally produced some solid evidence for the theory, in the form of a detailed underwater map of the English Channel produced by sonar sounding. The survey revealed a network of valleys extending out into the Channel from the mouths of the rivers Seine and Somme on the French side and the Solent on the English side. There are also two substantial 'trunk valleys' running along the main axis of the Channel. Unlike normal river valleys, they are unusually straight and broad – more than 10km (6 miles) wide, with long grooves in the floor that are typical hallmarks of a megaflood.

'To qualify for the name "megaflood", a flood must have a flow rate of at least 1 million cubic metres (220 million gallons) of water per second.'

The evidence suggests the English Channel was carved out by two separate floods, between 450,000 and 180,000 years ago. The source of the floodwaters was a vast lake of glacial meltwater to the north of the narrow Straits of Dover, held back by a barrier of bedrock some 30km (19 miles) long that connected Britain to France. Forming a natural rock dam, the barrier originally rose some 30m (100ft) above present-day sea level. Nevertheless, the rising glacial meltwater overwhelmed it around 450,000 years ago, allowing the waters to pour through and cut the first of the megaflood channels.

A subsequent warm period of deglaciation around 200,000 years ago triggered a repeat performance, in which the main northerly megaflood channel was further eroded and deepened by the floodwaters. During peak flow, this channel may have carried half the river and meltwater drainage from western Europe into the Atlantic. These megafloods could have been among the largest on Earth, and they have separated the British Isles from continental Europe ever since.

79

The Holocene rebound

DEFINITION A SLOW REBALANCING OF EARTH'S CRUST AS THE
WEIGHT OF ICE FROM THE LAST GLACIAL IS REMOVED

DISCOVERY FIRST IDENTIFIED FROM SCANDINAVIAN COASTAL
MEASUREMENTS IN THE 1760s

KEY BREAKTHROUGH SWEDISH GEOLOGIST GERARD JACOB DE GEER
LINKED THE UPLIFT TO 'ISOSTATIC' REBOUND IN THE 1890s

IMPORTANCE INEXORABLE CHANGES TO LAND LEVELS, COUPLED
WITH RISING SEA LEVELS, THREATEN MAJOR POPULATION CENTRES

Since the end of the last ice age and the beginning of the Holocene Epoch around 12,000 years ago, the high-latitude landmasses of North America, northern Europe and Asia have been slowly rising above sea level due to an effect called glacial or isostatic rebound.

The phenomenon of land's slow rise above sea level was first noted in Sweden during the 18th century. During a mapping survey, scientist and geographer Anders Celsius (1701–44) suggested placing markers at various points along the coastline, and within a few decades these showed what appeared to be a slow, uneven 'drop' in sea levels. At the time, the effect was attributed to evaporation of water from the sea – the true cause was only found with the discovery of evidence for an ice age across most of Europe, around a century later (see page 293). In 1890, Swedish geologist Gerard Jacob De Geer (1858–1943) published the first detailed measurements proving that the uplift was connected to the widespread glaciation.

Today, we understand that during the glacial depths of the recent Quaternary Ice Age, ice sheets several kilometres deep covered much of North America, Greenland, northern Europe, Arctic Russia, Asia and Antarctica. Their immense weight pushed the rocks of the crust many hundreds of metres down into the underlying mantle, squeezing the hot and 'plastic' mantle rocks out of the way.

Since the end of the last glacial around 12,000 years ago and the onset of a warmer climate, many of these ice sheets have melted away completely. The removal of their weight has allowed the crust to slowly push upwards as the mantle flows back into place, in a process known as isostatic rebound.

OPPOSITE The mountains of Scandinavia and northern Britain are remnants of ancient tectonic belts whose present upland topography, such as this steep-sided Norwegian waterfall, is partly the result of isostatic rebound following the retreat of ice age glaciers.

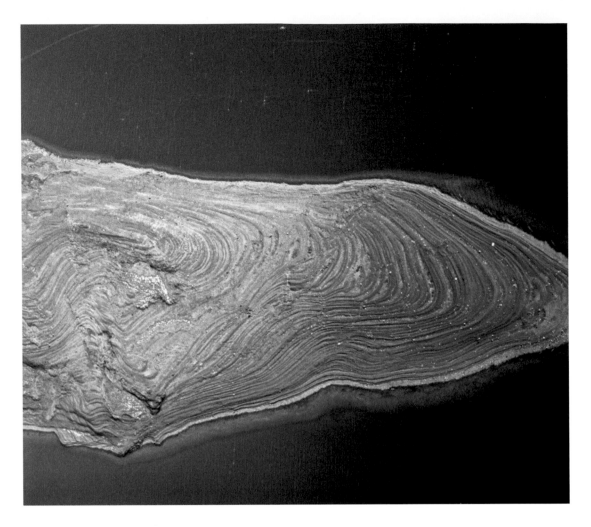

Data from the Global Positioning System (GPS) satellite network shows that the rebound typically causes land to rise at rates of around 1cm (0.4in) per year. This suggests that the process will continue for perhaps another 10,000 years, elevating parts of Scandinavia, Scotland and northern Canada by some 100m (330ft) above their current levels.

On the up

Perhaps the most significant economic effects of the uplift have been experienced in Scandinavia. In historic times, many ports, harbours and seaside settlements have been left literally high and dry as the sea has receded. Marine lagoons and bays all around the Baltic, but especially in Finland, have been cut off from the sea and transformed into freshwater lakes, marshes and peat fens, while some have silted up altogether. As the rebound continues, geologists predict that the northernmost part of the Gulf of Bothnia (the northern arm of the Baltic Sea) will become cut off from the rest of the Gulf and the Baltic. Water depths around the 5,600 islands of the Kvarken Archipelago are already no greater than around 25m (82ft), and as the islands gradually emerge from the sea, they will form a land connection between Finland and Sweden and a huge 250km (155-mile) lake. One 7,000-year-old Stone Age site, originally at the shoreline, already stands 150m (500ft) above sea level.

Altogether, the rise of the land here has already amounted to some 285m (935ft), and since 2006 the locality has been listed as a World Heritage Site showcasing the process of post-glacial rebound. Coastal sites in Norway, western Scotland and Ireland, with their deeply indented and glacially 'overdeepened' fjords have not been so drastically affected because they typically have much deeper offshore waters.

What goes up...

However, isostatic rebound quite literally has a downside – as the warped but still stiff crust flexes upwards in some places, it also produces a surrounding region of 'downwarp'. This is particularly noticeable in the British Isles, where Scotland, northern England and Wales and Northern Ireland are rising, but southern England, Wales and the majority of the island of Ireland are sinking, separated by a 'hinge' axis that runs roughly from northeast to southwest.

Across southern Britain the subsidence ranges from 5 to 12cm (2 to 5in) every century – a significant amount that must be added to predictions of sea-level rise caused by global warming. Around the coast, and especially in eastern England, remnants of drowned peatlands and woods now lie below sea level, but the growth of some other coastal marshes and peatlands has managed to keep pace with rising sea levels.

'The rebound typically causes land to rise at rates of around 1cm (0.4in) per year. This suggests that the process will continue for perhaps another 10,000 years, elevating parts of Scandinavia, Scotland and northern Canada by some 100m (330ft) above their current levels.'

The economic and environmental impact is likely to be keenly felt in highly populated regions that are already close to sea level – notably London and the Thames Estuary. In 1953, a high 'spring' tide coinciding with a major storm surge pushed North Sea waters more than 5m (17ft) above their normal levels, flooding low-lying coastal regions of the Netherlands, Belgium and eastern England. Altogether, some 2,400 people were killed, leading to the introduction of widespread flood control measures including a barrier to protect London.

Work on the 520m (0.3-mile) Thames Barrier, which has rotating gates to allow shipping through into the Port of London, began in 1974 and was completed in 1982, at a cost of £534 million (around a billion dollars). The barrier was closed due to flood risks 35 times in the 1990s, but has been raised some 75 times in the first decade of the 21st century, showing the increasing threat from rising sea levels and subsiding land. Although the barrier should remain effective until 2060, another much longer replacement will ultimately be needed in order to prevent future flooding.

80 Domestication of plants and animals

DEFINITION THE ADAPTATION OF CERTAIN SPECIES TO HUMAN NEEDS THROUGH SELECTIVE BREEDING

DISCOVERY THE ARCHAEOLOGICAL RECORD TRACES THE HISTORY OF DOMESTICATION BACK TO DOGS, AROUND 17,000 YEARS AGO

KEY BREAKTHROUGH DMITRI BELYAEV REPLICATED THE PROCESS TO PRODUCE A DOMESTICATED FOX IN THE 1950S

IMPORTANCE LIVESTOCK, WORKING ANIMALS AND FOOD CROPS ARE VITAL TO SUPPORTING OUR HUGE HUMAN POPULATION

The domestication of plants and animals is an essential element of human success, and ironically began at the same time that our species was abetting the extinction of the ice age megafauna. Our present population of close to 7 billion people could not survive without domesticated crops and livestock.

Media reports frequently suggest that 'genetic modification' is a new phenomenon, but the reality is that it has been with us since the very beginnings of human civilization. Successful domestication is, in effect, the genetic adaptation of an organism through selective breeding to suit the needs of humans. Over the millennia, our ancestors domesticated a huge range of organisms from rice to goldfish, for many purposes, ranging from food and materials to companionship and ornamentation.

Scientist are still puzzling over the exact nature of the domestication process, and why some species are more susceptible to our genetic manipulation than others. Experiments have attempted to domesticate a number of living wild animals such as the wolf, zebra and big cats without achieving any real success, and yet there is good genetic evidence, for instance, that all domesticated dogs are ultimately descended from wolves. One of the longest and most intensive studies was conducted by Russian scientist Dmitri Belyaev (1917–85) in the 1950s. By selecting silver fox cubs that showed the least antagonism towards humans and breeding them through numerous generations, he produced a population of foxes whose behaviour and appearance had changed. Their response to humans was no longer one of immediate fear or aggression, but more like that of a domesticated dog wagging the tail and licking at a proffered hand. In appearance, meanwhile, these foxes tended to have curly tails, floppy ears and smaller skulls.

OPPOSITE Today, domesticated rice is the second most cultivated plant after maize, and a dietary staple for millions of people. It was first cultivated around 10,000 years ago in the Yangtze River Valley of China, following a long period of harvesting as a wild crop from perhaps as far back as 12,000 years ago.

Puzzles of domestication

In contrast, attempts at domesticating big cats using similar techniques have been unsuccessful. Programmes to selectively breed zebras for domestication have also notably failed, even though their close wild horse relatives were first domesticated some 6,000 years ago and have played an important role in the development of human society ever since. Equally, the successful domestication of the Indian elephant contrasts starkly with the impossibility of domesticating the African elephant, showing that despite apparent similarities, some animals just cannot be bent to our will. So far as a pattern can be established, the successful domestication of animals generally seems to make use of some of the following characteristics: food requirements that can be easily satisfied, such as grass; fast growth and a short reproductive cycle coupled with ease of breeding in captivity; a relatively even temperament; and a social hierarchy that can be modified to suit human needs, often involving behaviours such as herding and pack leadership.

The earliest animal to become domesticated seems to have been the dog, and at least 17,000 years of archaeological evidence chronicle the story of canines living in close association with humans. The first livestock animals all seem to have been domesticated at around the same time, including sheep (13,000–11,000 years ago), boars and pigs (11,000 years ago), and goats and cattle (around 10,000 years ago). The domestication of cats, meanwhile, is particularly hard to trace, but there is evidence that they were introduced to the island of Cyprus around 9,500 years ago. Chickens were the first domestic birds, bred from red junglefowl around 8,000 years ago, and donkeys were the first pack animals, around 7,000 years ago. The Chinese, meanwhile, achieved a notable first by domesticating the silkworm some 5,000 years ago, while the Asian elephant has lived alongside people for around 4,000 years. There have been many more recent domestications, ranging from the European rabbit, 1,600 years ago, to Belyaev's silver foxes.

'The successful domestication of the Indian elephant contrasts starkly with the impossibility of domesticating the African elephant, showing that despite apparent similarities, some animals just cannot be bent to our will.'

Cultivating plants

Plants lie at the base of the food chain and their domestication and cultivation was fundamental to the agricultural revolution associated with the first permanent human settlements. The discovery that nutritious plants could be grown by cultivating their seeds was almost certainly accidental – perhaps a result of our ancestors seeing plants sprouting from piles of dung left by animals that had eaten similar plants. At first, there must have been small-scale planting of preferred 'annual' plants with large and easily identified seeds (such as squashes, legumes, grains and those with edible tubers).

Archaeological records suggest that plant domestication first occurred in western Asia, and preserved grains of rye showing some domestic traits have been found in Syria and dated to around 11,000 years ago. These grains were probably cultivated in small stands for domestic use and ease of gathering, rather than being selectively bred, but it would have been no great step to notice that some stands had better yields than others and to select those more successful grains for future planting.

Intriguingly, though, archaeologists have recently recovered the remnants of seeds (grains such as barley), legumes (peas and beans) and dates from the teeth of Neanderthal remains dating as far back as 46,000 years ago. Although these plants were certainly *not* domesticated, their discovery has vastly extended the record of such plant use back in time, and indeed beyond our own species for the first time.

Within Asia and Europe, there is little doubt that the full domestication of important cereals such as wheat, alongside legumes such as peas, originated in the so-called Fertile Crescent around 11,000 years ago. From the lower Nile Valley in Egypt to the broad Mesopotamian valleys of the Tigris and Euphrates, the nutrient-rich river floodplains and the climate of a region with hot, dry summers, were ideally suited to the early development of agriculture.

Disappearing megafauna

DEFINITION A WIDESPREAD EXTINCTION OF LARGE ANIMALS IN THE LAST 50,000 YEARS

DISCOVERY EXTINCT ANIMALS SUCH AS MAMMOTHS HAVE BEEN KNOWN SINCE PREHISTORIC TIMES, BUT THE TIMING OF THEIR EXTINCTIONS HAS ONLY RECENTLY BEEN ESTABLISHED

KEY BREAKTHROUGH NEW STUDIES SHOW A STRONG LINK BETWEEN LOCAL EXTINCTIONS AND THE SPREAD OF *HOMO SAPIENS*

IMPORTANCE EXTINCT SPECIES REVEAL THE DEVASTATING IMPACT THAT HUMANITY HAS HAD ON ITS ENVIRONMENT

Beyond the frozen wastes of ice and permafrost, the world of the Quaternary Ice Age was home to large herds of plant-eating 'megafauna', preyed upon by much smaller groups of carnivores and scavengers, including humans and our close relatives. What drove many of these creatures to extinction?

Today, except for their domestic descendants, most of the great ice age megafauna have largely disappeared – except in Africa, where they are increasingly endangered. Both climate change and human overkill have been blamed for this extinction, but which is the true culprit?

Megafauna (a name meaning simply 'large animals') is a term that generally refers to any animal over 40kg (88lb) in weight, and therefore includes many animals that are not particularly large, such as deer, kangaroo and humans. During the Quaternary Ice Age, the exact composition of each region's megafauna varied wildly, influenced by factors such as climate, vegetation, location and evolution. For example, the Australasian megafauna (including the giant wombat *Diprotodon*, the marsupial 'tiger' *Thylacoleo* and the huge lizard *Megalania*) was very different to that of South America (which included the giant sloth *Megatherium* and the ungulate *Toxodon*), and this in turn was largely distinct from the megafauna of North America (with the mastodon *Mammut*, the bison *Bison antiquus*, the 'sabre-toothed tiger' *Smilodon* and the short-faced bear *Arctodus*). Eurasian megafauna, meanwhile, included the famous woolly mammoth *Mammuthus*, the giant 'Irish elk' *Megaloceros* and the woolly rhinoceros *Coelodonta*, while in Africa, today's surviving species lived alongside the strange elephant relative *Deinotherium*, the moose-like giraffid *Sivatherium* and the giant wild ox *Pelorovis*.

OPPOSITE The dehydrated body of this eight-month-old baby mammoth, nicknamed 'Dima', was recovered from the permafrost in a Siberian gold mine in 1977. Originally, the body was covered in hair, but most of this was pulled away when the body was recovered. Radiocarbon dating shows that it died around 40,000 years ago.

What is more, the megafauna in a particular location could also change considerably between cold glacial periods and warmer interglacial times, with mammoths and woolly rhinos alternating with more familiar elephant and rhino species less suited to cold climates. Changing sea levels meanwhile, allowed animals to move in significant numbers across land connections between Africa and Eurasia, Asia and North America, and North and South America.

Overall, the ice age megafauna was complex and dynamic, involving several hundred mostly herbivorous species worldwide, of which some 85 percent became extinct by the end of the last glacial 11,500 years ago. It's significant that by that time, as global climates became warmer, *Homo sapiens* had colonized all major continents apart from Antarctica.

'According to recent research human entry into the various landmasses coincides suspiciously well with extinctions of the local megafauna.'

Extinction and climate change

The range and behaviour of most large plant-eating mammals can be constrained by a number of factors, including the availability of food, competition from other herbivores, pressure from predators and, above all, climate. Rapid changes in prevailing conditions during the Quaternary would have had a particular effect on high-latitude landscapes: during warm phases hippos wallowed among plant-rich waters in the east of England, while during cold phases, cool temperate forests extended down to southern Europe's Mediterranean shores.

Such changes in plant cover impacted upon both the plant eaters and the predators that depended upon them, including human hunters. Some predators (humans among them) were adaptable, so that if their normal prey became rare, they could change their choice of prey with a little time and practice. Other species, in contrast, were highly specialized, so that when their prey came under pressure, they did too.

But many of the megafauna, both herbivores and predators, seem to have survived through many cycles of dramatic climate change, suggesting that on each occasion, they were restocked by survivors from scattered refugia where climate change had not been so drastic. If so, then the final demise of the megafauna at the end of the last glacial must be due to some other cause – and humans are the main suspects.

The human factor

The *Homo sapiens* people that spread beyond Africa around 60,000 years ago were mobile hunters who evolved technologies that allowed them to bring down whatever game animals they came across. Their prey varied from bison and even mammoths to horses, deer and wild cattle, along with smaller game and aquatic animals. The spoils of the hunt supplied not only food, but also clothing and tools.

The global spread of humans took thousands of years, from the early colonization of Australia around 50,000 years ago, into Europe around 48,000 years ago, the Americas around 15,000 years ago and New Zealand a mere 800 years ago. According to recent research by David Burney of Hawaii's National Tropical Botanical Garden and Timothy Flannery of the South Australia Museum in Adelaide, human entry into the various landmasses coincides suspiciously well with extinctions of the local megafauna. Some 35 out of 45 genera went extinct in North America and 45 of 58 genera disappeared in South America, mostly between 12,000 and 10,000 years ago, coinciding with both climate change and the arrival of humans. In 2005, University of Florida researchers confirmed that the giant ground sloth survived in the West Indies until just 5,000 years ago, when humans first arrived on the islands.

In Eurasia, some 21 genera disappeared by around 11,000 years ago, although again there were a few late survivors in remote places. For instance, the giant deer of the Isle of Man survived until 7,700 years ago and the mammoths of Wrangel Island in the Arctic Ocean lasted until just 3,700 years ago. Many of these survivors, especially island species such as the dwarf elephants of Malta and the *Homo floresiensis* people (see page 309) were small in size as a result of limited food resources.

Australia is an interesting test case where post-glacial climate change caused an increasingly arid climate, but even so a dozen or so now-extinct members of the megafauna, ranging from *Diprotodon*, through giant kangaroos (*Macropus*) to the giant monitor lizard (*Megalania*) and the giant flightless bird (*Genyornis*), flourished until shortly after the arrival of humans. In Madagascar, the megafaunal collapse began 2,000 years ago, again following human arrival, and in New Zealand the megafauna, especially the flightless moas, survived right up until historical times when humans arrived.

Of course, there are arguments about the exact timing of extinctions and human arrivals, but modelling shows that any large animal species requires a large enough and genetically varied population with a broad enough range and food supply. Both climate change and human hunting fragmented populations and reduced food supplies until populations collapsed below a viable and renewable size.

ABOVE Excavations in North America have revealed a number of kill sites where prehistoric hunters drove large numbers of bison and even mammoths into natural topographic traps, where they slaughtered them for meat and other materials such as hides, bone and sinews. One of the most famous is the Olsen-Chubbuck kill site in Colorado, where some 200 bison were slaughtered around 10,000 years ago.

Biodiversity

DEFINITION THE OVERALL VARIETY OF LIFE IN AN ENVIRONMENT, WITH EACH SPECIES PLAYING ITS ROLE IN AN OVERALL ECOSYSTEM

DISCOVERY OUTLINED BY CHARLES ELTON IN THE 1920S AND EXPANDED BY WILSON AND MAY IN THE 1960S

KEY BREAKTHROUGH RECENT STUDIES CONFIRM THE INTRICATE WEB OF INTERDEPENDENCE WITHIN INDIVIDUAL ECOSYSTEMS

IMPORTANCE HUMAN UNDERSTANDING, MAINTENANCE AND RESPONSIBILITY FOR GLOBAL BIODIVERSITY IS ESSENTIAL FOR THE WELL-BEING OF THE ECOSYSTEM AND THE SURVIVAL OF LIFE

Recent scientific studies have confirmed the old saying that 'variety is the spice of life'. It seems that biodiversity – the variety of life in any particular natural environment – is of considerable importance in maintaining the environment's stability as a habitat for all forms of life.

Global differences in the relative abundance of life and diversity of organisms have fascinated biologists ever since records began. For European naturalists the extraordinary biodiversity of the tropics came as a considerable shock. For both Darwin and Alfred Russel Wallace (1823–1913), who developed the theory of evolution independently of one another, experience of life in the tropics was seminal to the development of their ideas.

The fundamental principles of ecology were clearly outlined in the late 1920s by scientists such as English zoologist Charles Elton (1900–1991), but it was not until the 1940s that the concept of biodiversity itself was fully developed. The following decades saw scientists such as US ecologists Eugene Odum and Edward O. Wilson and Australia's Robert May investigate the theoretical implications of biodiversity. However, it is only recently that scientists have begun to test their ideas in the laboratory and the field, establishing that diversity of life plays a variety of roles in maintaining complete ecosystems – supplying gases and nutrients required for the metabolisms of all the inhabitants, providing water, decomposing and recycling organic waste matter, and even stabilizing climate and controlling pests.

Today, scientists have realized that biodiversity, on levels ranging from small ecosystems such as an urban garden, through forests and coral reefs, to our entire planet, plays a vital role in the continuity and prosperity of life.

OPPOSITE While tropical rainforests are perhaps the most famous of Earth's biodiversity hotspots there are many other environments that act as hotspots, ranging from tropical coral reefs to the mountains of South Africa. Overall, some 50 percent of the world's plant species and 42 percent of terrestrial vertebrates are endemic to just 34 biodiversity hotspots around the world.

The need for improved conservation has led to new research into the patterns of biodiversity and the factors controlling them.

Species diversity

According to most estimates, Earth today is home to between 5 and 10 million species, some 2 million of which have been scientifically described. However, when all microbial life is taken into account, some think the global total could be well over 30 million species. But all this variety is not equally distributed – some environments support a much greater diversity than others. In general, biodiversity is at its lowest in cold and arid regions, and greatest where temperatures are high and water is plentiful.

In some tropical regions, biodiversity and the number of localized (endemic) species is significantly greater than surrounding areas. For example, the 17,000 or so islands of the Indonesian archipelago and their surrounding waters are home to 10 percent of the world's flowering plants, 17 percent of reptiles, amphibians and birds and 12 percent of mammals, and some 240 million people who are ultimately dependent on the diversity of life in this region. Known as biodiversity hotspots, such areas are particularly vulnerable to environmental change.

'It is only recently that scientists have begun to test their ideas in the laboratory and the field, establishing that diversity of life plays a variety of roles in maintaining complete ecosystems.'

Interactions between life and environment are both complex and dynamic. On land, the foundation is sometimes called the geosphere – the physical substrate, ranging from rocks and minerals to the complex mix of organic and inorganic components in sediments and soil. In order to be suitable for colonization by life, the substrate must interact with both the hydrosphere and atmosphere. Furthermore, almost all forms of life require solar energy either directly or indirectly, in the form of heat and light. The dynamic interactions between all these physical and chemical factors provide the basis for the biodiversity of any one habitat or ecosystem.

Marine and land ecosystems

The fossil record shows that life took some 3 billion years to evolve beyond simple microbes into more complex communities of multicelled organisms. Modern aquatic ecosystems are generally dependent on the sunlight in surface waters to provide energy for 'primary producers', most of which are microscopic phytoplankton. These in turn provide food for consumers that are mostly small, free-swimming animals. Dominated by the larvae of bottom-living shellfish, these primary consumers also include giant plankton – food for many whales and some sharks. Between these extremes there is a hierarchy of predators from small to large. Additionally, there are the bottom-dwelling creatures both on and within the sediment substrate. Some of these live at relatively shallow depths, building huge reefs in tropical waters that provide habitats for a range of different organisms. Others, living at greater depths, feed on food particles that drift down from above.

A recent global survey led by a team from Canada's Dalhousie University has uncovered previously undiscovered differences in patterns of marine biodiversity. Two distinct distributions emerged, with a separation between coastal and oceanic organisms. There are marked peaks in the diversity of coastal species in the Western Pacific between 45 degrees North and 30 degrees South, with gradients along the continental coastlines. Warm-blooded seals and other pinnipeds are the exception, with a low diversity in the tropics and peaks at higher latitudes. In contrast, ocean-dwelling organisms have tropical distributions that peak between 20 degrees and 40 degrees north and south of the equator in all oceans.

Terrestrial biodiversity, meanwhile, only began to develop around 450 million years ago, with the colonization of the land in late Ordovician times, and took a further 50 million years to extend beyond wetland environments. However, since late Devonian times, and the evolution of the first tree-sized plants and land communities, terrestrial biodiversity has expanded exponentially (despite suffering occasional catastrophic extinctions such as those at the end of Permian and Cretaceous times – see pages 209 and 261). Today, tropical rainforests are famously some of the most diverse land ecosystems.

ABOVE A remarkable abundance and variety of animals have recently been discovered on the deep Antarctic seabed at depths of 600m (2,000ft). Bryozoans and sponges live fixed to the seabed, with clams and other sessile animals between them, and more mobile organisms including numerous brittle stars foraging among them for food.

Extremophiles

DEFINITION ORGANISMS CAPABLE OF EXISTING IN EXTREME
ENVIRONMENTS BEYOND THE TOLERANCES OF 'NORMAL' LIFE

DISCOVERY FIRST HOT SPRING BACTERIA WERE IDENTIFIED IN THE
1960s, FOLLOWED BY DEEP-SEA VENT COMMUNITIES IN 1977

KEY BREAKTHROUGH SCIENTISTS HAVE RECOGNIZED THAT
EXTREMOPHILE ORGANISMS VASTLY EXPAND THE RANGE OF
HABITABLE ENVIRONMENTS FOR LIFE ON EARTH

IMPORTANCE EXTREMOPHILES HAVE IMPLICATIONS FOR THE ORIGINS
OF LIFE ON EARTH, AND THE CHANCES OF LIFE ON OTHER WORLDS

In the past decade, the study of extremophile organisms has taken off to
such an extent that there is now an academic journal devoted to their study.
Extremophiles are of great interest, particularly for the insights they provide
into the evolution of life on Earth, and the potential for life on other planets.

One of the 20th century's most remarkable discoveries was made in 1977, by
oceanographers using the deep submersible *Alvin* to explore the mid-ocean
ridge off the Galapagos Islands on the East Pacific Rise. Here, a team led
by Jack Corliss of Oregon State University found the first of many deep-
ocean sites where, despite the high pressure, low temperature and absence of
light, life thrives around hydrothermal vents that spout brines with
temperatures in excess of 90°C (194°F). These vents are generally associated
with ocean-floor spreading processes and the formation of oceanic crust
(see page 61). Ocean-bottom waters penetrate deeply fissured volcanic
rocks and circulate below ground, where they are heated from the ambient
temperature of 2°C (36°F) and enriched with minerals from the surrounding
rocks, before being spewed back into the ocean depths. Quite independent
of sunlight for photosynthesis and energy, the sea floor communities around
these vents rely instead on 'chemosynthesis', carried out by archaea and
bacteria that thrive on sulphur compounds (especially hydrogen sulphide)
from the escaping brines. These sulphur-loving organisms are just some of
the most remarkable among many 'extremophile' organisms that have been
found over the past few decades.

The range of extremophiles

As their name suggests, extremophiles are organisms that are well adapted
to life under extreme conditions – beyond the limits for most life on Earth.

OPPOSITE Hot springs
such as Yellowstone's
Grand Prismatic
Spring have attracted
the attention of
biologists for several
decades, since it was
first discovered that,
despite the apparently
hostile conditions,
their boiling, mineral-
enriched waters
are home to simple
extremophile life.

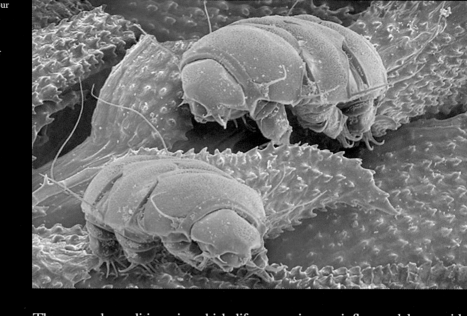

The normal conditions in which life can exist are influenced by a wide range of physical and chemical factors, such as temperature, pH (acidity or alkalinity) and the presence of water. Extreme conditions can cause the amino acid building blocks of life itself to break down, and so until the 1960s biologists understandably assumed that environments that were too hot, cold, arid, acidic or alkaline would be devoid of life.

Today, however, we know that extreme conditions and environments harbour a great variety of life, ranging from the deep-sea hydrothermal vents to the uniquely cold and arid McMurdo Dry Valleys of Antarctica. Various different extremophiles can tolerate extremes of pH, from the acidity of vinegar to the alkalinity of baking soda. They can cope with extremes of radiation (especially ultraviolet light and ionizing radiation), pressure, aridity and salinity, and temperatures from above 80°C (176°F) to below –15°C (5°F). Most extremophiles are micro-organisms, especially simple prokaryotic Archaea and Bacteria (see page 117), but recently a number of more complex, eukaryote extremophiles have been discovered, including some fungi, a few flowering plants, polychaete worms and brine shrimps.

Perhaps the strangest of all, though, are the water bears or tardigrades. These tiny, 1.5mm (1/16in) long relatives of the arthropods have been experimentally subjected to the extremes of outer space, where they have survived desiccation, temperatures in excess of 150°C (300°F) and doses of gamma rays that would be lethal to other organisms. Others have survived extreme cold as low as –272°C (–458 °F) (just 1°C/1.8°F above absolute zero, the coldest possible temperature) and still others have coped with pressures as high as 5,000 atmospheres – five times greater than the pressure in the deepest ocean trenches.

Extremophile environments

Some of the most intensely studied extremophiles are those found in terrestrial hot springs, such as those at Yellowstone National Park, USA. Here, waters are 'superheated' to above their normal boiling point in hot volcanic rocks, forming steam-driven geysers. Discovered in 1969, the archaean *Methanopyrus kandleri* can survive 122°C (252°F) and has an unusual metabolism that generates methane. Hot spring waters can be enriched with a variety of minerals, most notably sulphur, and are inhabited by sulphur-oxidizing extremophiles such as *Sulfurihydrogenibium rodmanii*, found at Hveragerdi, Iceland. Other springs, such as those at Rotokawa, New Zealand are notable for hot and highly acidic waters where bacteria such as *Thermoanaerobacterium aotearoensis* tolerate both low pH and high temperature.

Even more remarkably, a new organism found at California's Mono Lake is apparently capable of not only tolerating arsenic-rich conditions, but of putting this toxic element to use in its metabolism. This newly discovered bacterium seems to have a different biochemistry from all other life forms, and may even be using arsenic as a substitute for phosphorus, one of the essential ingredients of DNA and RNA. If confirmed, this discovery has huge implications, though it is still controversial.

'Extreme conditions can cause the amino acid building blocks of life itself to break down, and so until the 1960s biologists understandably assumed that environments that were too hot, cold, arid, acidic or alkaline would be devoid of life

Lessons from extremophiles

Much of the recent interest in extremophiles has been fired by their implications for our ideas about the origins of life, and the possibility that life might exist on other worlds with similarly extreme conditions. The recognition that our planet's oxygen-rich environment is a relatively recent development, and that microbial life has a long history stretching back for billions of years in a world of far less oxygen, has inspired some scientists to suggest that life may in fact have begun around deep-ocean vents. If true, then the first organisms were themselves extremophiles, relying on energy and nutrients from Earth's interior rather than, as has usually been assumed, sunlight in relatively shallow waters. They would then have evolved to take advantage of the more temperate conditions of shallow tropical seas once conditions on the surface had stabilized.

Meanwhile, as we learn more about the other planets and moons of our solar system, we are discovering tantalizing similarities to extremophile environments on Earth. The arid wastes of Mars are believed to have deposits of ice (and perhaps even liquid water) just below the surface, making them somewhat analogous to the dry valleys of Antarctica. Similarly, several of the moons of the distant giant planets seem to have oceans of liquid water sealed beneath icy outer crusts, kept warm by tidal activity that probably also generates volcanism. Jupiter's moon Europa, in particular, is thought to have sulphurous volcanic vents on its ocean floor, making it perhaps the most likely place in the solar system to find life beyond Earth.

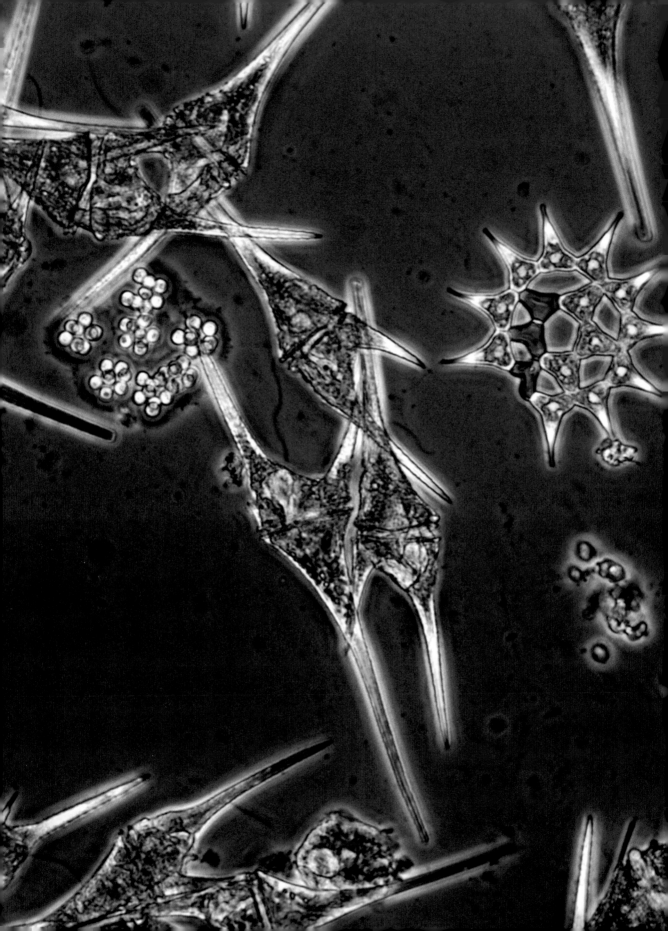

The Census of Marine Life

DEFINITION A DECADE-LONG, INTERNATIONAL EFFORT TO EXPAND
OUR KNOWLEDGE OF LIFE IN THE SEAS AND OCEANS

DISCOVERY FIRST HINTS OF THE VARIETY OF MARINE LIFE CAME IN
OCEANOGRAPHIC SURVEYS OF THE 18TH AND 19TH CENTURIES

KEY BREAKTHROUGH THE NEW MARINE CENSUS COMBINES
TRADITIONAL OCEANOGRAPHIC METHODS WITH GENETIC SCIENCE

IMPORTANCE AN UNDERSTANDING OF THE ABUNDANCE AND
DIVERSITY OF MARINE LIFE IS VITAL TO THE HEALTH OF THE
MARINE WATERS UPON WHICH WE ALL DEPEND

Uniquely within our solar system, vast amounts of Earth's surface are covered by seawater. Furthermore, some 99 percent of Earth's biota is thought to live in or under marine waters – yet only 5 percent of this ocean realm has so far been systematically explored.

Science fiction author and visionary Arthur C. Clarke once suggested that the planet we know as Earth should really be called Ocean. Water covers more than 70 percent of our planet's surface and has been heavily exploited by humans for food and mineral resources – many commercial fish stocks are now seriously depleted, and some may have already diminished beyond the point of recovery.

Part of the reason behind this crisis in the seas is our essential ignorance – we simply do not have an adequate understanding of the variety and abundance of ocean life in order to understand how our actions affect it. Recently, however, a ten-year Census of Marine Life has attempted to establish a scientific baseline for marine biodiversity. More than 5,000 new species have already been identified as a result, across habitats that range from coral reefs to ocean depths.

Historical investigation

Scientific exploration of the world's oceans began more than 250 years ago, when late 18th-century oceanographic expeditions set out to plumb the depths and investigate the life of the seas. They soon discovered that their crude lead-weighted ropes were rarely long enough to reach the seabed and that the ocean waters teemed with strange creatures new to science, many of which were microscopic and biologically hard to place.

OPPOSITE The surface waters of the world's oceans are home to a wealth of microscopic life without which the rest of the marine ecosystem could not be sustained. This life consists of two very different types of organism – the photosynthesizing phytoplankton that lie at the bottom of the food chain, and the zooplankton that feed upon them and are in turn consumed by larger animals.

Some of Charles Darwin's first investigations as a naturalist during his voyage aboard HMS *Beagle* (1831–36) involved sampling these microscopic inhabitants of the upper layers of the sea – creatures that we now call zooplankton. More than 6,800 such species are now known, and they are widely recognized as vital to the marine food chain and the biodiversity of the oceans.

In the mid-18th century, no more than 300 fish species were known, but throughout the 19th century large numbers of new ones were discovered each decade. Even so, the deep-ocean floor was thought to be a marine 'desert', virtually devoid of life. This is now known to be a complete misconception and remarkable levels of biodiversity have recently been discovered (see page 335). Today, more than 30,000 fish species are known, of which some 16,754 live in the seas, according to the new Marine Census. What's more, an estimated 5,000 more marine fish species still await discovery. But even as they are being discovered, we are learning that some species are in danger.

RIGHT Coral reefs are normally biodiversity hotspots, but in recent decades some of the world's major reef complexes, including Australia's Great Barrier Reef, have suffered 'bleaching' due to the death of the symbiotic micro-organisms that live in the coral tissues and are essential to the survival of the corals. Bleaching can be caused by numerous factors, but particularly by changes in water chemistry and temperature.

In 2006, some 1,173 fish species were placed on the International Union for Conservation of Nature (IUCN) 'Red List' of endangered species, and this number will undoubtedly increase.

What lives in the sea?

Launched in 1996 on a $650 million budget, the decade-long Census of Marine Life involved some 2,700 scientists from more than 80 countries in a wide-ranging study encompassing species from plankton and seabed invertebrates to marine mammals and seabirds. Some simple questions needed to be answered – what lives in the sea, where does it live and how abundant is it? Such is the vastness of the marine realm that despite more than 540 expeditions, 2,500 academic papers and 30 million records of species distribution, this survey is still literally a 'drop in the ocean' – for more than 20 percent of the marine environment is still entirely unrecorded.

But the census at least provides a starting point for future work. The estimated diversity of marine species has been raised from about 230,000 to nearly 250,000 species, with potentially 6,000 more in the process of being identified and described. An average of about 1,650 new species were described each year between 2002 and 2006. Of these, crustaceans and molluscs were by far the most numerous, but new fish were not far behind, much to the surprise of the experts.

'Late 18th-century oceanographic expeditions soon discovered that their crude lead-weighted ropes were rarely long enough to reach the seabed and that the ocean waters teemed with strange creatures new to science.'

One major innovation is the cataloguing of DNA sequences from some 35,000 species in a wide variety of different groups. This so-called 'genetic barcoding' can reveal previous errors in classification – for instance, some previously separated species have proved to be genetically identical, while other apparently identical organisms (especially microbes) have proved to have hidden genetic variation and therefore have to be reclassified as separate species. Microbes are notoriously hard to classify on the basis of appearance alone, and it is quite possible that the marine realm is home to tens of millions, perhaps even hundreds of millions, of species. What's more, most marine life seems to be microbial, comprising some 90 percent by weight of the ocean's total 'biomass'.

Wherever it looked, the census has found life – from the boiling waters of hydrothermal vents to the frozen depths of polar seas. Tracking of migratory animals, using a variety of techniques from sonar and satellite mapping to electronic tagging, has revealed countless new 'highways', meeting, feeding and breeding sites. What's more, this enormous scientific bounty has been made freely accessible to the public – anyone visiting the census website can find information about marine species and their distribution. Hopefully the availability of this information will lead to a revised assessment of our impact on the oceans and better stewardship of their bounty in the future.

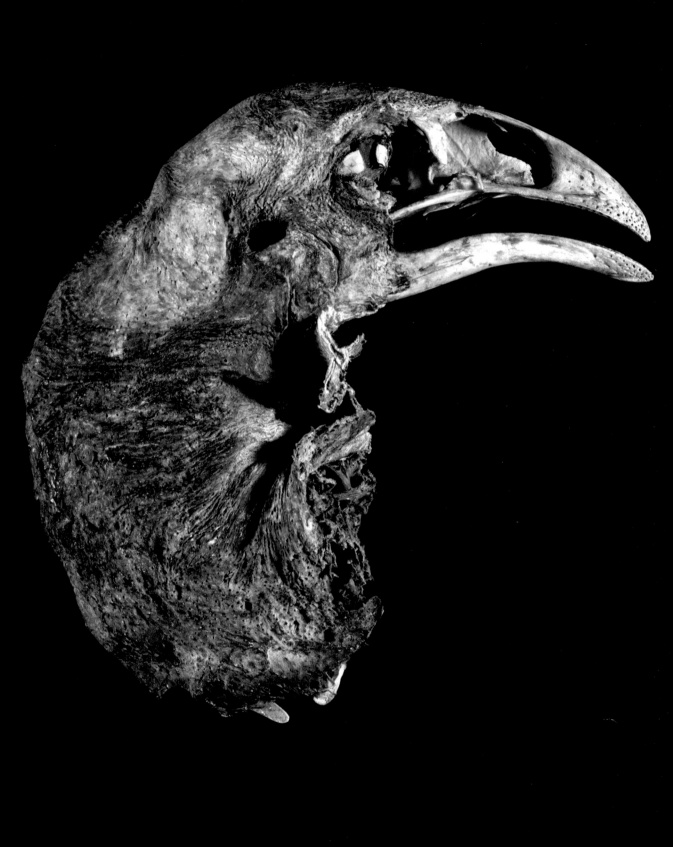

Human overkill

DEFINITION THE CURRENT WAVE OF EXTINCTION CAUSED BY HUMAN
EXPLOITATION OF BOTH INDIVIDUAL SPECIES AND THE WIDER
ENVIRONMENT

DISCOVERY FIRST WIDELY PUBLICIZED THROUGH THE STORY OF THE
DODO, THOUGHT TO HAVE GONE EXTINCT AROUND 1680

KEY BREAKTHROUGH THE CONSERVATION MOVEMENT HAS
DEVELOPED SINCE THE 1960S

IMPORTANCE INDIVIDUAL 'LANDMARK' EXTINCTIONS ARE
EMBLEMATIC OF OUR WIDER EFFECT ON GLOBAL BIODIVERSITY

Since the 16th century, humans have driven well over 780 species to
extinction. Often, we have been quick to blame such disasters on the
shortcomings of our victims, but it's only recently that we have begun to
wake up to our terrible toll on other species.

It has taken a long time, but gradually we humans have become aware that
our activities have made serious impacts upon Earth's wildlife. Historical
records and archaeological investigations both confirm that, whether
knowingly or not, we have driven animals and plants large and small, marine
and terrestrial, to extinction. For the mammoth, the thylacine and the quagga
it is too late, despite belated efforts to conjure them back into existence
from partially preserved DNA. Many others, such as the tiger and the giant
panda, stand on the brink, and only survive as icons of conservation through
intensive human intervention. In normal circumstances, their fragmented
populations are no longer biologically viable, and they too would almost
certainly become extinct.

One of the earliest, and certainly the most iconic, victims of human overkill
was a large flightless pigeon from the island of Mauritius – the dodo. The
last of these harmless birds, which had grown larger in size and abandoned
flight after finding themselves on a predator-free island paradise, was hunted
to extinction by humans in the late 17th century, despite the fact that its flesh
was widely considered unpalatable. Even today, the phrase 'dead as a dodo'
is commonly used to signify something that is not only dead but extinct,
possibly as a result of its own stupidity. The dodo's continued fame is due in
large part to its inclusion in Lewis Carroll's famous book *Alice's Adventures in
Wonderland*, and misleading depiction as an awkward, comical creature.

OPPOSITE The isolated
islands of New
Zealand were home to
11 species of flightless
plant-eating birds
generally known as
moas. The largest,
Dinornis robustus,
grew to around 3.7m
(12ft) tall. Moas had
few natural enemies
until humans arrived
around 1300. Within
about a century,
they were all
extinct, leaving only
occasional mummified
remains behind.

Extinction processes

Of course, it is worth remembering that some 99.999 percent of all life that has ever existed on Earth is now extinct – in general, extinction is just as normal a process as the appearance of new species, and there are a number of reasons why it occurs.

Organisms may be outcompeted and driven to extinction by newly evolved and better-adapted species (and indeed it could be argued that this is exactly what we humans are currently doing). Drastic changes in environmental factors, particularly climate and sea level, can also cause extinctions, particularly of highly specialized species that suddenly find their way of life is no longer viable. More sudden and extreme external events, such as extraterrestrial impacts (see pages 261 and 393) and volcanic eruptions (see pages 109 and 209) can have an even more widespread effect, culling species, genera and entire families more or less at random. The fossil and rock records tell us that such major extinctions, threatening a significant proportion of life on Earth, have happened in the past and will inevitably happen again. Smaller-scale extinctions are far more common and a 'normal' consequence of living on a dynamic planet whose climate, sea level and geography are in a state of continuous change – oceans open and close,

BELOW This huge pile of bison skulls was collected in America during the 1870s, when bison were being hunted on a massive scale – mostly for their hides, but also to clear the Great Plains for cattle and railroads, and to deprive Native American tribes of a mainstay of their livelihood. By 1884 only a few hundred bison were left.

mountain ranges rise and wear away, and there are occasional cataclysms such as volcanic eruptions, earthquakes and tsunami waves.

The evolution of one species into another, meanwhile, results in a 'pseudoextinction'. For instance, our immediate ancestor, *Homo heidelbergensis* (extinct around 400,000 years ago), evolved into at least two new species – *Homo neanderthalensis* (extinct 28,000 years ago) and *Homo sapiens*.

Reasons for the overkill

The unique feature of our current human predicament is that we are the first species that has not only the capacity to cause global extinctions across a wide range of organisms, but also the awareness that, in doing so, we often threaten our own well-being. Yet despite this awareness, we seem unable or perhaps unwilling to do much about it. We have ruthlessly decimated species ranging from marine mammals and fish to exotic birds and many smaller organisms, right down to viruses. Our reasons for doing this have ranged from our insatiable food requirements to demands for material goods and personal ornamentation. Sometimes, the motivation is one of self-preservation, ranging from so-called 'folk medicines' that use the body parts of endangered animals, to the scientifically justified drive to eradicate life-threatening pests and disease vectors.

'Historical records and archaeological investigations both confirm that, whether knowingly or not, we have driven animals and plants large and small, marine and terrestrial, to extinction.'

Historic extinctions

Of course, extinctions often result from more than just direct human action – the demise of the dodo may have been due as much to our introduction of animals such as dogs, cats, pigs and rats into its island habitat, as it was to deliberate human hunting. Other historic extinctions include the last of the zebra-like quaggas, which died in an Amsterdam zoo in 1883, and the thylacine, a wolf-like marsupial that was given protected status by the Tasmanian Government just 59 days before the last survivor died in Hobart Zoo in 1936. The North American bison was almost hunted to extinction in the late 19th century, when vast prairie herds of between 60 and 100 million were reduced to a few hundred animals by 1884. Latterly, they were mostly hunted just for their skins, but also to clear the prairie for cattle and to drive the Native Americans who depended on them into reservations. Fortunately, the bison population has been nurtured back from the grave, and currently stands at around 350,000 animals.

But many other animals are still threatened with extinction. So many whales were killed in the 19th and 20th centuries that commercial hunting was banned in 1986, but others are not so lucky. High-profile recent victims include the Chinese freshwater river dolphin and the Western Black Rhinoceros of Cameroon (both extinct in 2006). Unfortunately, like many smaller and less well-known species, the loss of these animals is largely due to habitat destruction rather than direct persecution.

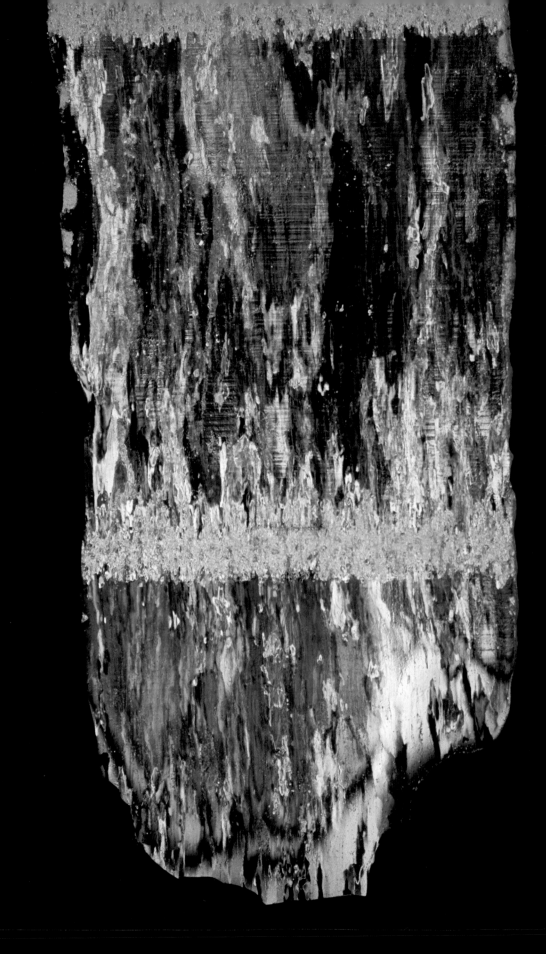

The carbon dioxide record

DEFINITION A RECORD OF THE GREENHOUSE GAS CARBON DIOXIDE EXTRACTED FROM ICE CORES AND OTHER SOURCES, WHICH CAN BE LINKED TO FLUCTUATIONS IN GLOBAL CLIMATE

DISCOVERY THE 'GREENHOUSE' EFFECT OF ATMOSPHERIC CARBON DIOXIDE WAS FIRST ESTABLISHED BY SVANTE ARRHENIUS IN 1896

KEY BREAKTHROUGH ANTARCTIC ICE DRILLED BY THE EPICA PROJECT RECORDS 800,000 YEARS' WORTH OF CARBON DIOXIDE LEVELS

IMPORTANCE THE LINK BETWEEN CARBON DIOXIDE AND CLIMATE HAS MAJOR IMPLICATIONS FOR THE FUTURE GLOBAL ENVIRONMENT

Little bubbles of gas deep inside the Earth's ice caps reveal that carbon dioxide levels in the atmosphere were at their highest during notably warm periods in the past. Today, carbon dioxide levels are rocketing again, due to fossil fuel use, and threaten to cause global warming at unprecedented rates.

Earth's atmosphere is mostly nitrogen (78 percent by volume) and oxygen (21 percent). Carbon dioxide (CO_2) currently comprises just 0.039 percent, or 390 parts per million (ppm). This might seem a trivial amount, but it plays a vital role for life on Earth. Plants depend on CO_2 for photosynthesis, and without them, there would not be enough atmospheric oxygen for us to breathe and many of the food chains that sustain life wouldn't exist.

Heated by sunlight, the Earth's surface emits long-wavelength thermal radiation at infrared wavelengths. Carbon dioxide is one of several so-called 'greenhouse' gases that absorb this radiation, effectively trapping energy that would otherwise escape to space. This greenhouse effect heats the atmosphere to a much greater extent than solar radiation alone would, raising global temperatures by some 30°C (54°F) and making Earth's surface habitable.

The geological record

The existence of a greenhouse effect caused by atmospheric gases was proposed in the 1820s by French mathematician Joseph Fourier (1768–1830), but the insulating effect of carbon dioxide was only established conclusively by Swedish scientist Svante Arrhenius (1859–1927) in 1897. Ever since then, scientists have speculated about the gas's role in regulating climate. Measuring CO_2 levels from the geological past is

OPPOSITE Ice sheets in Earth's polar regions have built up over many thousands of years, compressing stratified annual layers of snow into ice and trapping bubbles of air from the time of deposition within them. Ice cores such as this one, extracted from the glaciers, can be analysed to reveal subtle changes in past atmosphere and climate.

notoriously difficult, but there is indirect evidence that several warming events (around 183, 120 and 55 million years ago) were linked to major injections of carbon into the atmosphere. Sediments linked to the most recent of these increases suggest that between 1,500 and 2,000 billion tonnes of carbon were released first into the ocean and then into the atmosphere.

The most likely source of this carbon was methane hydrates buried within seabed sediments. Warming caused by their release was then superimposed on a pre-existing temperature rise. Warmer ocean waters would also have generated another greenhouse gas – water vapour. The net result was a sudden acceleration in warming that saw average global temperatures soar by around 6°C (11°F), and polar ones by about 20°C (36°F) – the so-called Palaeocene-Eocene Thermal Maximum (PETM – see page 273).

Evidence from ice bubbles

The firmest evidence for a relationship between carbon dioxide levels and atmospheric temperatures in relatively recent times comes from polar ice cores. An international collaboration called the European Project for Ice Coring in Antarctica (Epica) has drilled the longest, deepest ice column extracted so far. The core is 3.2km (2 miles) long and contains air bubbles trapped in the ice as snow became compacted. These bubbles record levels of carbon dioxide and air temperatures going back 800,000 years.

The Epica team have sampled and analysed this 'fossil air', providing a very useful direct measure of changes in greenhouse gases, including CO_2, that can be compared with independent measures of changing temperature over the same time interval. This revealed that as Earth has swung from cold glacials into warmer interglacials during the ongoing Quaternary Ice Age, average CO_2 levels have varied from around 170ppm and about 300ppm.

BELOW This illustration overlays a record of changing atmospheric carbon dioxide levels – measured at Mauna Loa, Hawaii, over the past decade – onto a map of the gas's distribution in the atmosphere. The map is derived from observations by the Atmospheric Infrared Sounder (AIRS) instrument carried aboard NASA's Aqua satellite.

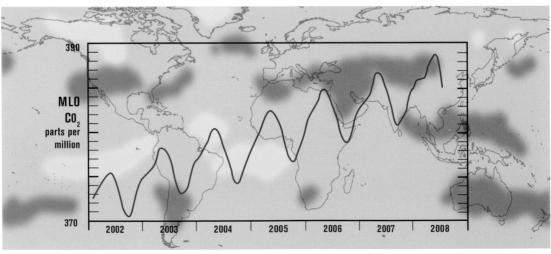

So was the rise in CO_2 the cause of warming, or the upshot? Scientists believe that ancient climate cycles were triggered by a variety of natural mechanisms, including subtle changes in the Earth's axial tilt and orbit that altered the amount of energy received from the Sun (see page 295). Other possible triggers are volcanic activity and changes to ocean currents, which transport a great deal of heat from the tropics to high latitudes.

Once any of these mechanisms initiated warming, a series of positive feedbacks would follow. As atmosphere and ocean warm, CO_2 is more readily released from ocean waters, further warming the air. Increased evaporation from the oceans would lead to more water vapour in the atmosphere, again promoting temperature rise. Additionally, sea ice would decrease as the ocean waters warmed, reflecting less solar energy back into space.

The modern CO_2 story

Today, atmospheric CO_2 levels are rocketing for a different reason. The level has risen by more than a third over the past 200 years since the beginning of the Industrial Revolution, and currently grows by around 1.9ppm each year. In 2010 the level reached 390ppm – higher than at any time in the history recorded by Epica's 800,000-year ice core.

'Epica has drilled the longest, deepest ice column extracted so far. The core is 3.2km (2 miles) long and contains air bubbles that record levels of carbon dioxide and air temperatures going back 800,000 years.'

The most obvious cause of this sharp increase is human activity, which is estimated to have released more than 500 billion tonnes of carbon (equivalent to more than 1,850 billion tonnes of CO_2) into the atmosphere since 1750. Around 65 percent of that total has come from the burning of fossil fuels (wood, coal and more recently oil and gas hydrocarbons), which powered the Industrial Revolution and still supply our energy-thirsty civilization today.

The UN Intergovernmental Panel on Climate Change concludes that the climate warmed by about 0.74°C (1.33°F) during the 20th century, largely as a result of human activity. This is comparatively little considering that the increase in CO_2 is already close to that measured in the PETM of 55 million years ago. Clearly, either some of the figures are wrong, or there are mitigating circumstances at work – one possibility is a greater rate of weathering and erosion during our current ice age (see page 353).

However, large temperature increases could eventually make the poles ice-free, while global sea levels could rise by well over 30m (100ft), flooding some of the most densely populated and cultivated land on Earth, and rendering the world's ports and harbours useless. Scientists predict that there would also be dramatic increases in extreme weather systems such as hurricanes.

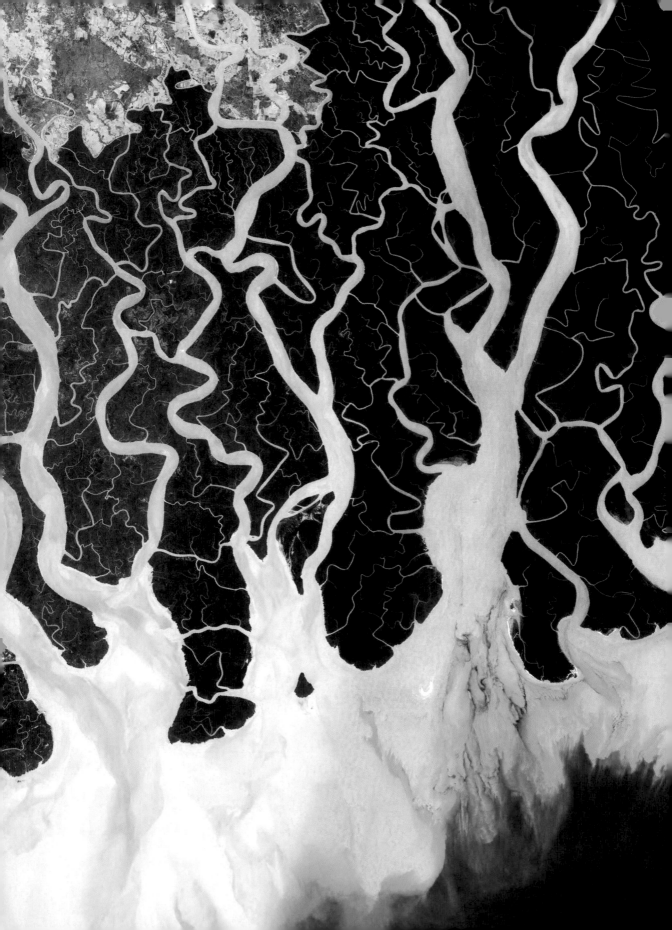

Climate change and mountain erosion

87

DEFINITION A LONG-STANDING THEORY LINKS THE INCREASED EROSION CAUSED BY MOUNTAIN BUILDING WITH REDUCTIONS OF ATMOSPHERIC CARBON DIOXIDE AND COOLING GLOBAL CLIMATES

DISCOVERY THE ABILITY OF CHEMICAL WEATHERING TO ACT AS A 'CARBON SINK' HAS BEEN KNOWN SINCE THE 19TH CENTURY

KEY BREAKTHROUGH RECENT RESEARCH SUGGESTS THAT THE AMOUNT OF RECYCLED MATERIALS IN NEWLY RAISED MOUNTAINS REDUCES THEIR POTENTIAL AS CARBON SINKS

IMPORTANCE THIS DISCOVERY UNDERMINES A FAVOURED EXPLANATION FOR THE ONSET OF THE QUATERNARY ICE AGE

Did the rise of major mountain ranges such as the Alps, Himalayas and Andes, beginning around 40 million years ago, tip our planet into its recent ice ages? And could tectonic changes today be linked to the current phase of global warming?

Chemical weathering is the widespread process that transforms silicate rocks on the Earth's surface into carbonate minerals, through physical erosion coupled with chemical reactions that involve rainwater and carbon dioxide (CO_2) from Earth's atmosphere. Ever since the 18th century, when the processes involved were first discovered, geologists have understood that weathering, along with the burial of organic carbon in seabed sediments, helps to remove carbon dioxide from the atmosphere for extended periods.

This long-term storage mechanism is an important 'carbon sink' – it helps to regulate the levels of CO_2 in the atmosphere and prevents the development of a runaway 'greenhouse effect', thus keeping Earth's environment hospitable for life as we know it today. However, just as too much CO_2 presents a threat to Earth's balanced climate, so does too little – a slight greenhouse effect plays a vital role in keeping our planet warmer than we might otherwise expect, and if the CO_2 levels fall too much, then Earth can cool significantly.

However, since the discovery of plate tectonics in the mid-20th century, it's become clear that the tectonic uplift associated with mountain building can increase rates of rainfall and erosion – both key elements in the weathering process. So if raising a major mountain range can reduce the amount of carbon dioxide in the atmosphere, can it also reduce the greenhouse effect and global temperatures?

OPPOSITE The removal and transport of eroded and weathered rock debris from upland areas downriver to the sea, as seen in this spectacular satellite image of the Bay of Bengal, is far more than just a simple relocation of material. The chemical weathering process also helps remove carbon dioxide from the atmosphere, while the eventual deposition and compression of weathered sediments helps to lock it up in seabed deposits.

ABOVE The formation
and uplift of mountain
chains exposes vast
new surface areas
of rock material
to weathering and
erosion, as seen here
in the Himalayan
foothills. Chemical
weathering removes
carbon dioxide from
the atmosphere and
locks it up in the
formation of new
carbonate minerals.

Testing the hypothesis

The latter part of the Cenozoic Era has seen several important mountain-building events, including the ongoing formation of the Himalayas, Andes and Alps. Over roughly the same period, there has been a long-term cooling trend that saw permanent glaciers begin to form in Antarctica around 15 million years ago and plunged the entire planet into the Quaternary Ice Age around 2.75 million years ago.

Were the two processes linked? In support of the hypothesis, some scientists have pointed to evidence of a fourfold increase in the volumes of sediment supplied to the world's oceans over the last 5 million years, indicating a massively increased rate of erosion that should, in theory, have involved chemical weathering. Weighed against this, however, is evidence that atmospheric CO_2 levels do not seem to have changed much over the last 20 million years, dropping only slightly even at the peak of the ice age.

The processes that transport sedimentary particles from highlands through rivers and ultimately to sea floors appear to show a strong link between physical erosion and chemical weathering. Erosion and fragmentation of rock into ever-smaller pieces through the actions of atmosphere, water, heat and gravity create smaller fragments with greater surface areas that are more susceptible to chemical reactions. The eventual deposition of these fragments into sediments traps and buries both the carbonate minerals, and carbon-rich organic particles caught up with them.

Carbon-neutral mountains?

Recently, however, the idea that increased mountain erosion leads to climate cooling has been challenged by a new study of eroded sediments supplied to the oceans during Earth's recent cooling. Instead of the previously estimated fourfold increase in erosion over the last 5 million years, this study, carried out by scientists from the German Research Centre for Geosciences at Potsdam, suggests that the *net* rate of erosion and sedimentation has been much more constant because, while mountain uplift has indeed increased erosion, much of that erosion has simply reworked previously deposited sediments.

'If raising a major mountain range can reduce the amount of carbon dioxide in the atmosphere, can it also reduce the greenhouse effect and global temperatures?'

But what about the rate of chemical weathering and carbon removal? The Potsdam researchers came up with a way of independently testing this by measuring ratios of beryllium isotopes from deep-ocean sediment cores. Cosmic rays (high-energy particles from space entering Earth's atmosphere) produce small but measurable quantities of 'heavy' beryllium (^{10}Be) that rain down on Earth's surface at a steady rate and mix with lighter ^{9}Be isotopes released by erosion and weathering of rock material. Together, the isotopes are transported and eventually buried in ocean sediment. Numerous measures of the isotope 'mix' from the Pacific, Atlantic and Arctic oceans show that they have been more or less constant for the last 10 million years, so it seems that the rate of weathering (and of carbon removal) must also have been more or less stable over this period.

If these measurements are borne out, then the uplift of the recent mountain ranges cannot have played any significant role in the cooling of late Cenozoic climates, but despite this, there is no doubt that the uplift of major mountain ranges does lead to increased erosion. At present three great rivers, the Ganges, Brahmaputra and Amazon, carry more than 20 percent of the total sediment load delivered to the world's oceans, all of which is essentially derived from the Himalayas and Andes. So, how could the sediment supply have remained constant in the face of major uplift events?

Possible explanations include the time limitations of the beryllium record, which does not extend back beyond 10 million years ago. There is good geological evidence that Himalayan uplift began at least 35 million years ago and intensified around 20 million years ago. The monsoon climate system (see page 243) began at this time, resulting in hugely increased rainfall that accelerated erosion on the southern flank of the Himalayas. All of this happened some 10 million years before the beryllium record begins, so we may just be missing an important older record of higher erosion rates.

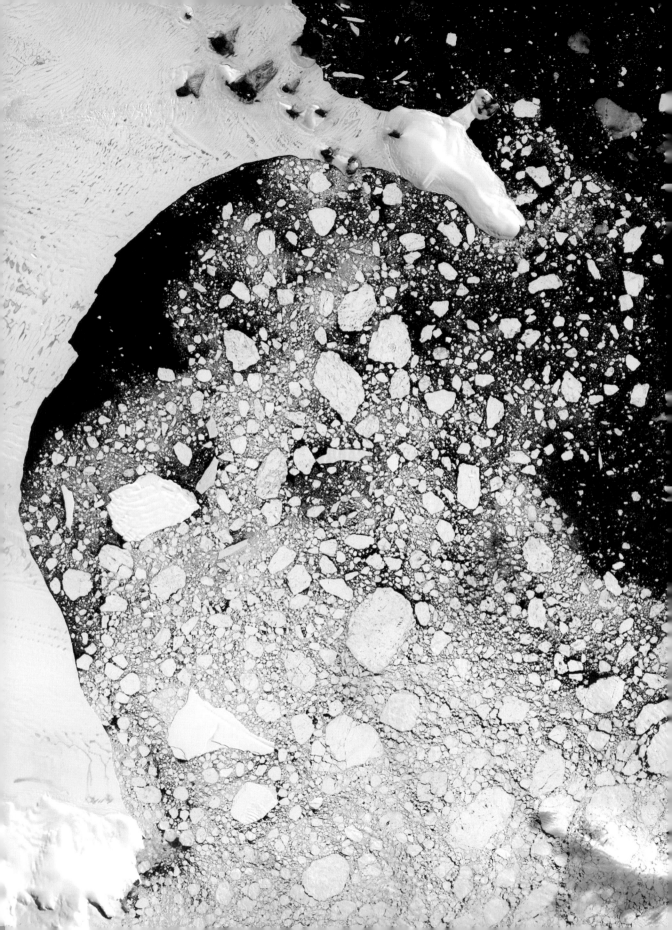

Changing sea levels

88

DEFINITION SEA-LEVEL CHANGES ON A SCALE OF METRES, LINKED TO
CLIMATE CHANGE AND GLOBAL WARMING

DISCOVERY CURRENT INCREASES AMOUNT TO AROUND 3MM (0.1IN)
PER YEAR – ROUGHLY TWICE THAT SEEN IN THE LAST CENTURY

KEY BREAKTHROUGH ROBERT KOPP'S TEAM SHOWED THAT SEA LEVELS
DURING THE LAST INTERGLACIAL WERE SIGNIFICANTLY HIGHER
THAN PREVIOUS ESTIMATES

IMPORTANCE RISING SEA LEVELS HAVE THE POTENTIAL TO
DISPLACE TENS OF MILLIONS OF PEOPLE, AS WELL AS DESTROYING
DELICATELY BALANCED ENVIRONMENTS

There is no question that sea levels are currently on the rise. But how far could they go if we continue to guzzle fossil fuels? Looking back to a slightly warmer period on Earth around 125,000 years ago reveals that calamitous rises of 6m (20ft) or more are possible.

In 2007, the Intergovernmental Panel on Climate Change (IPCC) concluded that global warming from human activities such as burning fossil fuels will cause a rise in sea levels of 18–58cm (7–23in) by the year 2100. However, rises so far measured suggest this prediction could be an overly conservative figure, and it might be better to plan for rises of about 1m (40in) by the end of the century. But what about the longer-term future? A new assessment, based on events during the last interglacial 125,000 years ago, suggests that sea levels could rise much higher over the next few centuries.

Rises on this scale are potentially catastrophic in both human and economic terms. Other factors such as tidal cycles and weather patterns will conspire to trigger rare events such as storm surges coinciding with high spring tides, producing significantly higher flood risks. Many of the world's most populous cities have coastal or riverside sites, and the most productive agricultural land, especially in the tropics, is also relatively low-lying. An estimated 145 million people live no more than 1m (40in) above present sea levels, and so risk losing their livelihoods and possibly their lives. Many more will suffer from economic disruption in the face of the rising seas.

Current sea-level rise
Global sea-level rise is currently around 3mm (0.12 in) per year, up from 1mm (0.04in) to 2mm (0.08in) per year in the last century, although there

OPPOSITE The current break-up of polar ice sheets and glaciers releases water back to the oceans as part of the natural ocean–atmosphere hydrological cycle, as shown spectacularly by the fragmentation of Antarctica's Larsen A and B ice shelves in the late 1990s and early 2000s. The wider concern is that global warming may alter the balance between accumulation and melting of ice and lead to rising sea levels.

are local and regional variations. The main contributing factors are global warming of the atmosphere and ocean, leading to a physical expansion of the ocean waters, and melting of glaciers and ice sheets, mostly in Greenland and Antarctica.

While the rise expected from thermal expansion is relatively easy to predict, the contributions from melting ice are not. Adding water from melting land ice to the ocean is not like pouring water into a bathtub – melting ice has domino effects for land elevations, for instance, as well as the large-scale distribution of ice, water and sediment. Relatively stiff crustal rocks take time to unload and rebound following the melting of glacial ice, whose weight physically depresses the crust (see page 321).

Lessons from the past

In an attempt to estimate the sea-level rise from human-generated warming, scientists have looked to the last interglacial period, 125,000 years ago, when global average temperatures were around 1.5–2°C (2.7–3.6°F) warmer than during the pre-industrial era. Until recently, they thought that the sea level during this interglacial was about 4–6m (13–20ft) higher than today. However, in 2009 a team led by Robert Kopp from Princeton University in New Jersey reported a more complex analysis. Their comprehensive study suggests that sea levels were at least 6.6m (22ft) higher, and perhaps as much as 9.4m (31ft).

'Taking a conservative and pragmatic sea-level rise of 0.5–2m (20–80in), there is a real risk that up to 2.4 percent of the global population, some 187 million people, will be displaced by such a rise in sea level.'

The calculations show that melting ice from Greenland and Antarctica each contributed at least 2.5m (8ft) to the rise, while thermal expansion of water and meltwater from mountain glaciers contributed up to 1m (40in), giving a total of *at least* 6m (20ft). If this last interglacial period provides a good comparison for the centuries to come, then global atmospheric warming of 2°C (3.6°F) above pre-industrial levels, which seems likely to happen, could result in sea-level rises of several metres – much more than scientists have typically estimated – and we should expect to see substantial melting of both the Greenland and Antarctic ice sheets. However, it is not certain that the last interglacial period is a good analogue for our fossil fuel-burning present – the Earth of 125,000 years ago was not warmed by an increase in atmospheric greenhouse gases, but by slightly greater summer heating of the northern hemisphere due to periodic variations in Earth's orbit (so-called Milankovitch cycles – see page 295).

Into the future

In 2011, an international group of scientists published their assessment of sea-level rise and its possible impacts in the 21st century, given a 4°C (7°F) rise in global climates. They were at pains to point out the uncertainties inherent in such predictions, particularly because of the problems associated with the response of the West Antarctic and Greenland ice sheets to global

warming. But taking a conservative and pragmatic rise of 0.5–2m (20–80in), there is a real risk that up to 2.4 percent of the global population, some 187 million people, could be driven from their lands.

However, scientists argue that it is wrong to assume that all these people need necessarily become displaced. Some particularly badly exposed regions such as coastal Africa, densely populated Asian river deltas and small low-lying islands such as the Maldives (the lowest-lying nation on Earth, with a high-point just 2.3m/7.6 ft above sea level) will indeed be severely challenged and are most likely to be abandoned, but the possibilities for adapting to rising sea levels cannot yet be fully appraised.

In general, though, experts agree that the response will require a mixture of adaptations, coupled with an effort to stabilize global warming in order to do the same for sea levels. Upgrading protection measures such as tidal barrages and levees could cost around 0.02 percent of the global Gross Domestic Product according to some estimates, but some nations will face a disproportionately high burden. The cost of transition to a more environmentally friendly energy regime will be far higher, but is likely to be more evenly spread. But even with such measures in place, it is also inevitable that sea levels will continue to rise for centuries.

ABOVE Many tropical islands, including the Maldives of the Indian Ocean, are essentially coral atolls with elevations of at most a few metres. Global warming and rising sea levels are likely to have an immediate impact on the survival of these islands and their populations.

Habitats on the move

DEFINITION A SHIFT IN CLIMATIC CONDITIONS ACROSS THE SURFACE OF THE EARTH, FROM WHICH SOME SPECIES, AT LEAST, BENEFIT

DISCOVERY US GEOLOGICAL SURVEY RESEARCH CONFIRMS A NORTHWARD SHIFT IN AVERAGE TEMPERATURES IN CALIFORNIA

KEY BREAKTHROUGH RESEARCH INTO A POPULATION BOOM AMONG MARMOTS HAS SHOWN THAT THEY ARE THRIVING DUE TO MILDER AND SHORTER WINTERS

IMPORTANCE UNEXPECTED SHIFTS IN ANIMAL AND PLANT POPULATIONS MUST BE ACCOUNTED FOR IN FUTURE PLANS TO MANAGE ENDANGERED SPECIES

Although we hear a great deal of bad news about the damage global warming is doing to animal and plant life, there is some good news at least – a number of creatures, including the yellow-bellied marmot, *Marmota flaviventris*, are thriving in the changing conditions.

All plants and animals are affected by climate change, whether from season to season, or on a more long-term basis. Such changes, especially in temperature, control the timing of events and processes such as plant budding, flowering and fruiting, and animal reproduction, migration and hibernation. Climate change can even affect an organism's body form and size along with the structure, distribution and dynamics of its population.

However, while most organisms have life cycles that take account of normal seasonal climate variations, and can often adapt to relatively slow trends, rapid global climate change threatens to have a much greater impact. Such threats, both to wild species and domesticated ones, are more difficult to predict thanks to a multitude of interacting environmental and biological factors. What's more, changes in population sizes can be influenced by many factors other than climate, and it can be difficult to distinguish between the various possible causes.

Shifting climates

A 2010 survey of some 500 protected reserves in the San Francisco Bay area of California predicted that even with moderate climate warming, all but a handful will have significantly different summer temperatures in the future. The study, carried out by scientists from the US Geological Survey, calculated that for any given average summer temperature, the

OPPOSITE Most organisms have a fixed tolerance for certain environmental conditions. When those conditions change, animals may be able to move immediately to a more suitable habitat, but plants such as these towering Californian sequoias can only do this through dispersal over successive generations. In the face of rapid climate change, human intervention may be needed to conserve both the plants and the ecosystems that they support.

locations experiencing that temperature will shift north by about 5km (3 miles) per year and will therefore move across hundreds of kilometres within the next few decades.

As a result, most nature reserves will have summer temperatures unsuitable for any of their present plant species. In order to survive, the plants and the animals that depend on them will have to move north, either by their own means of propagation and dispersal, or with human assistance. This kind of dynamic change within the reserves will require whole new management strategies – instead of being treated as living herbariums or museums of certain plant and animal species, they will have to be seen as temporary 'staging posts' for a biota on the move.

The first step in developing this new strategy will be to gather more detailed information about the plant and animal species affected, and a recent in-depth survey, using data gathered in Colorado's Upper East River Valley since 1976, demonstrates the kind of information needed to assess the impact of climate change on the life history and population of a single mammal.

Life on the edge

The yellow-bellied marmot (*Marmota flaviventris*) lives on the front line of climate change, in a mountainous subalpine habitat where it survives cold winters by hibernating. Species such as this, living in relatively extreme environments, are the first to be affected by climate change, since their habitats are likely to change first. By studying individual marmots throughout their lives, an international team led by Arpat Ozgul of London's Imperial College has revealed how climate change is indeed shifting their patterns of reproduction and hibernation.

'Instead of being treated as herbariums or museums of certain plant and animal species, reserves will have to be seen as temporary "staging posts" for a biota on the move.'

As winters have warmed over the last decade and more, the summers have effectively become longer. Consequently, the feeding and growing season has been extended, so the marmots are measurably heavier by the time they begin their delayed hibernation. Because they are better fed by the time they hibernate, more animals survive winter hibernation and are in better condition when they emerge in the spring. This increases their chances of reproduction and has led to a threefold rise in numbers from 2000–2010. In this instance, the direct connection between climate change and population increase is clear. Over such a short period, the marmots are not changing genetically – instead, it is the changing environment around them that has led to increases in their body size and population.

Hibernation problems

But there are still interesting questions around the marmots increasing size. Hibernation requires that an animal's body temperature falls as the ambient air temperature drops. The energy requirements of a body entering

hibernation also diminish and the animal survives by drawing on fat stored during its growing and feeding season. Herbivores like marmots make use of particular plant compounds, such as polyunsaturated fatty acids, to see their chilled bodies through hibernation.

But the plants on which marmots feed have also been influenced by climate change, so that, for instance, far fewer of their favoured bluebells (*Mertensia ciliata*) are now flowering. It is just possible that dietary changes imposed on the marmots by climate change may actually be responsible for their weight gain. A better understanding of physiological mechanisms of hibernation, such as dietary and energy requirements and temperature control, will be needed to get a better understanding of the marmots' population explosion.

This study is an excellent example of how evolutionary processes such as environmental change can influence a particular aspect of an organism's biology without genetic change. However, it will be necessary to combine such physiological studies with genetic research to get a fuller picture of the potential impact of rapid climate change upon population survival and biodiversity – not just for the marmot, but for many species.

ABOVE So far, the yellow-bellied marmot (*Marmota flaviventris*) is one of the few creatures that definitely seems to be benefiting from climate change. A recent survey in the Upper East River Valley of Colorado, USA, shows that they are surviving winter hibernation in greater numbers and emerging in better condition to breed, leading to an increase in their population.

90 | Danger from the soil?

DEFINITION THE POTENTIAL THREAT FROM VAST AMOUNTS OF
GREENHOUSE GASES RELEASED BY THAWING ARCTIC PERMAFROST

DISCOVERY THE FULL EXTENT OF PERMAFROST AND ITS CARBON
CONTENT WAS MAPPED FOR THE FIRST TIME IN 2008

KEY BREAKTHROUGH TESTS ON SMALL AREAS OF ALASKAN
PERMAFROST CONFIRMED THEIR POTENTIAL TO RELEASE VASTLY
MORE CARBON WITH JUST A SMALL INCREASE IN TEMPERATURE

IMPORTANCE BURIED CARBON IN THE PERMAFROST HAS THE
POTENTIAL TO SIGNIFICANTLY AMPLIFY GLOBAL WARMING RISKS

Melting glaciers and ice sheets that calve country-sized icebergs into the ocean often grab the headlines in media coverage of global warming. But the thawing of permafrost soils across the Arctic could play a far more devastating role in climate change.

From Greenland through Arctic Canada and Siberia to Scandinavia, Earth's northern continents are ringed by vast swathes of tundra landscape where the soil is permanently frozen except for an uppermost layer that can briefly thaw in summer. This 'permafrost' comprises about 50 percent of all Canadian soils, 60 percent of Russian soils and an enormous 90 percent of Alaskan soils. In total, it contains an estimated 500 billion tonnes of organic carbon – more than twice the total mass of carbon in the atmosphere.

Today, global warming is threatening to thaw much of the permafrost, which could then release its trapped greenhouse gases to accelerate global warming in an effect that spirals out of control. An experiment to deliberately warm Alaskan permafrost has shown that it does indeed release excess carbon dioxide, so this is a genuine threat we should be concerned about. If thawed permafrost is kept wet by rain, it might develop into peatlands that keep the carbon stores 'locked up'. But if the permafrost thaws and stays dry, then the release of just a small fraction of its carbon as the organic matter within it decomposes could drastically amplify the rate of global warming.

Unique habitats

The permafrost regions are remarkable environments that are among the few untouched wilderness areas left on Earth. Even aside from the extreme cold, people have found them difficult to occupy – construction of roads,

OPPOSITE The permafrost and tundra landscapes of polar regions are one of the last relatively unspoilt environments on Earth, and their peaty soils lock up a significant volume of organic carbon. As global warming threatens to melt the permafrost, there is a risk that more of this carbon will be released into the atmosphere, accelerating the warming process.

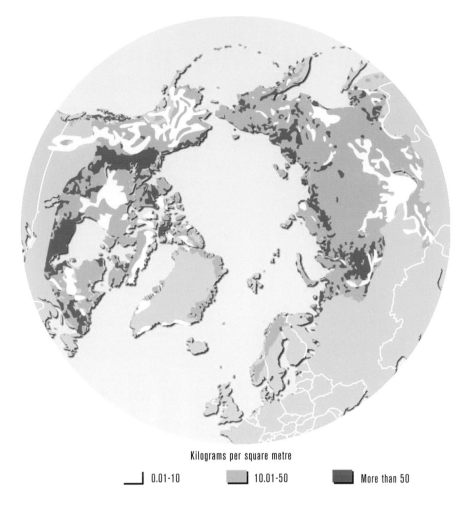

Kilograms per square metre

0.01-10   10.01-50   More than 50

buildings and pipelines is hindered by a land surface that alters drastically from season to season. Large structures tend to sink during the summer thaw unless they are supported on deep piles that reach the permanently frozen levels, where temperatures remains around a constant –5°C (23°F).

Where seasonal temperatures rise above freezing, the surface of the permafrost may thaw sufficiently to form an 'active layer' 0.5–4m (1.6–13ft) deep, in which hardy plants such as black spruce can survive. Ice is a key component of permafrost, and its deposits form a significant proportion of Earth's fresh water. Annual thawing and refreezing of the active layer forms structures such as ice wedges several metres deep, which interconnect over vast regions in polygonal patterns, while ice-covered mounds called 'pingos' grow several metres high.

When the ice thaws, water-filled cracks and ponds in the tundra generate lethal traps for animals tempted by the lush plant growth on their margins. Permafrost at high latitudes has preserved the remains of many ice age plant eaters that foraged there, such as the mammoth and woolly rhinoceros. However, until recently, even the full global extent of the permafrost was largely unrecognized. It was only in 2008 that the results of a three-year international initiative were published in the form of a comprehensive map of permafrost and its associated soils across the entire Arctic region above latitude 50 degrees North.

As atmospheric temperatures rise, the summer thaw of the uppermost permafrost lasts longer, penetrates deeper and extends over a greater region. As it does so, an ever-larger mass of organic matter in the soil starts to decompose and release increasing volumes of greenhouse gases, including methane, carbon dioxide and nitrogen dioxide, into the atmosphere.

Just south of the permafrost at high northern latitudes, vast peatlands also store huge amounts of carbon. If these were to dry out, there is the risk of increased decomposition of organic material, as well as wildfires that could release even more carbon dioxide. Limited loss of a relatively small area of permafrost and its associated peatlands could lead to global warming that causes more permafrost and peatlands to disappear, creating a snowball effect.

'In total, the permafrost contains an estimated 500 billion tonnes of organic carbon – more than twice the total mass of carbon in the atmosphere.'

Testing the mechanism

However, another effect might counteract this. Increased growth of photosynthesizing plants during warmer northern summers would increase CO_2 uptake, offsetting the release of the gas. So what would be the overall outcome? To test this, a team led by Susan Natali from the University of Florida in Gainsville set out to artificially simulate warmer winters and summers on a series of experimental plots in Alaska. The team planted fences around the plots to retain snow cover, creating an insulating effect that increased the temperature of the buried soil by about 1.5°C (2.7°F) during the winter of 2008–09. They later removed the excess snow and set up open-air chambers to increase the summer soil temperature.

Natali's team discovered that, even when a 20 percent increase in CO_2 uptake due to enhanced summer growth was taken into account, the warmer winter soil temperatures led to a doubling of emissions over the course of a year – conclusive proof that warming tundra could well have a dramatic positive feedback effect on global warming.

Estimates from computer models for the global average temperature increase during the 21st century typically range from about 1.8°C (3.2°F) to 4.0°C (7.2°F). However, these models do not factor in the possible feedback from thawing permafrost, which could eventually push temperatures up by more than this. If average temperatures rise by 7°C (12.6°F), many regions of the Earth could start to become uninhabitable because people would suffer fatal heat stress. Needless to say, with such potentially catastrophic consequences, scientists are closely monitoring the permafrost to see if they can detect the first signs of any such large-scale thawing.

91

Freshwater challenges

DEFINITION A RANGE OF THREATS CAUSED BY HUMAN INTERVENTION IN THE PLANET'S SUPPLIES OF FRESH WATER

DISCOVERY THE FIRST GLOBAL SURVEY OF FRESHWATER THREATS WAS COMPLETED IN 2010

KEY BREAKTHROUGH THE SURVEY REVEALS THAT WATER SUPPLIES TO 80 PERCENT OF THE GLOBAL POPULATION ARE THREATENED, AND BIODIVERSITY IS AT RISK IN 65 PERCENT OF FRESHWATER HABITATS

IMPORTANCE THREATS TO WATER SUPPLIES ENDANGER HUGE SWATHES OF THE HUMAN POPULATION, AS WELL AS PUTTING THE WIDER ENVIRONMENT AT RISK

The conclusions of the first global survey of the Earth's fresh waters are bleak. The water supplies of nearly 80 percent of the world's population are threatened by problems like pollution or overconsumption, which also jeopardize the wildlife and biodiversity of most of the rivers.

Salty oceans cover about two-thirds of our blue planet Earth. Warm air evaporates moisture from the sea surface to form clouds that rain fresh life-sustaining water down onto the land, creating small streams that feed lakes or join into branching river networks, including the vast Amazon and Nile systems. The plants and animals that live in natural fresh waters, such as Africa's lake fish, constitute vital food supplies in many parts of the world.

People have always settled at sites with plentiful water, but often populations burgeoned until supplies became overstretched and contaminated. Over the centuries, humans have developed a range of techniques to overcome problems of water supply and waste disposal, from simple wells to aqueducts, dams, reservoirs, recycling and sewage works. Today, this enables populations of several million people to live together in bustling cities.

To stay alive, people need to drink about 2 litres (3.5 pints) of water every day. In the developed world, each person uses more than 50 litres (88 pints) a day on average for drinking and hygiene. But in the rest of the world, less than 10 litres (17.6 pints) are available per person per day.

Degradation of fresh waters

Many countries have constructed dams and other infrastructure, redirecting water to supply their freshwater needs. But this has come at a severe cost

OPPOSITE A spectacular view of the Amazon Basin from the International Space Station shows the extent of fresh water covering this famously abundant region. But a new global survey suggests that fresh waters in the Amazon and around the world are under threat.

to the environment. Dams divert water needed for healthy river systems downstream, preventing the flow of nutrients and impeding wildlife migration. Agricultural practices are another major problem. Farmers often apply nitrogen fertilizers to crops, because nitrogen is an essential nutrient for healthy plant growth. But in excess, as run-off from farmland, nitrogen quickly becomes a pollutant in rivers, causing a decrease in oxygen levels in the water that has severe effects on fish and other animals.

The problems in river systems and habitats vary enormously even within individual countries. For instance, the Amazon suffers its greatest pollution near its Peruvian source as a result of human activity, but most of this pollution is naturally removed as the water flows downstream through the Brazilian tropical rainforest. However, slash-and-burn deforestation for agriculture and livestock grazing is introducing new threats downstream and could pose a serious threat to water supplies in future.

The problem is already severe on the Nile, which supports the livelihoods of more than 180 million people in Africa, many of whom are already on the brink of poverty and starvation. Human activities are seriously degrading the river's headwaters at Lake Victoria, and the problem is exacerbated by mid-stream settlements such as Khartoum in Sudan, as well as the dense human populations of Cairo and the Nile Delta further downstream.

A global survey

Until recently, assessments of water resources around the world relied on patchy data that seriously hampered efforts to organize their management and protection. But in 2010, scientists reported the results of the first worldwide synthesis of detailed information about water resources. Charles Vörösmarty of the City University of New York and Peter McIntyre of the University of Michigan led a team that created a sophisticated computer model of the Earth's river networks and incorporated available information about 23 'stressors' that affect water quality, including regional human and livestock density, pollution, overfishing and the locations of dams.

'The results show that multiple environmental stressors, such as agricultural run-off, pollution and invasive species, threaten the rivers that serve nearly 80 percent of the world's population.'

The results show that multiple environmental stressors, such as agricultural run-off, pollution and invasive species, threaten the rivers that serve nearly 80 percent of the world's population – more than 4.8 billion people. What's more, the model suggests that biodiversity is also under considerable threat in 65 percent of habitats associated with continental river discharge.

Not surprisingly, the report revealed that rivers in regions of intensive agriculture and high population densities are highly threatened. This includes much of the United States and virtually all of Europe (excluding Scandinavia and

northern Russia), as well as large portions of central Asia, the Middle East, the Indian subcontinent and eastern China. Smaller areas of high threat appear in various countries including Nigeria, South Africa, Cuba, Mexico, Korea and Japan.

Water scarcity

The report also highlighted regions where water scarcity accentuates the problems, including the desert-belt transition zones in Argentina, the Sahel in Africa, Australia and central Asia. And while the quality of fresh water was excellent in some very cold or unsettled warm regions – including northern Siberia, Canada and Alaska as well as Amazonia and northern Australia – a mere 0.16 percent of the Earth's land surface had low scores for all of the 23 stressors that Vörösmarty and McIntyre considered in their study.

Global climate change and population growth are bound to make things worse. The human population is expected to reach about 8 or 9 billion by the year 2050, and scientists predict that global warming will increase extreme weather events including floods. Alongside increasing demand for drinking water, this will increase pressure on countries to redirect the flow of their rivers, threatening the global freshwater supply even more.

ABOVE Freshwaters and the plants and animals (including humans) that depend upon them are particularly vulnerable to pollution from human activity. In October 2010, a toxic red sludge from an aluminium works in Hungary flowed into a local river and several kilometers downstream. Heavy metals in the sludge may contaminate local groundwater for decades.

92

Does coal have a future?

DEFINITION COAL IS A VITAL FOSSIL FUEL FOR GLOBAL INDUSTRY,
BUT ALSO A MAJOR POLLUTANT AND A FAST-DWINDLING RESOURCE

DISCOVERY M. KING HUBBERT DEVELOPED AN ACCURATE 'BELL
CURVE' TECHNIQUE FOR ESTIMATING PEAK PRODUCTION OF FOSSIL
FUELS IN THE 1950S

KEY BREAKTHROUGH MODIFIED VERSIONS OF HUBBERT'S TECHNIQUE
PREDICT THAT COAL PRODUCTION WILL PEAK IN THE NEXT DECADE

IMPORTANCE THE LIMITED FUTURE FOR COAL REINFORCES THE
URGENT NEED TO DEVELOP MORE SUSTAINABLE ENERGY SUPPLIES

The questions of how much coal is left, how expensive it will become and whether we should continue to exploit it, may not seem like pressing issues at the moment. But in fact they are vital to the whole issue of future energy supplies amid ongoing global economic and climate change.

Over the last 300 years or so, ever since the beginnings of the European Industrial Revolution, human industry and domestic use have burned up ever-increasing amounts of coal – the 'fossil fuel' laid down in huge amounts by the sedimentation and compression of ancient forests (see page 201). Since the 1950s, increased use of other fossil fuels, in the form of oil and gas, has triggered a sharp decline in production and use of coal in the industrialized Western world, but global coal production and consumption is still increasing, thanks largely to the insatiable demand from a rapidly developing China.

Although it is a daunting task, scientists and economists have made various attempts to estimate how much coal is left within economically accessible deposits, and how many years it may last given present and future levels of consumption. The results have generally been reassuring, agreeing on something like 2,500 billion tonnes or 150 years' worth of global reserves. By way of comparison, however, a 19th-century estimate for coal reserves within the British Isles alone put the total at around 200 billion tonnes, then thought to be enough to last the population and industry for 900 years.

Bell curve estimates

However, there is another independent method for assessing coal resources, analysing historical coal production data using techniques pioneered by US

OPPOSITE Humans have exploited fossil hydrocarbons, and especially coal, for centuries, but the Industrial Revolution saw an exponential rise in consumption. Estimates of hydrocarbon reserves, production and demand suggest that we have already consumed close to half the global reserve of accessible fossil fuels.

Geological Survey geophysicist M. King Hubbert (1903–89). Published in 1956, Hubbert's 'distorted bell curve' model of US oil production predicted a peak in the early 1970s followed by a gradual decline, and has been borne out by subsequent figures. Applied to global oil reserves, the same technique predicts that world production should peak around 2010–15 – and indeed global oil production has not significantly increased since 2005.

Hubbert's method has since been applied to coal production in the United Kingdom (the first nation to undergo major industrialization), the US and China. Both UK and US data show peak productions during the First World War followed by declines. The pattern was briefly disrupted by outside political events, with another smaller peak from the Second World War into the 1950s as these countries were forced back to reliance on domestic fuel sources, but it resumed with a steep decline thereafter. The estimated cumulative total of UK production amounts to around 28 billion tonnes – nowhere near the 19th-century geological estimate of around 200 billion tonnes, even though it was ultimately the increasing cost of extraction and competition from imports that saw the country's coal industry truncated. In this case, the cheaper imported coal is coming from relatively unexploited coalfields in Russia, China and Australia, and still from the United States.

'In reality, there are significant global coal reserves, but major questions over the quality of what remains and the rising costs of extraction in economic, human and environmental terms.'

An international commodity

The largest global coal exporters are currently Australia, Indonesia and South Africa. Russia also has major undeveloped reserves in Siberia that could be exported in the future. The United States still has the largest global reserves overall, though estimates of their useful lifetime have been cut from 400 years in 1975 to 240 years today. The United States also has the advantage that a considerable portion of its coal can be strip mined at the surface, though new environmental protection legislation may restrict this practice.

However, when modified versions of Hubbert's analytical methods are applied to the historical output from the world's major coal producers they give a global production total of 660 billion tonnes, about a quarter of the geologically based estimates. Estimates of peak production come out between 2010 and 2020, with a 30-year plateau followed by a steep decline.

What's more, global demand for coal is still increasing, at a rate of 3.8 percent per year since 2000. China is the biggest consumer, currently consuming about 3 billion tonnes a year, 80 percent of which is used for electricity generation. It also has the biggest production, at around 40 percent of the global annual total. In the 1990s, China revised its estimated reserves of economically viable coal downwards to 187 billion tonnes, one-sixth of that claimed in the late 1980s. A 2009 report predicted China's coal demand rising by between 700 million and 1 billion tonnes per year by 2020, which would reduce its

reserves to around 33 years. Even before the peak of production, the quality of the coal will be reduced, and since most of China's coal is in deep mines, the technical challenges and cost of extraction will only increase.

In reality, there are significant global coal reserves, but major questions over the quality of what remains and the rising costs of extraction in economic, human and environmental terms. Even with the most modern technology, coal mining is a dirty and dangerous business – as we are frequently reminded by accidents around the world. In China alone, there were 188 accidents, each with a death toll of more than ten, between January 2001 and October 2004. In November 2009, an accident at Heilongjiang killed at least 104 people. Even in the US, annual deaths from mining still averaged 93 in the 1990s. On 5 April 2010, 29 miners were killed by an underground explosion at Upper Big Branch Mine in West Virginia and on 29 November of the same year another 29 died in New Zealand.

Cheap and clean coal?

What is more, all these models fail to take environmental concerns into account. Today it's generally recognized that the burning of coal has made a major contribution to the recent increase in atmospheric carbon dioxide, and that this is leading to global climate change. Engineers and politicians have assumed that new technology will allow the development of 'clean coal' power, but there are considerable problems with this assumption – how much coal will be usable and will its price be economic?

ABOVE For a variety of geopolitical reasons, continuing supplies of coal are more reliable than those of oil and to a lesser extent natural gas. However, the use of coal for power generation is a major source of carbon dioxide emissions, with black coal emitting twice as much carbon dioxide as natural gas. Although the technology for capture and storage of these emissions is available, it is not yet economically viable.

93

Atmospheric pollution

DEFINITION HARMFUL MATERIALS, IN THE FORM OF GASES OR
PARTICLES, CONTAMINATING EARTH'S ATMOSPHERE

DISCOVERY THE INFLUENCE OF 'ACID RAIN' LINKED TO CERTAIN
CHEMICAL POLLUTANTS WAS IDENTIFIED AS EARLY AS THE 1850S

KEY BREAKTHROUGH DISCOVERY OF THE ANTARCTIC 'OZONE HOLE'
IN THE 1970S TRIGGERED A MAJOR SHIFT IN UNDERSTANDING OF
THE INFLUENCE WE CAN HAVE ON THE ATMOSPHERE

IMPORTANCE THE ATMOSPHERE IS A COMPLEX SYSTEM THAT WE
STILL UNDERSTAND ONLY POORLY, AND POLLUTE AT OUR PERIL

Since the Industrial Revolution, the quantity of man-made pollutants in the
atmosphere has increased on a scale that threatens to seriously endanger
living organisms and the environment. But there is also growing evidence
that 'natural pollutants' can have a major influence on our planet.

Simply defined, pollutants are harmful materials that contaminate an
environment where they would not normally be found. In the atmosphere,
they range from volcanic ash to radioactive isotopes.

Typically, we tend to think of pollution as a result of human activity. Ever
since people started smelting metals and using fire to clear forests we have
been contributing to atmospheric pollution, and today the major sources are
power plants (especially coal-fired ones), vehicle exhaust emissions, heating
fuels, chemical fumes from a variety of sources and waste disposal (which
generates methane). A host of other sources range from chemical aerosols and
dust generated by human activity such as farming and industry, to the use of
asbestos and other potentially dangerous fibrous minerals. Recognizing and
combating the effects of these various pollution risks has been something of
a running battle throughout the history of industrialized society.

Pollutant risks

When gases such as sulphur dioxide, nitrogen oxide and carbon dioxide are
introduced to the atmosphere, they can combine with atmospheric water to
form sulphuric, nitric and carbonic acids that can damage the environment
and life. Known as acid rain, this phenomenon was discovered and linked to
pollution as early as the 1850s by Scottish chemist Robert Smith (1817–84).
Of course, these acids can be formed by natural processes, such as volcanic

OPPOSITE One of
the largest holes in
Antarctic ozone was
recorded and mapped
in 2006. Reductions in
ozone concentration
over Antarctica were
first observed from
the late 1970s and
reported in 1985.
Since than there has
been a steady decline
of around 4 percent
per decade in the
total volume of
atmospheric ozone.

Ocean hypoxia

DEFINITION A SHORTAGE OF OXYGEN IN THE OCEANS THAT MAKES IT HARD FOR MOST ANIMALS TO SURVIVE

DISCOVERY HYPOXIC AND ANOXIC REGIONS IN THE DEEP OCEANS ARE A NATURAL PART OF THE OCEAN SYSTEM, IDENTIFIED BY OCEANOGRAPHIC SURVEYS IN THE 19TH CENTURY

KEY BREAKTHROUGH RECENT STUDIES HAVE SHOWN THAT ANOXIC EVENTS MAY BE LINKED TO MAJOR EXTINCTIONS OF MARINE LIFE

IMPORTANCE THE CURRENT SPREAD OF HYPOXIC 'DEAD ZONES' MAKES ENTIRE REGIONS OF OCEAN INHOSPITABLE FOR LIFE

Oxygenation of Earth's oceans has been essential for the evolution of marine life since the oceans first formed between 4.4 and 3.8 billion years ago. However, there is growing evidence that oxygen levels in seawater can sometimes dip low enough to plunge life into crisis.

Oxygen is necessary for the survival of life in the sea just as much as it is for life on land. But the sea's free oxygen is essentially derived from gases dissolved out of the atmosphere, and this dissolved oxygen doesn't easily travel deep into the oceans, with their average depth of more than 3.5km (2.2 miles). Oceanic oxygen levels diminish rapidly from an average of about 6.5ml (0.22fl oz) of oxygen per litre of water at the surface to less than 20 percent of that value at depths of around 50m (160ft).

Most fish cannot survive if the oxygen saturation in water is less than about 30 percent of the maximum value, so the waters below 50m (160ft) are harmful to them for all but short periods. When the concentration of dissolved oxygen in ocean waters falls from 30 percent to 1 percent, the condition is known as hypoxia, and is seriously deleterious to life. Normally, between 20 and 50 percent of the 'water column' is hypoxic. Below 1 percent oxygen saturation, the condition is called anoxia, and can only be tolerated by rare organisms adapted to life without oxygen.

The oxygenation of the oceans

More than 2.45 billion years ago, the levels of oxygen in the ocean were too low for the evolution of aerobic (oxygen-dependent) marine organisms. It was only during the so-called Great Oxidation Event (GOE), around 2.4 billion years ago, that photosynthesis by cyanobacteria dramatically

OPPOSITE Around 499 million years ago, a sudden drop in oxygen levels within Earth's oceans led to a considerable extinction among seabed-dwelling animals such as trilobites. 'Anoxic' episodes like this offer a warning of the potential risk to life from similar events triggered by human activity.

increased levels of oxygen in Earth's atmosphere and surface waters. Photosynthetic microbes were present and producing oxygen long before the GOE, but organic matter and dissolved iron chemically captured all the 'free' oxygen they produced. It was only during the GOE that these 'oxygen sinks' became saturated, allowing the excess free oxygen to accumulate in the atmosphere for the first time (see page 125).

During this period, rising oxygen levels were a disaster for many anaerobic life forms, perhaps causing the largest extinction in Earth's history. Conversely, they were a boon for the evolution of aerobic life. However, there are signs in the rock and fossil record that ocean oxygen has dipped very low on several occasions since the GOE, during periods known as 'anoxic events'.

A late Cambrian anoxic event

The fossil record from the Cambrian Period (542–488 million years ago) preserves a tumultuous period of crises and extinctions for marine life. While the Cambrian is generally viewed as a time of rapid diversification and flowering of animal life (see page 165), it seems that many groups, such as trilobites, diversified only for many of their families to go extinct a little later. Scientists suspected that ocean anoxia might explain the apparent volatility of these early marine ecosystems, but evidence has been sparse until now. Early in 2011, however, a team led by Benjamin Gill of the University of California at Riverside reported strong geochemical evidence that oxygen depletion triggered major crises.

'Oceanic oxygen levels diminish rapidly from an average of about 6.5ml (0.22fl oz) of oxygen per litre of water at the surface to less than 20 percent of that value at depths of around 50m (160ft)'.

Cambrian rocks preserve marked changes in carbon isotope levels that are brief in geological terms, occurring over timescales of about a million years, but reflect significant disturbances in the global carbon cycle. Gill's team analysed one of the largest of these, from the late Cambrian around 499 million years ago. It lasted for about 2–4 million years, and coincides with a known trilobite extinction.

Comparison of carbon and sulphur isotope data shows that there was a large-scale anoxic event, involving increased burial of both the sulphate mineral iron pyrites and organic carbon in ocean-floor sediments. The organic carbon built up because an anoxic environment was preventing its decay, while iron pyrites has a known tendency to occur beneath anoxic water columns that are rich in hydrogen sulphide. This is the strongest evidence yet that drops in the oxygen levels of the oceans caused serious crises for Cambrian marine life, and it has important implications for our understanding of today's oceans.

Hypoxic 'dead zones'

In recent years, scientists have become increasingly concerned about the growth of oxygen-deficient or hypoxic 'dead zones' in the world's marine

waters. Since 2004, the documented number of dead zones has increased from 146 to well over 400. Some occur naturally, but many are caused by the influx of chemical nutrients from fertilizers, especially nitrogen and phosphorus, running off agricultural land. Fertilizer spills create large increases in algal 'phytoplankton', which in turn deplete oxygen levels in water. Recent demands for higher agricultural yields of grain crops – both for food and ethanol 'biofuels' – require greater use of fertilizers, but growing hypoxia is damaging coastal fisheries and local economies based upon them.

Notable dead zones occur in the Baltic, the North Sea, around the Japanese islands and in the Gulf of Mexico, as well as along the Atlantic and parts of the Pacific coasts of North America. In recent years hypoxic conditions have developed annually in important fishing waters off the coast of Oregon. Here, it seems that expansion of hypoxic waters from offshore 'oxygen minimum zones' (OMZs) is causing the problem. These zones extend over some 8 percent of the world's oceans (more than 28 million square km/ 11 million square miles).

These offshore OMZs occur naturally, usually at water depths of 600–1,200m (2,000–4,000ft) where oxygen levels are permanently lowered to less than 10 percent of typical surface values. However, even lower levels of oxygen have been measured in some equatorial OMZs of the Atlantic, Indian and Pacific Oceans. As a result, the toxic zone where fish cannot survive has now grown by more than 4 million square km (1.5 million square miles).

ABOVE When plentiful nutrients are available from either natural sources or human pollution, water-dwelling phytoplankton can reproduce very quickly, growing into extensive 'blooms' such as this one imaged off the Texas coast in 2000. The phytoplankton may be dense enough to deplete the oxygen in the surrounding water, causing hypoxia that can have a deadly impact on other organisms.

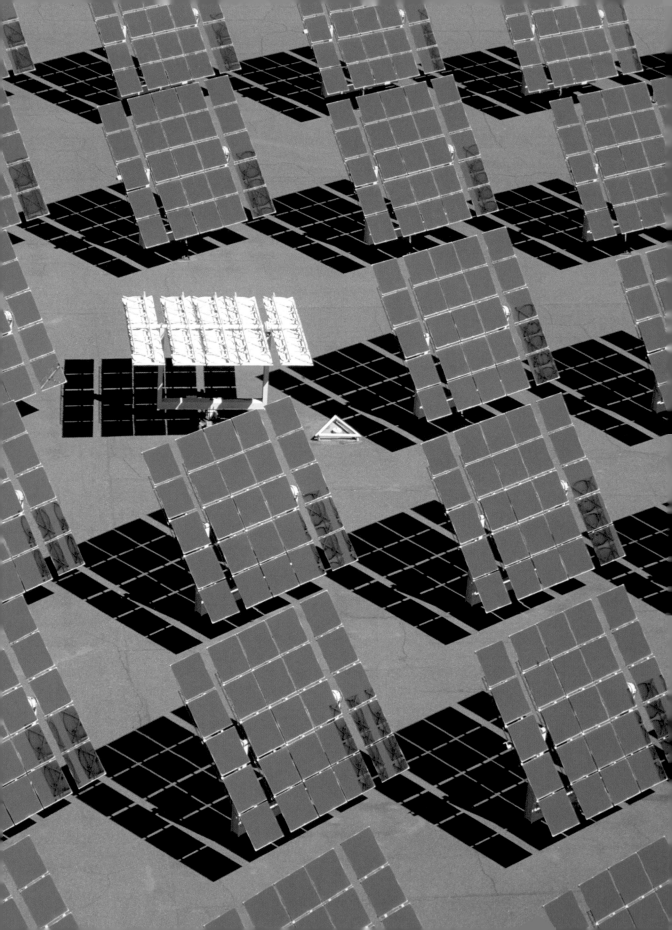

95

Energy futures

DEFINITION STRATEGIES FOR THE FUTURE DEVELOPMENT OF SOCIETY
BASED ON AVAILABLE ENERGY SOURCES

DISCOVERY THE FIRST A.C. HYDROELECTRIC PLANT WAS OPENED AT
NIAGARA FALLS IN THE 1890s

KEY BREAKTHROUGH CONTROLLED NUCLEAR FUSION FIRST ACHIEVED
BY BRITISH PHYSICIST JOHN COCKCROFT IN 1958

IMPORTANCE EXPLOITATION OF RENEWABLE ENERGY SOURCES NOT
ONLY HELPS THE ENVIRONMENT – IT IS VITAL IF OUR CIVILIZATION
IS TO BE SUSTAINED IN THE LONG TERM

According to the most recent figures available, more than 80 percent
of worldwide energy use is derived from fossil fuels – oil, gas and coal.
However, these are finite resources with enough global reserves for perhaps
another 40 years or so. How will we manage when they are gone?

The global human population is currently close to 7 billion, and will almost
certainly top 8 billion by around 2025. With increasing living standards,
the total consumption of mass-produced, 'marketed' energy is predicted to
rise by some 50 percent between 2007 and 2035, even in the face of a major
global recession. During 2007, energy use in less-developed nations outside
of the Organisation for Economic Co-operation and Development (OECD),
such as China and India, had already exceeded that of the 31 OECD nations
including the United States.

More than one-third of the global population lives in China and India,
where economies and energy demands are growing rapidly as industry
expands. Together, China and India accounted for just 10 percent of
global energy consumption in 1990, but by 2007, this consumption had
doubled to 20 percent, with a predicted rise to 30 percent by 2035. In
contrast, the United States, with just 4.5 percent of the global population,
consumes some 25 percent of world energy – the highest per-capita
consumption in the world.

Hooked on hydrocarbons

Ever since humans first discovered that burning hydrocarbons was one of
the easiest ways to release energy, these resources have been increasingly
exploited worldwide. The early 19th-century Industrial Revolution in

OPPOSITE Solar power
offers an efficient
and clean way of
generating energy, but
requires considerable
capital investment
and is dependent on
the availability of
adequate sunlight.
These reflectors,
known as heliostats,
form part of an
experimental solar
power station in
Albuquerque, New
Mexico. They
concentrate the
Sun's radiation on a
collector tower where
it is used to drive
a turbine.

northern Europe was located around major coalfields, but in an increasingly globalized society, energy resources and demand do not always match up so well – the global 'lottery' resulting from Earth's long history of environmental development has left a small number of countries with the bulk of the world's supply of oil and gas.

Some 65 percent of the world's proven reserves of crude oil are in the Middle East, mostly in Saudi Arabia. Oil and gas have been the major hydrocarbon fuels since the mid-1960s, coinciding with the peak of discovery of new sources. Global production of oil and gas is predicted to peak around 2015 (US production peaked in the 1970s) and to decline steeply thereafter, despite exploration and exploitation of controversial reserves in environmentally sensitive or inaccessible areas such as the deep oceans, poles and elsewhere (see page 389). Among today's main energy-using countries, the only ones that are self-sufficient in energy supply are Russia and Canada. India and China in particular are deficient in oil reserves and so resort instead to the large-scale exploitation of their massive coal deposits.

'A potential solution is the development of a "hydrogen economy", in which hydrogen (manufactured by splitting water molecules using electricity) is used to both store and transport surplus energy.'

Future energy scenarios

Most models or scenarios for future energy policy envisage an inevitable worldwide reduction in our reliance on finite and environmentally damaging energy resources, including all fossil fuels but especially hydrocarbons. In their place, they predict a rise in the exploitation of less damaging energy resources such as renewables, which at present supply around 20 percent of the global energy supply, and nuclear power.

The most highly developed among these alternative energy sources are hydroelectric and nuclear power. Hydroelectricity has been tried and tested ever since Serbian inventor Nikola Tesla (1856–1943) built the first A.C. hydroelectric plant at Niagara Falls in the 1890s. It is still the dominant form of alternative electricity supply, but is geographically limited by suitable sites and environmental pressures. Nuclear power, meanwhile, currently supplies about 16 percent of global electricity demand. Despite all the issues relating to safety, cost and disposal of nuclear waste, the actual energy production process involved is remarkably 'clean', and nuclear will certainly continue to be a major source into the future, especially in countries with traditions of hi-tech industry such as France.

Solving renewable problems

Other renewables, such as solar, wind, biomass and tidal power, have significant problems of cost, siting in relation to demand and storage of surplus energy. But despite these drawbacks, the urgency of the energy situation is leading to rapid development of technological solutions that allow these resources to be increasingly exploited.

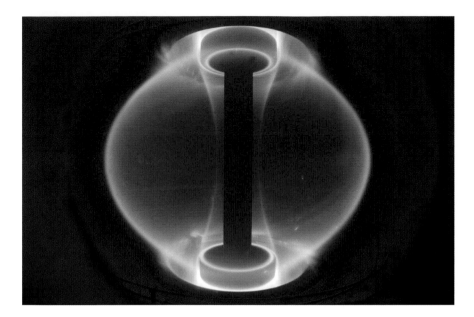

LEFT A spherical ball of superheated, electrically charged gas, known as a plasma, is contained within powerful magnetic fields at an experimental nuclear fusion facility in Oxfordshire, UK. The ability to create and manipulate such plasmas is a vital part of controlled nuclear fusion, but itself requires the harnessing of enormous amounts of energy.

The major problem for all these 'green' energy supplies is the difficulty of storing large amounts of surplus power generated by intermittent sources. A potential solution is the development of a 'hydrogen economy', in which hydrogen (manufactured by splitting water molecules using electricity) is used to both store and transport surplus energy. Hydrogen has the additional benefit that it is a relatively 'clean' fuel – recombination with oxygen in the atmosphere simply produces water. However, practical hydrogen storage on both large and small scales requires the widespread adoption of advanced 'fuel cell' technologies that are still in development.

Another potential technological 'fix' for our energy problems is nuclear fusion. In contrast to the widely used nuclear fission process, which generates energy by splitting heavy radioactive atoms into lighter ones, fusion releases energy by combining light atoms (the hydrogen isotopes deuterium and tritium) to form the slightly heavier element helium. Fusion is the process that powers the stars themselves, and is perhaps the ultimate source of 'clean' energy, since the hydrogen isotopes required can be extracted from water. However, initiating fusion requires that the atoms involved are transformed into a multimillion-degree 'plasma' contained and compressed in a powerful magnetic field – a process that requires enormous amounts of power and expensive equipment. As a result, while artificial controlled fusion is now a matter of routine at research reactors such as the Joint European Torus (JET), in Oxfordshire, UK, commercial development is some way off.

So it seems there is no single cure-all solution – instead, future development of energy sources in a world with a booming population and growing demand from countries such as India and China will almost certainly have to combine a number of strategies.

The Athabasca tar sands

96

DEFINITION A LARGE HYDROCARBON RESERVOIR EMBEDDED IN TAR
SANDS IN NORTHEASTERN ALBERTA, CANADA

DISCOVERY FIRST EXPLOITED AS A SOURCE OF BITUMEN BY NATIVE
PEOPLES, AND LATER BY EARLY EUROPEAN SETTLERS

KEY BREAKTHROUGH DEVELOPMENT OF EXTRACTION TECHNIQUES
ALLOWED COMMERCIAL PRODUCTION TO BEGIN IN THE 1960s

IMPORTANCE ATHABASCA AND OTHER SIMILAR DEPOSITS OFFER US A
WAY TO STAVE OFF THE IMPENDING HYDROCARBON CRISIS, BUT AT
WHAT COST?

Our energy-hungry world faces a huge dilemma over the exploitation of
'dirty' hydrocarbon resources such as oil shales and sands, which will lead to
ever-greater greenhouse gas emissions and environmental problems. One of
the most important of these is Canada's vast Athabasca tar sands deposit.

Crude bitumen is a sticky, tar-like form of petroleum that people have used
since ancient times – for instance, in waterproofing canoes and as mortar
for bricks. In the 1700s, European fur traders came across remarkably rich
reserves in the sands of northeast Alberta, Canada, along the Athabasca
River. By the late 1800s, scientists recognized the Athabasca tar sands as the
most extensive reserve of crude bitumen in the world.

The tar sands cover an area of more than 140,200 square km (54,100 square
miles) of the Canadian Shield. Experts estimate that around 10 percent of
their content can be economically recovered with current technologies, with
the potential to provide an estimated 174 billion barrels of crude oil – second
only to the reserves of Saudi Arabia. But their exploitation is proving highly
controversial – 'gung-ho' exploiters are keen to cash in on an economic
windfall, while green campaigners are set against a process that will have
huge impacts on native forests and wildlife, as well as increasing greenhouse
gas emissions from energy-intensive mining.

The geological background

Geologists first appraised the tar sands in the middle of the 19th century
and later showed that the bitumen could be successfully separated from the
sediment with hot water. Up to 60m (200ft) thick, the oil sands date back to
the early Cretaceous Period, about 128–118 million years ago. They were

OPPOSITE Exploitation
of Canada's Athabasca
tar sands involves strip
mining to remove
vegetation, soils and
thin superficial deposits
prior to extraction
of hydrocarbon
deposits. This is
having a considerable
environmental impact
on the 'virgin' muskeg
terrain of the region.

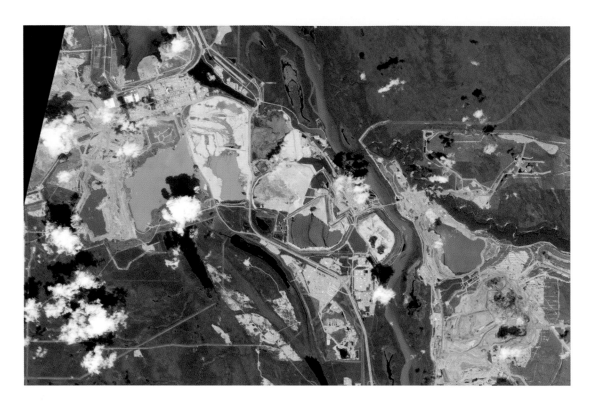

originally deposited as part of a vast wedge of sedimentary strata about 6km (3.7 miles) thick, which thins eastwards from the Rocky Mountains onto the Canadian Shield. The sediments were deposited by rivers and then shallow seas that flooded the region.

The warm and humid ancient climates of early Cretaceous times stimulated prolonged development of lush vegetation. Rapid and eventually deep burial of the organic plant debris produced an abundance of various hydrocarbons including coals, gas and bitumen. Altogether, the strata contain huge reserves of coal, as well as the tar sands and one of the world's largest reserves of natural gas, which today supplies much of the North American market at a rate of around 450 million cubic m (98 billion gallons) per day.

Exploitation techniques

By the 1960s, improved technology and the rising price of oil made extraction of bitumen from the tar sands economically viable. Because about 20 percent of the oil sand deposit lies at or very close to the surface, it can be easily scooped up once the soil and vegetation have been stripped away. Consequently, the mining process is much safer than that required to extract deep coal or ore deposits.

Sticky bitumen is separated from the surrounding sediment using vast amounts of hot water – about two to five barrels of water for every barrel of oil extracted. This requires large amounts of natural gas to heat the water to between 50 and 80°C (122 and 176°F). The process yields a bitumen froth that separates with the help of chemical surfactants to yield 60 percent bitumen, 30 percent waste water and 10 percent solids by weight. The waste products are stored in huge tailing ponds, while coke-fired upgraders convert the raw bitumen into synthetic crude oil for export.

Recently, new techniques have allowed exploitation of subsurface deposits, by a process called steam-assisted gravity drainage. Engineers drill two horizontal wells into the deposit and pass low-pressure steam into the tar sands through the upper pipe. The steam heats the bitumen and reduces its viscosity, allowing it to drain into the lower pipe and be pumped out.

Exploitation of the Athabasca deposits has fluctuated in relation to oil prices since production began in 1967. But recently mining and production has increased significantly, from 760,000 barrels a day in 2005 to 1.3 million barrels a day in 2006. Estimates suggest production will reach 3.3 million barrels a day by 2020. At present, Canada is a major supplier of oil to the US, sending 1.4 million barrels of oil across the border each day.

Environmental concerns

Green campaigners are concerned about the way that bitumen mining has led to the destruction of native forests and the region's peaty muskeg soil. Although the industry says that the land will be restored, little work towards this has taken place so far. The mining also guzzles a lot of natural gas. By 2020, this energy use will generate about 108 million tonnes of carbon dioxide each year – about 20 percent of Canada's current total emissions.

'The warm and humid ancient climates stimulated prolonged development of lush vegetation. Rapid and eventually deep burial of the debris produced an abundance of various hydrocarbons.'

The tailing ponds that store mining waste are also a threat to wildlife, since they replace natural wetlands. The toxic water poisons migratory birds and threatens to seriously contaminate local groundwaters, as well as the Athabasca River itself, which drains into Lake Athabasca and flows on from there into the Beaufort Sea.

Campaigners have also voiced concerns that there is not enough oversight of the environmental and health impacts of the oil sand exploitation. During 2009 to 2010, a Standing Committee on Environment of the Canadian House of Commons failed to issue a consensual public report on the matter, but the minority parties condemned the monitoring programme. What's more, in November 2010 Canada's Senate defeated a climate change bill, quashing legislation to cut greenhouse gas emissions to 25 percent below 1990 levels by 2020, with a long-term target of a 50 percent cut by 2050. It seems that Canada's environmentalists have a tough fight on their hands.

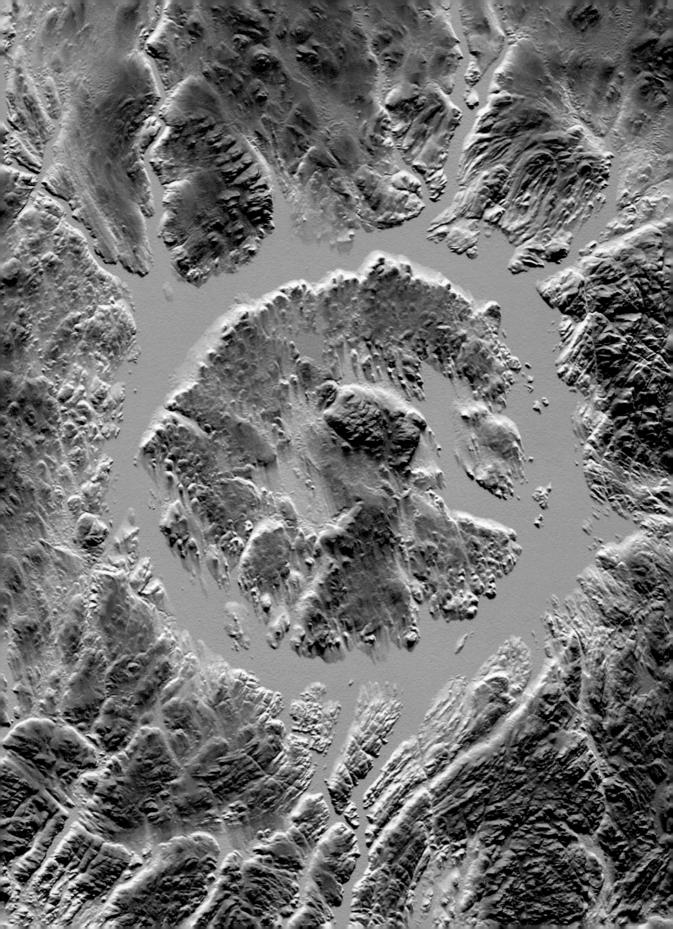

97

Threat from the skies

DEFINITION A MAJOR THREAT TO OUR PLANET PRESENTED BY
IMPACTS OF ASTEROID OR COMET FRAGMENTS FROM SPACE

DISCOVERY FIRST IDENTIFIED IN 1980 THROUGH THE DISCOVERY
OF THE END-CRETACEOUS IMPACT

KEY BREAKTHROUGH DISCOVERY OF SEVERAL RECENT, SMALL-SCALE
CRATERS HAS EMPHASIZED THE THREAT THESE IMPACTS PRESENT

IMPORTANCE FUTURE IMPACTS OF THIS KIND ARE INEVITABLE, AND
HAVE THE POTENTIAL TO INFLICT WIDESPREAD DESTRUCTION

Time and again during the Earth's history, comets and asteroids have smashed into our planet, sometimes with catastrophic effects on the environment and life. Such impacts will happen again in future – it's not a question of if, but when.

The Moon's scarred face bears testimony to the fact that our solar system is a violent place. Asteroids and comets have bombarded all the planets and moons throughout history, sometimes leaving vast craters. These are well preserved on the Moon, but on Earth, the constant recycling of ocean floor crust by plate tectonics has erased many crater structures, while others have been buried deep beneath thick layers of younger strata.

Even today, the Earth is constantly bombarded by objects from space. Thankfully, most of them are tiny fragments of rock and dust that burn up as they enter the atmosphere. But over the past few decades it has become increasingly clear that some impacts have been utterly devastating to the environment. Most notorious of all is the Chicxulub impact that devastated the Gulf of Mexico at the end of the Cretaceous Period around 65 million years ago. Current estimates suggest that an asteroid roughly 11km (7 miles) wide hit the Earth, releasing energy equivalent to 100 million megatonnes of TNT.

The impact would have raised a vast hot cloud of dust, ash and steam, while colossal shock waves would have triggered global earthquakes and volcanic eruptions. The ash-laden skies would have blocked sunlight for years, cooling the environment dramatically. Today, the consensus is that this caused the mass extinction of many plants and animals, including the dinosaurs,

OPPOSITE The vast Manicouagan impact crater in the Canadian Shield is clearly visible in this map produced by the Space Shuttle Radar Topography Mission of 2000. The impact happened around 215 million years ago, when a 5km (3-mile) asteroid ploughed through the atmosphere, creating a 100km (60-mile) scar in the landscape.

ABOVE Arizona's
famous 'Meteor
Crater' was the first
structure on Earth to
be recognized as the
result of an impact
from space. Geologist
Daniel Barringer
(1860–1929) and
others subsequently
spent 25 years drilling
in search of the iron-
rich meteorite they
believed was buried
beneath the crater, but
in fact the impacting
body was vaporized
on impact.

although there is still some debate about whether just one impact caused
the mass extinction, or whether several were responsible. Nevertheless, the
impact and the environmental havoc that followed are thought to have led
to the extinction of around 65 percent of all living creatures (see page 261).

How big, how frequent?

While the devastating effects of the Gulf of Mexico impact are now beyond
doubt, the big question for astronomers is how often such impacts occur, and
how large they can be. Recent satellite-based radar mapping has detected
several hundred significant impact craters on Earth's surface, but many of
these date back for tens, even hundreds of millions of years. Only a few,
such as Manicouagan Lake in Quebec, Canada, are obvious today.

The largest identified impact from space in modern times was the 'Tunguska
event' of 1908. This blast is believed to have been caused by the 'air burst'
explosion of a large meteoroid or comet fragment, tens of metres across,
several kilometres above the Earth's surface, which felled 2,150 square km
(830 square miles) of Siberian forest.

In 1994, however, astronomers had a 'ringside seat' for a much larger
explosion – the impact of comet Shoemaker-Levy 9 into the atmosphere
of the giant planet Jupiter. This substantial comet had already been ripped
apart by Jupiter's gravity, and the largest of its fragments released energy
600 times greater than the world's entire nuclear arsenal.

But however spectacular such events are, we could at least comfort ourselves with the thought that explosions such as Tunguska are rare. However, the recent discoveries of a series of craters around the world, with diameters ranging from tens of metres to kilometres, and ages counted in centuries rather than millions of years, shows that the threat from impacts is very real. According to Eugene Shoemaker (1928–97), co-discoverer of Shoemaker-Levy 9 and a geologist with the US Geological Survey, incoming fragments of rock and ice trigger roughly one Hiroshima-scale explosion in Earth's atmosphere every year, but these almost always go unnoticed.

'Incoming fragments of rock and ice trigger roughly one Hiroshima-scale explosion in Earth's atmosphere every year, but these almost always go unnoticed.'

Tracking future threats

Since the late 1990s, astronomers have discovered thousands of Near-Earth Objects (NEOs) that could pose a threat in future. For instance, in January 2002, a 300m (1,000ft) asteroid called 2001 YB missed Earth by little more than twice the distance to the Moon. The most worrying aspect of the event was that the asteroid was not detected until 12 days before its closest flyby.

Collision threats prompted Apollo 17 astronaut Harrison Schmitt and a group of influential scientists to write to the US Congress in 2003, urging it to make preparations for dealing with such a threat. In 2005, Congress tasked NASA with cataloguing 90 percent of an estimated 20,000 NEOs with diameters larger than 140m (460ft) by 2020.

But the rate of discovery of new and potentially threatening objects has vastly outstripped the agency's ability to track their paths with precision. In 2009, NASA told Congress that it lacked the funds to track large NEOs accurately, so the US is currently considering how best to respond to the threat and whether to establish technologies for effective planetary defence. Unfortunately, all the suggested means of deflecting an incoming asteroid are prohibitively expensive and carry no guarantee of success.

Many scientists believe the immediate priority should be for better asteroid detection and tracking systems. One proposal calls for the launch of a telescope into orbit around the Sun, where it would detect and track asteroids down to 140m (460ft) in diameter and cover a blind spot that terrestrial instruments cannot monitor. However, a telescope with these capabilities would cost an estimated $600 million.

Canada and Germany also have plans for NEO survey satellites, but whether they will come to fruition in the current economic climate is uncertain. Many experts argue that new terrestrial telescopes, such as the Large Synoptic Survey Telescope planned for Cerro Pachon in Chile, offer better value for asteroid monitoring because they are cheaper, have a much longer working life and can be up and running more quickly.

98

Future sea levels

DEFINITION LONG-TERM FLUCTUATIONS IN SEA LEVELS TRIGGERED BY TECTONIC PROCESSES

DISCOVERY HIGH SEA LEVELS IN THE CRETACEOUS WERE RECENTLY CONFIRMED BY DIETMAR MÜLLER AS REACHING AROUND 170M (560FT) ABOVE PRESENT-DAY LEVELS

KEY BREAKTHROUGH THE DISCOVERY OF SUBSIDENCE ALONG THE NEW JERSEY MARGIN RESOLVED CONFLICTING EVIDENCE

IMPORTANCE THE LONG-TERM TREND OF FALLING SEA LEVELS IS UNLIKELY TO HELP IN THE FACE OF THE PRESENT RISE

While today's concern about changing sea levels focuses on the potential devastation wrought by rises of a metre or so, the geological record shows that both past and future hold far more drastic changes in Earth's sea level, with peaks more than 150m (500ft) higher than today.

Ancient rocks record global changes in sea level over long timescales of tens or hundreds of millions of years. During the late Cretaceous Period, for instance (100–65 million years ago), levels peaked at more than 150m (500ft) above present levels, flooding low-lying landmasses with shallow seas. At other times, seas have fallen by 50m (160ft) or more below their current average, leaving shallow continental shelves dried out and exposed to the atmosphere as happened in early Triassic times (251–245 million years ago).

Such changes have had radical effects on the environment and shaped the evolution of life. But what drives such huge changes in sea level? Today, we know that global warming is causing sea-level rise. But the much larger changes of the past, which are likely to recur in the future, are difficult to explain due to a large number of potential factors that gradually change the lie of the land. These include sedimentation, the eruption of Large Igneous Provinces (see page 109), mantle convection, ocean spreading rates and changes in the lengths of the mid-ocean ridges (see page 61).

Fluctuating basins

Scientists generally accept that changes in the size and volume of the ocean basins themselves are the most likely cause of the large long-term changes in sea level. Ocean basin size is driven by heat flow in the mantle and expressed

OPPOSITE Global sea levels are governed by a number of factors. While there are very real concerns about comparatively short-term changes due to global warming, geological processes can and have caused much greater changes in the past. During the Cretaceous Period, for example, high sea levels submerged much of North America beneath the 'Western Interior Seaway', where marine reptiles, including the giant sea turtle *Archelon*, thrived.

at the surface by the formation and growth of mid-ocean spreading ridges and the eruption of Large Igneous Provinces. Mid-ocean ridges often rise some 2km (1.2 miles) above the deep-ocean floor, and are typically tens of thousands of kilometres long. New volcanic rocks on the ocean floor slowly move away from the ridge, cooling and subsiding as the ocean crust thickens. As a result, their formation displaces significant and changing volumes of ocean water over time.

Some of the best evidence for long-term changes in sea level comes from the Cretaceous Period (145–65 million years ago). In Europe, the late Cretaceous saw the formation of chalk-rich seas over all continental Europe. Other continents, such as Africa and North America, were also inundated by shallow seas around 80 million years ago, showing that the sea-level rise was

RIGHT Raised coral reefs, such as these on Niue Island in the South Pacific, are witness to relative height changes between land and sea. Some of these changes are due to tectonic uplift, while others are due to varying sea levels.

a global phenomenon. Shortly after the end of the Cretaceous, the floods receded and continental landmasses reappeared.

The New Jersey continental margin

One major mystery about the Cretaceous sea-level rise has always been the way that measurements from different locations have thrown up wildly different figures – in some areas seas appeared to be up to 250m (820ft) higher than today, while analysis of sediment deposits on the New Jersey margin of North America suggested a rise of no more than 40m (130ft). Recently, however, a team led by Dietmar Müller at the University of Sydney in Australia has resolved this discrepancy using new geophysical data to create an accurate reconstruction of the history of ocean basins over the past 140 million years.

The team used a simulation of mantle convection to show that the whole New Jersey region has actually subsided by more than 100m (330ft) as North America has moved westwards during the past 70 million years. When this mechanism is taken into account, it explains the anomalously low Cretaceous sea levels previously measured in the region. The model also takes account of a wealth of factors including crustal subduction of the ocean floor north of Africa, India and Australia since Cretaceous times. Eventually, Müller's team was able to reconstruct the historical development of the mid-ocean ridges, as well as sedimentation rates, at 5 million year intervals, to produce the most detailed picture yet of the way that ocean volumes and surface areas have changed over time.

'If such a change suddenly occurred now, we would find Russia connected to Alaska by land across what is currently the Bering Strait, and it would be possible to walk from France to Britain.'

The results, published in 2008, show that at their late Cretaceous peak, seas were roughly 170m (560ft) above today's levels. Later, they dropped dramatically, largely due to a sharp fall in the production of new ocean crust at spreading ridges in the Pacific. The existing oceanic crust has sunk as it grows older and denser.

Predictions for the future

While our warming world is likely to see higher sea levels over the next few centuries, the longer-term trend is probably downwards. Müller's model predicts that, on the largest scales, the oceanic crust will continue to subside for the foreseeable future. Specifically, it predicts that around 80 million years into the future, sea levels could be about 120m (390ft) lower than today.

If such a change suddenly occurred now, we would find Russia connected to Alaska by land across what is currently the Bering Strait, and it would be possible to walk from France to Britain. Even if all the ice on the polar caps were to melt, causing a sea-level rise of about 50m (160ft), the net result 80 million years from now would still be a 70m (230ft) fall, due to the unstoppable increase in the depths of ocean basins.

99

Future tectonics

DEFINITION FUTURE REARRANGEMENT OF EARTH'S TECTONIC PLATES, WITH HUGE CONSEQUENCES FOR GEOGRAPHY, GEOLOGY AND LIFE

DISCOVERY THE FIT OF OPPOSING CONTINENTS WAS PROPOSED AS EVIDENCE FOR CONTINENTAL DRIFT BY ALFRED WEGENER

KEY BREAKTHROUGH DISCOVERY OF THE OCEAN-FLOOR SPREADING MECHANISM THAT GENERATES FRESH CRUST AND MOVES EARTH'S TECTONIC PLATES

IMPORTANCE TECTONICS ARE A KEY PROCESS IN THE DEVELOPMENT OF OCEANS, CONTINENTS, MOUNTAINS AND EARTH HISTORY

As long as there is enough heat in the Earth's core, the tectonic plates that make up our planet's crust will continue to drift across the surface at a stately pace. Some oceans will expand while others will contract, but where are the continents headed?

Most of Earth's crustal plates are constantly on the move, creeping slowly but inexorably by 1–100mm (0.04–4in) each year, with an average speed similar to the rate at which our fingernails grow. These movements are so slow that any significant motion on a planetary scale – over thousands of kilometres or tens of degrees of latitude – takes tens of millions of years.

In the much shorter term, however (though still over centuries), opposing pressures build up across faults in the crust where they are suddenly relieved by earthquakes and surface displacements, typically of a few centimetres or metres. Over millions of years, meanwhile, regional pressures build up, which can only be resolved by combinations of folding and faulting of the rocks. Through tens of millions of years, rocks can be crumpled into new mountain ranges, or pulled apart into rifts, accompanied by volcanic eruptions and the formation of depressions that will ultimately grow to become new ocean basins.

All these plate movements produce erratic surface effects, which are often catastrophic and still largely unpredictable. But in our planet's future, the longer-term effects will not only reconfigure the global jigsaw pattern of plates, but also radically alter ocean circulation patterns and climates. So how long will it take for there to be a significant transformation of the global map, and what will that future world look like?

OPPOSITE The Red Sea, which separates Africa from Arabia, also marks a major geological boundary between diverging tectonic plates. The sea's long parallel coasts, seen in this spectacular Space Shuttle view, mark the boundary faults of a complex rift valley system.

Predicting plate movement

Plate dynamics are driven by heat that is lost from Earth's central core, through convection in the mantle and the formation and motion of crustal plates, whose strength and thickness control the dissipation of heat from the surface. But there are still many uncertainties about the forces that both drive and resist plate motion.

The upshot is that trying to predict exactly how the continents will move over the next 100 million years or so is far from easy. However, geologists have developed reasonable models of the present direction and rate of plate motion from satellite and ground-based measurements, and consequently they can at least make some informed guesses about short-term plate motions over the next few million years.

There is strong evidence, for instance, that Africa is currently rotating clockwise and moving to the northeast. As a result, it is converging on the massive Eurasian Plate, which lies to its north separated by the Mediterranean and Red seas. The ongoing collision between these two plates is a complex process. Subduction zones and small fragmentary plates (generally referred to as microplates) have developed complex patterns that seem to be influenced by forces other than the main convergence. Recently a group of Chinese researchers have argued that small-scale convection in the mantle may be influencing their behaviour.

'New subduction zones will open up along their eastern coasts as they begin to drift back to the east, shrinking the Atlantic Ocean once again and ultimately forming a new supercontinent, "Pangea Ultima", in around 250 million years.'

However, the long-term prognosis for the Mediterranean is not good. The basin will close up as its oceanic crust is subducted, and will eventually become the site of a major new mountain belt. Northern Europe will experience the tectonic repercussions and be pushed polewards over tens of millions of years.

Varied predictions

Other geologists have gone further in their predictions of the global-scale changes that may happen in Earth's future. Although the complexities of all the jostling tectonic plates and the mysteries of Earth's inner mantle mean that all such predictions must depend on certain basic assumptions, and small variations in those initial assumptions can produce wildly different outcomes, it is still possible to find some agreement.

Some of the best predictions have come from Christopher R. Scotese of the University of Texas at Arlington, whose Paleomap project has set out to chart the influence of Earth's geography and climate from the distant past to the limits of the foreseeable future. According to the work of Scotese and others, likely future changes include the break-up of Africa, as the Great Rift Valley (see page 277) continues to stretch and widen, eventually opening up to form a new ocean basin. The splinter of eastern Africa and the island of Madagascar will drift away to the east, subducting the Indian Ocean Plate

beneath them as they go. Meanwhile, the Atlantic will continue to widen and the Pacific continue to shrink, as the Americas move westwards. Parts of California will ultimately separate from the rest of North America, sliding north along the line of the San Andreas Fault.

One of the most dramatic plate motions, however, is that of the Indian-Australian Plate, which is moving north across the equator as its northern margin is subducted beneath the Indonesian archipelago. Eventually, Australia and the Indonesian islands will collide with Southeast Asia.

By around 100 million years in the future, the bulk of Africa and Eurasia will have merged completely, while Antarctica, moving north from its present position, will reunite with Australia. When it comes to the fate of the Americas, however, predictions diverge. Scotese's work predicts that new subduction zones will open up along their eastern coasts as they begin to drift back to the east, shrinking the Atlantic Ocean once again and ultimately forming a new supercontinent, 'Pangea Ultima', in around 250 million years.

Other geologists, however, believe that the Americas will continue to drift west, finally uniting with the Afro-Eurasia-Australia continent on its extreme eastern side to form a supercontinent that has been dubbed 'Melonia'. In this scenario, North America moves towards Australia and Asia, while South America rotates and collides with the eastern margin of North America.

100

The fate of the Earth

DEFINITION THE DISTANT FUTURE OF THE EARTH, AND THE ULTIMATE
FATE OF OUR HOME PLANET

DISCOVERY THE PATTERNS OF STELLAR EVOLUTION, AND THE SUN'S
LIKELY FATE AS A RED GIANT, WERE WORKED BY ASTROPHYSICIST
ARTHUR EDDINGTON AND OTHERS IN THE MID-20TH CENTURY

KEY BREAKTHROUGH JEFFREY S. KARGEL HAS PREDICTED THAT EARTH
MAY EVENTUALLY BECOME TIDALLY LOCKED IN ORBIT AROUND THE
RED GIANT SUN

IMPORTANCE THE EXAMPLES OF OTHER STARS REVEAL THAT EVEN
THE SUN HAS A LIMITED LIFESPAN

Earth formed as part of the solar system over 4.5 billion years ago and has been constantly changing ever since. These dynamic processes, and life as we know it, are dependent upon light energy from the Sun and internal energy from Earth's core – but how long will these energy sources last?

We might think of planet Earth and its rock material as the epitome of substance, durability and permanence, but our study of Earth's geology tells us that nothing is permanent, everything is in flux. Rocks are created, altered and recycled into other forms by continuing natural processes over timescales of many millions of years. Furthermore, the very processes of geology, such as plate tectonics and volcanism, which often endanger life, are a vital part of the planet's dynamic, driven by energy from the core. Without this internal heat source, Earth would become a geologically dead and lifeless world like the Moon. And when the core's energy is eventually used up, our planet will indeed 'die'. The living organisms on its surface, meanwhile, are largely dependent on energy from the Sun – another finite source that will eventually run out.

How long have we got?

According to most estimates, the major long-term threat to the survival of life on Earth comes from the Sun itself – the star whose surrounding nebulosity gave birth to our planet and which has nurtured us through 4.5 billion years of history. Developments in astronomy during the 20th century have show that the Sun is an unremarkable, mid-sized 'main sequence' star that is now about halfway through the stable phase of its life cycle. Despite this, its luminosity has increased steadily at a rate of roughly 1 percent every 110 million years, and this pattern will continue into the future.

OPPOSITE As stars like the Sun near the end of their lives, they go through a series of increasingly violent pulsations, shrugging off their outer layers in a series of shells that may be sculpted into fantastical shapes by the star's rotation, stellar winds and other factors. The result is a short-lived but beautiful 'planetary nebula' such as the Cat's Eye Nebula, photographed here by the Hubble Space Telescope.

Inevitably, this will impact upon Earth's climate, which will get slowly hotter and probably eventually overcome any natural mitigating processes of the atmosphere, ocean and biomass. In a billion years or so Earth's biosphere, as we know it, will become unsustainable and collapse, with life gradually retreating to refugia deep in the oceans or underground. Once the Sun's luminosity has increased by 10 percent, global surface temperatures will reach 47°C (117°F) and most, if not all, of the oceans will evaporate. As water molecules enter the atmosphere, they will probably be broken apart by solar radiation, allowing lightweight hydrogen atoms to escape into space.

'Eventually, all that will remain of the Sun will be an incredibly dense and luminous body known as a white dwarf. Perhaps a hundred times more luminous than it is now, it will contain about half of its present mass packed into a body the size of Earth.'

Earth's continuing power source

Even with the loss of the oceans and atmosphere, Earth will still have plenty of energy in its core, and plate tectonic motion will continue with crustal stretching, rifting, volcanism, subduction and mountain building. The volcanic release of carbon dioxide and water vapour into the atmosphere may generate a 'supergreenhouse' state, but whether that will be a dry one (similar to that found on our nearest planetary neighbour, Venus, today) or a moist one is unclear. Whatever its eventual composition, the Earth will probably retain some form of atmosphere for another 5 billion years or so.

By this time, the Sun will have exhausted the hydrogen fuel in its core that has kept it shining for so long. As it begins to plunder other material in order to keep shining, its luminosity will increase perhaps a thousandfold, while its outer layers expand until it stabilizes with a radius around 250 times its present size. In the process, the Sun will consume Mercury and possibly Venus, but what happens to Earth is less clear.

Geologist Jeffrey S. Kargel of the University of Arizona, together with Bruce Fegley and Laura Schaefer of Washington University in St Louis, Missouri, have looked at one possibility, in which Earth by this time has become 'tidally locked' with one face permanently facing the Sun (just as the Moon currently has one face towards Earth). In this scenario, intense heating on Earth's daylit side will result in melting of the surface, restructuring of the interior and 'frosts' of various compounds forming on the night side.

Another possibility is that Earth's orbit moves further away as the Sun's gravitational pull decreases thanks to a slow but steady mass loss. The Moon's orbit, meanwhile, may reduce to some 18,000km (11,200 miles), where tidal interactions could tear it apart into a ring system, sending rocky debris raining down onto Earth.

Final throes

The final evolution of the Sun is now thought to be a remarkably complicated process of elemental change, involving phases of expansion and contraction

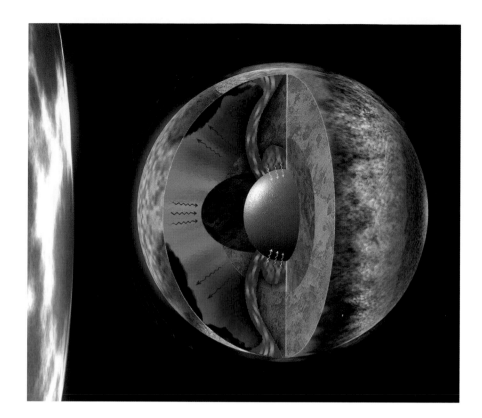

in size, temperature and mass. The loss of mass will weaken its gravitational pull upon Earth and the remaining planets in a somewhat unpredictable way. However, it is likely that, whatever state it is in, Earth will now drift away into a more remote orbit – darker, colder and completely lifeless.

Eventually, all that will remain of the Sun will be an incredibly dense and luminous body known as a white dwarf. Perhaps a hundred times more luminous than it is now, it will contain around half of its present mass packed into a body the size of Earth. Over billions of years, the white dwarf will continue to cool, with the remaining carbon and oxygen of its core crystallizing until it eventually loses all remaining energy and becomes a black dwarf, perhaps trillions of years from now.

Of course, the human race will not be around to see much, if any, of this distant future. Few species have ever survived for more than a few million years, especially those that have a record of rapid evolution like humankind. Despite all our current environmental woes and political concerns this does not necessarily mean that we will become extinct in the traditional sense – instead, like our immediate ancestors, *Homo sapiens* may speciate into new forms with new genetic templates. Perhaps some distant descendants of theirs, evolved into creatures we cannot begin to imagine, will have found a new home among the stars by this time, and may at least spare a thought for the dying embers of the world where they originated.

Glossary

Angiosperm
A member of a major group of mostly land-living plants that first appear as fossils in Cretaceous times but probably evolved before that in the Jurassic Period. They are characterized by a number of features, especially associated with the formation of their seeds within an ovary from flowers.

Asthenosphere
One of the two layers of Earth, the other being the outermost lithosphere, defined on the basis of the strength, rather than the composition, of their material. Heat and pressure make the minerals of the asthenosphere ductile, allowing them to slowly deform and flow.

Banded Iron Formation (BIF)
Layered, iron-rich sedimentary strata deposited in oxygen-poor ocean waters during Precambrian times.

Bilaterian
An animal with an axial symmetry to its embryonic body, which divides it into left- and right-hand sides.

Cambrian explosion
A rapid increase in the diversity of life in early Cambrian times, marked in the fossil record by the earliest known representatives of many animal groups.

Chemical fossil
The remains of an organism preserved in the form of organic chemicals derived from the breakdown of its cellular matter.

Chemical weathering
The breakdown of mineral components of rock through reaction with active solutions in the environment, producing new chemical compounds – for example, the creation of limestone (calcium carbonate) through the action of acidic rainwater.

Composite volcano
A typically steep-sloped volcano commonly associated with subduction zones. It is distinguished by a cone constructed of layers of lava, ash and other eruptive products, situated above a source of magma within the crust.

Core
Earth's innermost and densest layer, composed mostly of iron and nickel at exceedingly high temperatures and pressures, with an outer liquid section and a sold inner component. The core generates Earth's magnetic field, and its heat, conducted outwards through the overlying mantle, ultimately drives plate tectonics.

Cratons
Ancient Precambrian regions of rock lying at the heart of continental plates, which are thought to be the remnants of the first continental crust to form on Earth.

Crust
Earth's thin, cool and brittle outermost rocky layer, composed of relatively dense ocean-floor rocks lying mostly below sea level, and slightly less dense continental rocks that are mostly above sea level.

Cyanobacteria
A group of primitive single-celled bacterial micro-organisms, also known as blue-green algae, which obtain energy by photosynthesis.

Cyclothem
A sedimentary sequence of deposits that reflects environmental changes such as sea-level fluctuations, generally repeated through time and thus forming a recurring series of strata (such as coal measures).

Cynodont
A member of an extinct group of tetrapods that evolved from the reptiles and acquired a number of mammalian characters in their skeletons. Cynodonts include the ancestors of the mammals themselves.

Geological timeline

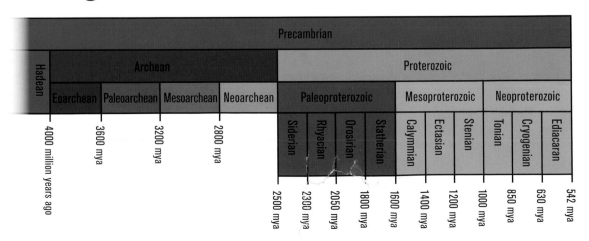

Dinosaur

A member of a large and diverse group of reptiles, characterized by the structure of the hip and pelvis, which flourished between the Triassic and Cretaceous Periods and are now mostly extinct (except for those that evolved into birds).

Dropstone

A boulder carried by floating ice and dropped onto the sea floor when the ice melts, often far from its place of origin.

Eon

A major division of geological time, comprising a number of eras – such as the Phanerozoic Eon, containing the Palaeozoic, Mesozoic and Cenozoic eras.

Epoch

A relatively short division of geological time lasting perhaps tens of millions of years, and itself comprising a number of stages – for example, the Paleocene Epoch of the Paleogene Period.

Era

A major division of geological time, comprising a number of periods – for example, the Mesozoic Era with its Triassic, Jurassic and Cretaceous periods.

Erosion

The breakdown of rock material by physical and chemical processes on the Earth's surface, such as the action of wind, water or ice, and chemical weathering.

Eukaryote

A member of a major division of life, distinguished by the presence of a membrane-enclosed nucleus and mitochondria within each cell.

Fault

A dislocation and displacement of brittle rocks across a plane relative to one another – the result of a catastrophic failure of the rocks in response to a build-up of forces within the crust.

Foraminiferans

A group of single-celled, aquatic eukaryote organisms that generally produce small shells or other protective structures.

Fossil

The remains of an organism preserved in the sedimentary rock strata of Earth's crust, ranging from hard parts, such as shells and bones, to mineralized soft tissues, such as feathers and hair, chemical compounds or impressions, such as footprints.

Fossilization

The processes by which the remains of an organism are preserved after death within rock materials. Normally only hard parts, such as shells, bones and woody tissues, are preserved.

Genome

The complete series of inherited cellular instructions, encoded in the form of DNA, that allows an organism to function and reproduce.

Greenhouse gas

An atmospheric gas, such as carbon dioxide or methane, which both absorbs radiation from the Sun and emits radiation within the thermal or infrared range, warming the atmosphere as a result.

Hotspot

A region of Earth's crust where heat rising through the mantle is concentrated, resulting in partial melting of the crust and eruptions of magma at the surface.

Igneous rock

A major rock type formed by the cooling and crystallization of minerals from a melt.

The chart spread across the following pages shows the geological timeline as currently defined by the International Commission on Stratigraphy. The largest formal divisions of time are the Eons – Archean, Proterozoic, and Phanerozoic (the overarching term 'Precambrian' is frequently used to describe all time before the Phanerozoic). Within the eons lie the shorter geological eras, which from the Proterozoic onwards are subdivided into periods, epochs (shown here only for the recent Cenozoic Period) and even briefer stages, omitted for clarity. Numbers marked at the divisions indicate time in millions of years before the present.

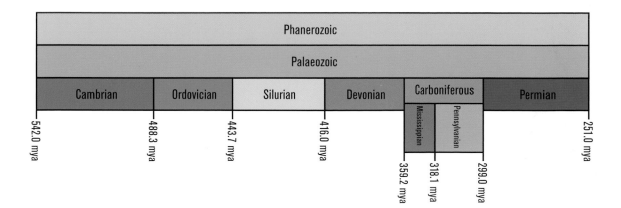

Isotope

An atom of an element that contains the same number of electrons and protons as the other atoms, but carries a different number of neutrons. As a result, isotopes of an individual element have essentially identical chemistries, but different atomic weights.

K/T boundary

The traditional scientific name for the transition between the Cretaceous Period (K for kreide, German for chalk) and the Tertiary Era (today known as the Cenozoic), dated to 65.5 million years ago. The boundary is marked by one of Earth's major extinction events and the impact of a large extraterrestrial body.

Large Igneous Province (LIP)

One of several regions of Earth's crust where massive volumes of lava have erupted from fissures and volcanoes over a relatively short period of time, creating extensive 'plateau basalts' such as those at Deccan, India, and the Columbia River Plateau, USA.

Laurentia

The geological name for the ancient North American continent (comprising most of the United States, Greenland and Canada).

Limestone

An aquatic sedimentary rock composed of calcium carbonate derived from organic remains such as shells and/or from the chemical precipitation of calcium carbonate.

Lithosphere

The outermost cool, brittle and rocky layer of Earth, defined by the strength of its rock material rather than by composition, which lies above the hot and ductile asthenosphere. The lithosphere consists of the crust and the upper part of the mantle.

Mantle

An immensely deep intermediate layer within the Earth, lying beneath the crust and above the core, and composed almost entirely of silicate minerals.

Mass extinction

A catastrophic event in the history of life, when a significant proportion of life forms are killed off in a relatively short time. Five particularly severe events are now recognized, during each of which more than 30 percent of life became extinct.

Metamorphism

Transformations experienced by rocks and minerals that are exposed to pressures and temperatures different from those in which they were formed, giving rise to 'metamorphic' rocks.

Metazoan

A eukaryotic organism whose body is composed of many cells organized into tissues.

Mineral

A naturally occurring solid chemical compound, normally crystalline in structure, that forms the substance of Earth's rock material.

Mitochondrial DNA (mtDNA)

The genetic material that encodes the mitochondria, cellular structures that are inherited directly from the mother. Mitochondrial DNA therefore survive unchanged from generation to generation (aside from 'genetic drift' – random changes that accumulate over time).

Molecular clock

A method for dating the evolutionary divergence of groups of organisms, based on estimated rates of 'genetic drift' in their mtDNA and comparison of the genes of their living representatives.

Ornithischian

One of two major divisions within the dinosaurs (the other being the saurischians). Ornithischians were all plant eaters and are characterized by the bird-like structure of their hip bones.

Peridotite

An iron-magnesium silicate rock that comprises much of the upper mantle, from which many basalt lavas are derived.

Period

A major interval of geological time, lasting for tens of millions of years and made up of subdivisions known as epochs.

Phytoplankton

A wide-ranging collection of mostly microscopic organisms that obtain energy by photosynthesis and live in the upper sunlit (photic) layer of water bodies such as oceans and lakes.

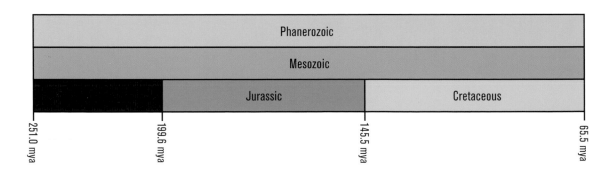

Prokaryote

A member of a major division of living organisms, composed of primitive unicellular microbes that lack a defined cell nucleus and instead have genetic material dispersed throughout their cells (see also eukaryote).

Protist

An abbreviation of protistan, a large group of single-celled and mostly aquatic eukaryotes, containing a wide range of otherwise unrelated organisms.

Quaternary Ice Age

A recent interval in Earth history, beginning around 2.8 million years ago, during which a succession of cold glacial and warmer interglacial climates had a dramatic effect on life and environments, especially in the major landmasses of the northern hemisphere.

Radioactivity

The property of certain chemical elements whose atoms spontaneously emit ionizing radiation and transform naturally over time into a series of isotopes.

Radiocarbon dating

A radiometric technique by which an age of growth or formation can be provided for organic materials, such as bone, teeth and wood, based on the presence of radioactive isotopes of carbon.

Radiometric clock

A dating technique that uses the known rates of decay of natural radioactive elements and isotopes. By measuring the ratio of isotopes, an estimate can be made of when a mineral first formed.

Rifting

The process of fracture by which Earth's crust is broken into parallel faults with a central depression, during the stretching, thinning and rupture of brittle rocks.

Rock

A solid material made from an aggregate of minerals. The rocks that make up Earth's crust can be generally classified as igneous, metamorphic or sedimentary.

Saurischian

One of two major divisions within the dinosaurs (the other being the ornthischians). Saurischians were characterized by the lizard-like structure of their hip bones, and include giant four-footed sauropods and bipedal theropods (including their modern bird descendants).

Sauropods

A major group of essentially plant-eating saurischian dinosaurs, some of which reached very large size.

Sedimentary rock

One of the major divisions of rock material, produced by the deposition of weathered and eroded rock and organic material into layers upon Earth's surface, and their subsequent compression.

Snowball Earth

A supposed glacial condition of Earth during Precambrian times, in which the entire surface from the poles to the equator was frozen over with ice sheets, glaciers and sea ice.

Stratification

A result of layering of sediments one upon another in the process of deposition by natural processes, so that the relatively younger stratum is laid above and on top of a relatively older one.

Subduction

The tectonic process triggered when two plates converge on each other and one plate is forced beneath the other, descending into the mantle.

Supercontinent

An agglomeration of continents brought together by plate tectonic processes – for example the ancient supercontinent of Gondwana.

Tetrapod

A four-limbed vertebrate animal, such as amphibians, reptiles, birds, mammals and their extinct ancestors.

Theropod

A member of a major group of bipedal saurischian dinosaurs, most of which were carnivorous. Most theropods are extinct, but modern birds are descended from one group.

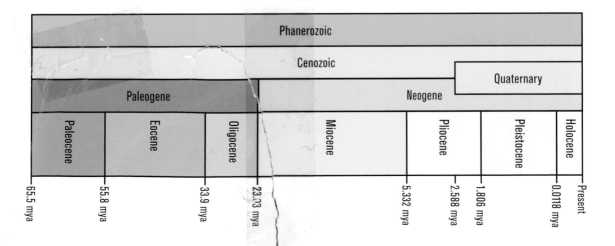

Index

Page numbers in **bold** denote an
illustration/caption text
Page numbers in *italics* denote caption text only

A FIREFLY BOOK

Published by Firefly Books Ltd. 2011

First printing

Publisher Cataloging-in-Publication Data   (U.S.)

Palmer, Douglas.
A history of Earth in 100 groundbreaking
discoveries / Douglas Palmer.
[416] p. : col. photos. ; cm.
Includes index.
Summary: A journey back through the
evolutionary history of our planet, from the
origins and formation of the earth.
ISBN-13: 978-1-55407-807-3 (pbk.)
1. Geology--History. 2. Earth sciences--
History. I. Title.
550.9 dc22   QE11.P356 2011

Library and Archives Canada Cataloguing in
Publication

Palmer, Douglas
A history of Earth in 100 groundbreaking
discoveries / Douglas Palmer.
Includes index.
ISBN 978-1-55407-807-3
1. Earth--Popular works. I. Title. II. Title: History
of the Earth in one hundred discoveries.
QB631.2.P35 2011   550   C2011-900024-5

Published in the United States by
Firefly Books (U.S.) Inc.
P.O. Box 1338, Ellicott Station
Buffalo, New York 14205

Published in Canada by
Firefly Books Ltd.
66 Leek Crescent
Richmond Hill, Ontario L4B 1H1

Printed in China

Douglas Palmer thanks Giles Sparrow and Hazel Muir
for their invaluable work on editing and improving
the text and to Tim Brown for all his efforts on the
design of the book. Any remaining errors remain
the responsibility of the author who has made every
effort to report the work of others as accurately as
possible and will seek to amend remaining mistakes in
any future edition of the book.

www.douglascpalmer.com